Semiconductors for Solar Cells

The Artech House Optoelectronics Library

Brian Culshaw, Alan Rogers, and Henry Taylor, *Series Editors*

Acousto-Optic Signal Processing: Fundamentals and Applications, Pankaj Das

Amorphous and Microcrystalline Semiconductor Devices, Optoelectronic Devices, Jerzy Kanicki, ed.

Electro-Optical Systems Performance Modeling, Gary Waldman and John Wootton

The Fiber-Optic Gyroscope, Hervé Lefèvre

Field Theory of Acousto-Optic Signal Processing Devices, Craig Scott

Highly Coherent Semiconductor Lasers, Motoichi Ohtsu

Introduction to Electro-Optical Imaging and Tracking Systems, Khalil Seyrafi and S. A. Hovanessian

Introduction to Glass Integrated Optics, S. Iraj Najafi

Optical Control of Microwave Devices, Rainee N. Simons

Optical Fiber Sensors, Volume I: Principles and Components, and *Volume II: Systems and Applicatons*, John Dakin and Brian Culshaw, eds.

Optical Network Theory, Yitzhak Weissman

Principles of Modern Optical Systems, Volumes I and II, I. Andonovic and D. Uttamchandani, eds.

Reliability and Degradation of LEDs and Semiconductor Lasers, Mitsuo Fukuda

Single-Mode Optical Fiber Measurements: Characterization and Sensing, Giovanni Cancellieri

For further information on these and other Artech House titles, contact:

Artech House
685 Canton Street
Norwood, MA 01602
(617) 769-9750
Fax:(617) 762-9230
Telex: 951-659

Artech House
6 Buckingham Gate
London SW1E6JP England
+44(0)71 630-0166
+44(0)71 630-0166
Telex-951-659

Semiconductors for Solar Cells

Hans Joachim Möller

Artech House
Boston • London

Library of Congress Cataloging-in-Publication Data

Möller, H. J. (Hans Joachim)
Semiconductors for Solar Cells/Hans Joachim Möller
Includes bibliographical references and index.
ISBN 0-89006-574-8
1. Solar cells—Materials. 2. Semiconductors. I. Title
TK2960.M65   1993   92-40718
621.31'244—dc20   CIP

British Library Cataloguing in Publication Data

Möller, H. J. (Hans Joachim)
Semiconductors for Solar Cells.
I. Title

ISBN 0-89006-574-8

© 1993 ARTECH HOUSE, INC.
685 Canton Street
Norwood, MA 02062

International Standard Book Number: 0-89006-574-8
Library of Congress Catalog Card Number: 92-40718

10 9 8 7 6 5 4 3 2 1

Contents

Preface

Semiconductors have emerged as the most promising class of materials that can convert sunlight directly into electrical energy. Historically, silicon was the first commercially used solar cell material and is today the most extensively studied semiconductor. The advantages of silicon are its mature processing technology (because of its use for microelectronic devices), the large abundance of it in the crust of the earth, and its nontoxicity, which is an important consideration from the environmental perspective. With high-purity silicon and optimized solar cell designs, efficiencies of 23% under normal sunlight can be achieved today in laboratory experiments. With polycrystalline silicon used in commercial cells and modules, efficiencies of about 10% to 13% are obtained now. The drawback of the silicon cell technology is that the production costs are still too high because of the sophisticated processing steps required. The energy costs of photovoltaic power are not yet competitive with other energy sources, and therefore a large driving force exists for the development of low-cost materials and fabrication processes.

Currently, a wide range of semiconductors is explored for their potential use in photovoltaic applications. The general criteria determining the choice of a particular semiconductor are efficiency and cost considerations, the environmental conditions (e.g., terrestrial or space applications and duration of sunshine), and the availability and toxicity of the raw materials. Each material requires a particular solar cell device structure for optimum performance, and the choice is primarily determined by the available processing techniques and the photovoltaic properties that can be achieved under these conditions.

This book is divided into three parts addressing different aspects of the photovoltaic energy conversion. In the first part (Chapters 2 to 4), the fundamental principles of the solar energy conversion process and the most common cell technologies are discussed. Energy conversion in semiconductors uses the fundamental physical process in which a photon penetrating into the semiconductor can generate electron-hole pairs. Because of their opposite charges, the electrons and holes can be separated by an internal electrical field and collected at two contacts, thus giving rise to a voltage and photocurrent if the two contacts are connected externally. The most widely used device principle is the operation of a solar cell as a pn-homojunction diode. Alternative concepts are heterojunction devices, tandem structures, and concentrator cells, which will be addressed in Chapter 3. However, completely different technologies, such as electrochemical cells (which may become important in connection with the hydrogen technology of photoelectrolytic decomposition of water) are beyond the scope of this book.

In the second part (Chapter 5), the material science aspects of photovoltaic semiconductors are presented. The electronic and optical properties of semiconductors are determined to a large extent by point defects and extended lattice defects like grain boundaries, dislocations, and precipitates. Contrary to high-purity electronic grade material, solar cell semiconductors usually contain higher concentrations of impurities and lattice defects. This is a consequence of the application of low-cost techniques and the utilization of novel materials. The continued improvement and successful development of solar cell materials therefore requires a thorough understanding of the defect structure and the ramifications for the photovoltaic properties. A main emphasis in this chapter will therefore be to review the current knowledge of the physical properties of crystal defects in semiconductors and their process-related generation, as far as they are relevant for photovoltaic properties.

In the third part (Chapters 6 to 9), the currently most promising semiconductor materials will be presented. These chapters treat in some detail the practical issues that are typical for each material and important for the processing and fabrication of the solar cells.

Monocrystalline and polycrystalline silicon is still the most widely used semiconductor for solar cells, and its technology shall be discussed in Chapter 6. Compound semiconductors usually have a much higher absorption, and thin layers are only necessary for light absorption. This offers the possibility of preparing solar cells using thin-film technologies, which are already available in the microelectronics industry. The most promising materials for thin-film solar cells are the copper ternary semiconductors and some binary compounds such as cadmium telluride and gallium arsenide. They are discussed in Chapters 7 and 8. Thin-film solar cells can also use amorphous semiconductors as the active photovoltaic material; however, so far, only silicon and germanium have been used for commercial devices.

The entire area of the fabrication of modules, arrays, and solar power plants is omitted in this book. Although this is also an important part of a complete photovoltaic conversion system, the issues belong primarily to the field of engineering and are less related to the science and technology of semiconductors.

The improvement of the performance of existing solar cells or the development of new semiconductors requires the participation of scientists from different areas, such as electrical engineering, physics, chemistry, and material sciences. Traditionally, the scientific issues and the methods of approaching and solving them are different in these areas, which can hamper the communication and the mutual understanding of researchers with different backgrounds who work with photovoltaic materials. It is thus the intention of the book to cover the broad range of fundamental problems and the relationship between the physics, material science, and technology aspects of solar cell development. The book is primarily intended for scientists and engineers who are working in the field or are attracted by it, but is also suitable for students with some background in semiconductor physics and materials science.

I would like to acknowledge a number of people who have stimulated my interest in the field of solar cells over the last few years. In particular, I wish to thank Ingo Schwirtlich and Wolfgang Koch and a large number of scientists from the Bayer Co., too numerous to mention individually. Special thanks are also due to Ted Massalski and Peter Haasen who encouraged me in this venture.

The content of the book has been the subject of graduate courses over several years, and I would like to thank the many students who participated in the interesting discussions and made suggestions for improving the presentation of the subject.

During the fairly intense period in which the book was written, I received much encouragement and support from my wife Kerrin.

Finally, I would like to thank Mrs. Pamela Ahl and Mr. Mark Walsh from Artech House for their support and critical suggestions during the preparation of the book.

Hans J. Möller
Cleveland and Hamburg, September 1992

Chapter 1
Introduction

1.1 HISTORICAL BACKGROUND

The conversion of sunlight directly into electricity using the photovoltaic properties of suitable materials is the most elegant energy conversion process. A laboratory curiosity for more than a hundred years, solar cell technology has seen enormous development during the last three decades, initially in providing electrical power for spacecraft, and more recently for terrestrial systems. The driving force for the technological development is the realization that the traditional fossil energy resources, coal, oil, and gas, are not only limited, but they also contribute to unpredictable and possibly irreversible climate changes in the near future through the emission of carbon dioxide. The increasing concern for environmental pollution problems in industrialized countries has also discredited nuclear power as a long-term alternative energy concept. From this perspective, the use of sunlight offers a conceivable alternative to the worldwide energy problems.

Solar cells use the photovoltaic effect for their operation, which was discovered in 1839 by Becquerel, who studied the behavior of solids in electrolyte solutions. He observed that when metal plates immersed in a suitable electrolyte were exposed to sunlight, a small voltage and current were produced. The first solid-state materials that showed a significant light-dependent voltage between two contacts were selenium in 1876 and later cuprous oxide [1–4], which indicated already that semiconductors would eventually be the most promising class of materials for photovoltaic energy conversion. Technological development began with the development of a diffused silicon *pn*-junction in 1954 [5], a forerunner of the present silicon solar cell, which converted light into electricity with reasonable efficiency. Almost coincident with the beginning of silicon solar cell technology was the first development of cuprous sulfide/cadmium sulfide heterojunctions, which served as the basis for intense research on thin-film solar cell devices [6]. The initial applications were on a small scale, and the first real impact of solar cells was only realized with the advent of space exploration. The demand for a reliable, longlasting power source was the major reason for the application of solar cells, and by 1958 the first silicon solar cells were used in spacecraft.

Interest arose in solar cells as an alternative energy source for terrestrial applications in the mid-1970s after the political crisis in the Middle East and the oil embargo, and the realization that fossil fuel sources were limited. Increased research efforts resulted not only in further improvement of the efficiency of silicon solar cells and a considerable reduction of energy costs, but also in the development of new photovoltaic materials and devices. The increase in production volume and new cost-efficient solar cell technologies rendered this achievement possible. The cost target for electricity from a photovoltaic plant operating for 30 years was established in 1986 to be equal to about 0.06 US$/kWh. It was estimated that this requires module efficiencies in the range of 15% to 20% for a flat plate panel system and 25% to 30% for a system operating under concentrated sunlight. These correspond to module area costs of $45 to $80/m^2 and $60 to $100/m^2, respectively [7]. In a recent German study (1987), energy costs of about 0.20 US$/kWh (module area cost of about $500/m^2) were estimated for power plants in Europe with a

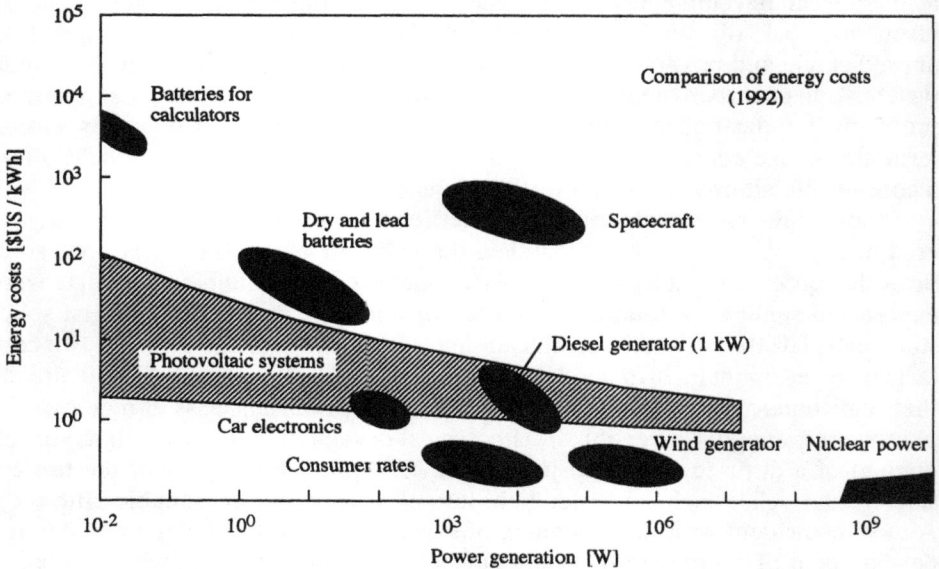

Figure 1.1 Comparison of energy costs and generated power for photovoltaic systems and various other electrical power sources.

capacity larger than 10 MW and a plant operating lifetime of 30 years (Figure 1.1). The calculation was based on the performance characteristics of polycrystalline silicon solar cells available today [8].

Since the beginning of the 1980s, new solar cell materials and innovative device concepts have been evaluated at the pilot production stage, and a number of promising options for future development are currently available. However, the expected breakthrough of solar technology as a competitive energy source could not be achieved, mainly because the required efficiencies could so far only be obtained for laboratory solar cells. Part of the barrier to more widespread use of photovoltaic technology is the inexpensiveness of fossil and nuclear power energy.

A comprehensive comparison of an alternative energy system in general is not trivial and has to take into account the full costs to society of the energy system, such as environmental and climatic effects, hidden subsidies, or production of intermediate goods. These external costs are only beginning to be considered, but already have some impact on the energy policies of several European countries. Though solar cell technology is still not mature and expectations are that it will not contribute more than 5% to 7% to the total energy consumption over the next two decades, it offers a realistic and promising energy technology for the future.

A photovoltaic system consists of several components, from the solar cell to a module, the array of modules, and the entire plant. If one examines the individual components, the major hurdle to technological feasibility remains the basic component, the solar cell itself. Therefore, research into photovoltaic alternatives is imperative to make the technology competitive. This includes the development of low-cost techniques, higher efficiency cells using new materials and cell concepts, and thin films that require less material. Inherent to any option is successful tailoring of the semiconductor material and the control of the electro-optical properties during each processing step.

1.2 TECHNICAL CONCEPTS FOR SOLAR ENERGY CONVERSION

Semiconductors use the fundamental physical process that a photon that penetrates into the crystal and is absorbed can generate electron-hole pairs. Because of their opposite charges, electrons and holes can be separated by an internal electrical field and collected at two contacts, which gives rise to a photocurrent if the two contacts are connected externally. In semiconductors, internal electric fields occur in connection with space charges at junctions, and a variety of technological concepts are used to produce a built-in electrical field. The most widely used device principle is the operation of a solar cell as a *pn*-homojunction diode. Other concepts are heterojunction devices, metal-semiconductor junctions (Schottky diodes) or *metal-insulator semiconductor* (MIS) structures. These are conventional device structures which are also used for other microelectronic devices, and the development of solar cells thus benefits from the rapid advancement of semiconductor technology.

Alternative solar cell concepts are also pursued, such as the development of the photoelectrochemical cells, where the internal electrical field is created at the surface of a semiconductor brought into contact with an electrolyte. This technology may become important in connection with the hydrogen technology, the photoelectrolytic decomposition of water. The electrochemical cell and other unconventional concepts, however, will not be considered in the following.

Silicon is the most extensively studied semiconductor and was the first commercially used solar cell material. Monocrystalline silicon as an indirect semiconductor has a rather low absorption coefficient, which requires a relatively large volume to absorb the light compared to the width of the space charge region where the charge carriers are separated. Recombination mechanisms that limit the lifetime of electrons or holes before they can diffuse to the junction region are therefore of great importance. High purity silicon is necessary for maximum-efficiency cells, and after 40 years of solar cell research, the optimization of material properties and of the cell design has resulted in efficiencies of 23% for silicon under normal sunlight [9]. Such high efficiencies, however, are only obtained in laboratory experiments on small cells with areas of a few square centimeters, and a large gap still exists between the lower efficiencies of about 10% to 13% for commercial cells in modules. Another drawback of high-efficiency concepts is that the production costs for the solar cells are still too high, mainly due to the more sophisticated processing technology, so that, despite the gain in efficiency, the energy costs of photovoltaic power probably cannot become competitive in the near future.

The cost problem for monocrystalline silicon has led to the development of low-cost technologies at the expense of the efficiency of the solar cell. So far, inexpensive processes only allow the production of low-purity, polycrystalline ("terrestrial solar grade") silicon. Nonetheless, this technology has been established commercially in particular areas of the energy market and has reduced the energy price to about $2/kWh, mainly due to the scaling-up of the production. Most commonly, polycrystalline silicon is produced by casting and solidification in a crucible, which produces blocks (ingots) of about 100 kg. This technique requires the subsequent sawing of the ingots into wafers of 10 by 10 cm^2 and a thickness of about 400 μm, and results in the waste of about 50% of the silicon. Efficiencies of about 17% have been obtained with cast polycrystalline silicon wafers by using high-efficiency concepts for the solar cell design [10].

A further reduction of costs is expected from technologies that avoid the slicing step by growing the silicon directly in the final shape in the form of sheets or ribbons. A number of continuous ribbon growth techniques are currently developed and will allow the mass production of solar cells in the future [11–15]. However, tremendous material problems still have to be solved before solar cell efficiencies can be achieved that are sufficient for a competitive commercial product.

Nonetheless, the technologies that have been developed for monocrystalline and polycrystalline silicon are the most advanced so far and continue to be improved.

Parallel with the development of solar cell technology based on crystalline silicon, other concepts are pursued that try to avoid some of the fundamental disadvantages of silicon. A wide variety of compound semiconductors, such as GaAs, CdTe, InP, and CdS, have a band gap that is more suitable to the sunlight spectrum and therefore has the potential for higher efficiency. In fact, recent developments of GaAs solar cells have demonstrated efficiencies of about 25% [16–21]. Despite this success, most of the semiconductor compounds cannot yet be produced inexpensively enough or with sufficient quality to be competitive in the near future.

Most of the compounds are direct semiconductors with high absorption coefficients; therefore, only thin layers of a few micrometers' thickness are necessary to absorb the light. The electron-hole pairs can thus be generated close to or in the junction region, and the loss of charge carriers due to recombination is greatly reduced. Because of the technological developments in the microelectronics industry, a number of thin-film deposition techniques such as *chemical vapor deposition* (CVD), *metal-organic chemical vapor deposition* (MOCVD), or *molecular beam epitaxy* (MBE) are available, which can be used for the growth of high-quality thin-film solar cells. Since the growth of high-quality films is a rather time-consuming and expensive technology, its application to large-scale solar cell production is still limited. In addition, some of the compounds contain heavy metals or toxic elements, which may prevent their widespread use because of environmental concerns. Nonetheless, thin-film solar cells have the advantage that they can be deposited (at least in principle) on different substrates (e.g., polymers, steel, or glass) and only a fraction of material is necessary compared to bulk solar cells. Since many of the processing techniques are also used in microelectronic device production, the integration of other electronic components is also feasible. Therefore, thin film solar cells may have considerable advantages in the future if the technology and the material properties can be further improved.

One strategy to cut down the production costs is to develop or apply less expensive deposition techniques and obtain thin films of lower quality. One of the most advanced technologies in this direction is the deposition of amorphous silicon. Compared to monocrystalline silicon, it has a different band structure and a higher absorption so that it can be used as a thin-film material. The photovoltaic properties have been explored for more than a decade now, and amorphous silicon solar cells are already used commercially in small electronic devices such as watches and calculators. Despite this success, fundamental problems with regard to the improvement of the efficiency and the long-term stability of the material have not yet been solved, and they hamper utilization on a larger scale.

A wide range of new semiconductor materials for bulk and thin-film devices have been opened up with the exploration of ternary and quaternary compounds. The most promising candidate is the compound $CuInSe_2$, which has the highest reported absorption coefficient and shows good photovoltaic properties as a polycrystalline thin film [22]. This offers a number of less expensive deposition techniques, such as vacuum evaporation, spraying methods, or electrodeposition. Efficiencies of about 12% have already been reported for $CuInSe_2$ solar cells [23], and the theoretical limit is estimated to be about 20%.

None of the photovoltaic semiconductors has a band gap that can utilize the entire spectral distribution of sunlight. The selection of the optimum band gap for the solar cell material can be considered a compromise between choosing a band gap wide enough so that not too many electrons are wasted, and yet narrow enough so that enough photons can create electron-hole pairs. The strategy to overcome the problem is to use semiconductors with different band gaps optimized for different wavelength ranges of the spectrum. Two concepts for directing the sunlight with different wavelengths have been developed: spectrum splitting and the tandem cell approach. In the first case, the sunlight is directed on the appropriate cell by spectrally sensitive mirrors, whereas, in the second case, two or more different semiconductors are stacked upon each other. For instance, a wider band gap material on top absorbs the blue part of the sun spectrum, whereas the red part with longer wavelengths passes through and is absorbed by the lower cell with a smaller band gap. This tandem concept has proven to be very efficient in some cases and has recently yielded the record efficiency of 30% for a GaAs-Si tandem cell. This strategy also offers a large number of cell concepts for multinary compound semiconductors where band gaps can be modified by the composition.

Another possible approach to reduce the costs of photovoltaic systems is to concentrate the sunlight on the active area of the solar cell. In addition, this approach requires concentrating elements such as lenses or mirrors and, mostly, sun tracking systems. The higher the concentration ratio, the smaller is the range of angles of light rays that the system can accept. For ratios greater than 10, only direct sunlight can be used, and the sun must be tracked in its path across the sky. Maximum possible concentration ratios of about 45,000 are theoretically possible, but concentrating the sunlight will also increase the operating temperature, which decreases the efficiency or requires an additional cooling system.

The research in the last decade has greatly enhanced knowledge of the structural, electronic, and technological aspects of semiconductors, mainly because of their utilization in microelectronic applications. Further improvements of photovoltaic semiconductors can certainly be achieved if these advances are used and a better understanding of the electronic properties of the solar cell materials is obtained. The main emphasis in this book will therefore be to review the current understanding of electronic properties of semiconductors relevant to photovoltaic applications.

An introduction to the operation principles of solar cells and the different concepts can be found in Hovel [24] and Green [25,26]. Recent developments are given in the latest proceedings of the regular Photovoltaic Specialists conferences, the European Solar Energy conferences, or other solar cell conferences. Many of the advances and improvements of solar cells during the last decade have become possible because of innovative cell designs and the utilization of new process technologies. The material aspects of solar cells have been discussed by Bachmann [27] and Möller [28], and Pulfrey [29] and Johnston [30] have reviewed the photovoltaic power generation. Particular topics have also been reviewed in several papers and books and will be cited in later chapters.

REFERENCES

[1] Smith, W., *Nature*, Vol. 20, Feb. 1973.
[2] Fritts, C. E., *Lumiere Electr.*, Vol. 15, 1985, p. 226.
[3] Schottky, W., and W. Deutschmann, *Phys. Z.*, Vol. 30, 1928, p. 839.
[4] Schottky, W., *Z. Phys.*, Vol. 11, 1930, p. 460.
[5] Chapin, D. M., C. S. Fuller, and G. L. Pearson, *J. Appl. Phys.*, Vol. 25, 1954, p. 676.
[6] Raynolds, D. C., G. Leies, L. L. Antes, and R. E. Marburger, *Phys. Rev.*, Vol. 96, 1954, p. 533.
[7] Kazmerski, L. L., *Intern. Mat. Rev.*, Vol. 34, 1989, p. 185.
[8] Wagner, P,. *Microelectronics Journal,* Vol. 19, 1988, p. 37.
[9] Blakers, A. W., A. Wang, A. M. Milne, J. Zhao, and M. A. Green, *Appl. Phys. Letters,* Vol. 55, 1989, p. 1363.
[10] Green, M. A., A. W. Blakers, and S. R. Wenham, *Proc. 9th E.C. Photovolt. Solar Energy Conf.,* Dordrecht: Kluwer Academic Publ., 1989, p. 301.
[11] Surek, T., and B. Chalmers, *J. Cryst. Growth*, Vol. 29, 1975, p. 1.
[12] Ravi, K. V., R. C. Gonsiorawski, and A. R. Chaudhuri, *Proc. 18th IEEE Photovolt. Specialists Conf.*, New York: IEEE, 1985, p. 1222.
[13] Ciszek, T. F., and J. L. Hurd, *Proc. Symp. on Electr. and Opt. Properties of Polycrystalline or Impure Semiconductors and Novel Silicon Growth Methods*, K. V. Ravi and B. O'Mara, eds., Princeton: Electrochemical Soc., 1980, p. 213.
[14] Wald, F.V., *Crystals*, Vol. 5, Berlin: Springer, 1981, p. 147.
[15] Meier, D. L., J. S. Spitznagel, and R. B. Cambell, *Proc. 20th IEEE Photovolt. Specialists Conf,* New York: IEEE, 1988, p. 415.
[16] Kuribayashi, K., H. Matsumoto, H. Uda, Y. Komatsu, A. Nakano, and S. Ikegami, *Jpn. J. Appl. Phys.*, Vol. 22, 1983, p. 1828.
[17] Tyan, Y. S., S. Vazan, and T. S. Barge, *Proc. 17th Photovolt. Specialists Conf,* New York: IEEE, 1984, p. 840.
[18] Prince, M. B., *Proc. 8th E.C. Photovolt. Solar Energy Conf,* Dordrecht: Kluwer Academic Publ., 1988, p. 1632.
[19] MacMillan, H. F., N. R. Kaminar, M. S. Kuryla, M. J. Ladle, D. D Liu, and G. G. Virshup, *Proc. 20th IEEE Photovolt. Specialists Conf.*, New York: IEEE, 1988, p. 462.
[20] Weinberg, I., C. K. Swartz, R. E. Hart, and T. J. Coutts, *Proc. 20th IEEE Photovolt. Specialists Conf.*, New York: IEEE, 1988, p. 893.
[21] Coutts, T. J., and M. Yamaguchi, *Current Topics in Photovoltaics*, Vol. 3, T. J. Coutts, and J. D. Meakin, eds., New York: Academic Press, 1988, p. 79.
[22] Kazmerski, L. L., and S. Wagner, *Current Topics in Photovoltaics*, T. J. Coutts and J. D. Meakin, eds., New York: Academic Press, 1985, p. 41.

[23] Morel, D., *Solar Cells*, Vol. 24, 1988, p. 157.

[24] Hovel, H. J., *Semiconductors and Semimetals*, Vol. 11, R. K. Willardson and A. C. Beer, eds., New York: Academic Press, 1975.

[25] Green, M. A., *Solar Cells*, Englewood Cliffs, NJ: Prentice Hall, 1982.

[26] Green, M. A., *High Efficiency Silicon Solar Cells*, Aedermannsdorf, Switzerland: Trans. Tech. Publ., 1987.

[27] Bachmann, K. J., *Current Topics in Material Science*, Vol. 3, E. Kaldis, ed., Amsterdam: North Holland, 1979.

[28] Möller, H. J., "Semiconductors for Solar Cell Applications," *Progress in Materials Science*, Vol. 35, 1991, p. 205.

[29] Pulfrey, D. L., *Photovoltaic Power Generation*, New York: Van Nostrand Reinhold, 1978.

[30] Johnston, W. D., *Solar Voltaic Cells*, New York: Dekker, 1980.

Chapter 2
Physical Principles of Photovoltaic Energy Conversion

In this chapter, the fundamental physical principles of a typical *pn*-junction solar cell will be described and the electrical parameters that characterize the performance are introduced. The physical processes are light absorption, minority carrier transport, and recombination mechanisms, which are determined by the electro-optical properties of the material. The corresponding material parameters and the technical design of the solar cell finally determine and limit the efficiency of the energy conversion process. The relationship between material properties and the important electrical parameters will be discussed. To a large extent, the performance of a solar cell can also be improved by the technical cell design. Though this aspect is not the main emphasis here, a brief summary of different cell concepts will be given. For a more comprehensive discussion, see [1–4]. Solar cells are devices that are optimized with respect to sunlight, and therefore the solar spectrum will be presented first.

2.1 SOLAR SPECTRUM

The spectral distribution of the radiation emitted from the sun is determined by the temperature of the surface (photosphere) of the sun, which is about 6000K. The wavelength distribution of the sunlight (power per unit area and per unit wavelength) follows approximately the radiation distribution of a black body at this temperature, as can be seen in Figure 2.1. The deviations at certain wavelengths are due to absorption effects in the sun's atmosphere. For the comparison of the conversion efficiency of different solar cells, the knowledge of the exact spectral distribution is important, since they can respond differently to different wavelengths. The total energy per unit area integrated over the entire spectrum and measured outside the atmosphere perpendicular to the direction of the sun is

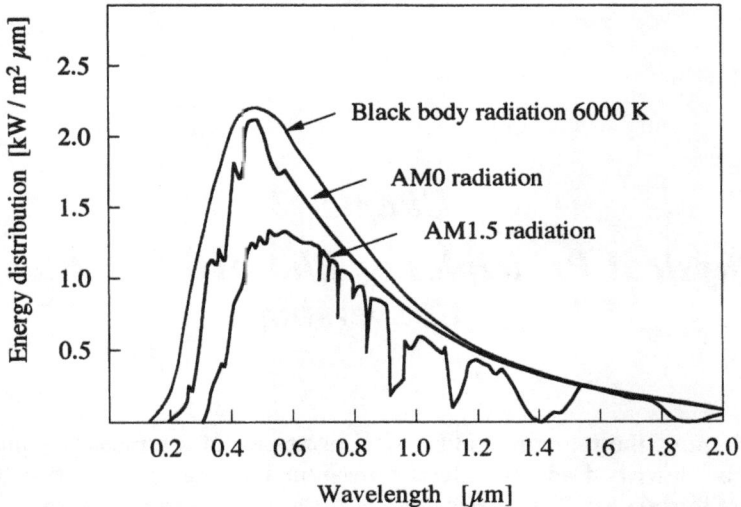

Figure 2.1 Spectral distribution of sunlight. Shown are the radiation outside the earth's atmosphere (AM0) and at the surface (AM1.5). The dashed line indicates the radiation distribution expected from the sun if it were a black body at a temperature of 6000K [7–9].

essentially constant. This radiation power is referred to as the *solar constant* or *air mass zero* (AM0) radiation. Measurements taken at high altitudes have yielded the currently accepted average value of 1.353 kW/m^2 [5,6].

The spectral distribution is changed considerably when the sunlight penetrates through the earth's atmosphere. Even for a clear sky the light intensity is attenuated by at least 30% because of scattering at molecules, aerosols, and dust particles, and absorption by the atmosphere's constituent gases, such as water vapor, ozone, or carbon monoxide. The attenuation mechanisms are wavelength-dependent, which explains the strong absorption bands in the spectral distribution measured at the earth's surface, given in Figure 2.1 [7–9]. The measurement of the radiant power (per unit area) is further complicated by the fact that the degree of attenuation is highly variable because of the constantly changing position of the sun and the corresponding change of the light path through the atmosphere. The ratio of the actual path length to its minimum value has been defined when the sun is directly overhead as *optical air mass* and the corresponding radiation as *air mass one* (AM1) radiation. For any given angle θ with respect to the overhead position, the atmospheric path length h relative to the minimum path length h_0 when the sun is directly overhead is given by $h = h_0/\cos \theta$. The optical air mass is thus increased by the factor (or air mass number) $m = 1/\cos \theta$ and takes the value AMm. It is evident that terrestrial sunlight varies greatly both in intensity and spectral distribution. Therefore it became necessary to define a terrestrial standard that allows a meaningful comparison of different solar cells tested at different locations. In the U.S. photovoltaic program, the spectral distribution for an AM1.5 radiation with a total power density of 0.844 kW/m^2 (for θ = 45 deg) has been adopted as a standard. Tabulated values of the spectral distribution are given in [9] and refer to sunlight in a cloudless sky. The AM1 spectrum is for an angle of θ = 0 deg and has an incident power of about 0.925 kW/m^2. The solar spectrum has contributions from direct and diffuse (or scattered) sunlight. The diffusive component can account for about 20% of the total radiation in a clear sky, but this fraction can increase considerably with the cloud coverage and the position of the sun. For instance, the mean daily global irradiation in sunny locations in Europe (Mediterranean countries) ranges from 3.2 to 5.2 kW/m^2 [8]. Since the spectral distribution for diffuse radiation is different from direct sunlight, there exists a great variability both in the intensity and the spectral composition at different locations on the earth (Table 2.1).

This is especially important for photovoltaic systems using concentrator solar cells (see Section 3.6) which can only accept radiation from a limited range of angles and require a tracking system to utilize the direct component of the sunlight. Since the diffusive component is wasted in this case, the design of a photovoltaic system requires detailed records of the solar insolation at the site of installation. Reasonable estimates of the global radiation received annually are available now for most regions of the world, though there are still uncertainties involved that can be caused by local geographical conditions. It is evident from these considerations that an optimized photovoltaic system has to use solar cells that are adjusted to the average spectral distribution of the local sunlight.

Table 2.1
Integrated Solar Energy (per Area) Received Annually in Various Countries. (Maximum values are for the most sunny locations in each country.)

Country	Annual Solar Energy [kW/m^2]
USA	1,100–2,200
Soviet Union	800–1,400
Europe	800–1,700
Japan	1,200–1,400
China	1,100–1,700
Saudi Arabia	2,200

2.2 ABSORPTION OF LIGHT IN SEMICONDUCTORS

The light-generated current I_L is determined by the absorption behavior of the semiconductor. The fraction of incident light $D = 1 - R$ that actually penetrates the absorbing material can be calculated from the complex refraction index $n_c = n - i\kappa$, where κ is the extinction coefficient and the reflectivity R given by

$$R = \frac{(n - 1)^2 + \kappa^2}{(n + 1)^2 + \kappa^2} \tag{2.1}$$

where $n(\lambda)$ and $\kappa(\lambda)$ are functions of the wavelength λ of the incident light. Substituting the appropriate values shows that for many semiconductors a considerable fraction of the light is reflected. It is therefore desirable from the perspective of making an efficient solar cell to decrease R. This can be achieved by antireflection coating or by a textured structure of the surface (Section 2.4.3).

The important process for photovoltaic conversion is the excitation of electrons from the valence into empty states of the conduction band, which can occur if the energy of the incident photons is larger than the band gap energy. The light passing through the material is absorbed then, and the number of generated electron-hole pairs depends on the number of incident photons $S_o(\nu)$ (per unit area, unit time, and unit energy) that can be calculated from the spectral distribution of the sunlight in Figure 2.1. The frequency ν or the photon energy $h\nu$ is related to the wavelength λ by the relation $\lambda[\mu m] = c/\nu = 1.24/h\nu$ [eV] where c is the speed of the light. Inside the crystal the photon flux $S(x, \nu)$ decreases exponentially according to

$$S(x, \nu) = S_0(\nu) \exp(-\alpha x) \quad \text{with} \quad \alpha(\nu) = \frac{4\pi\kappa\nu}{c} \tag{2.2}$$

The absorption coefficient $\alpha(\nu)$ is determined by the absorption process in the semiconductor and can be used to calculate the generation rate $G(x, \nu)$ of electron-hole pairs (per unit time, volume, and energy) at a distance x from the semiconductor surface. The fraction of photons that penetrate into the crystal is given by $S_0(\nu) (1 - R)$; therefore, the number of electron-hole pairs generated per unit time in the volume between x and $x + \Delta x$ can be calculated from the derivative of (2.2) with respect to x:

$$G(x, \nu) = \beta(\nu)\alpha(\nu)S_0(\nu)(1 - R) \exp(-\alpha(\nu)x) \tag{2.3}$$

The quantum efficiency $\beta(\nu)$ (of the internal photoeffect) takes into account that only a fraction of the absorbed photon energy generates electron-hole pairs. For many compound semiconductors it is observed that $\beta(\nu) \ll 1$ near the absorption edge. This is due to the formation of excitons or bound electron-hole pairs, which carry no charge and do not contribute to the conductivity. However, for the elemental semiconductors germanium and particularly silicon, and some III-V compounds, values of $\beta(\nu) \approx 1$ are measured at room temperature even near the absorption edge.

For photons with energies higher than the band gap energy, the electrons and holes carry excess (kinetic) energy that will be dissipated to the lattice until they occupy states near the band edges. If the kinetic energy is large enough, more electrons may be generated through impact ionization so that values $\beta(\nu) \gg 1$ can be observed. This process occurs in silicon for photon energies above $h\nu > 3\ E_g$ and is thus not important in the energy range where the visible light is absorbed. Therefore, the kinetic excess energy does not contribute to the photocurrent and is wasted from the perspective of the energy conversion.

The absorption coefficient $\alpha(\nu)$ depends on the band structure of the semiconductor. In direct band gap semiconductors, such as GaAs and many other compound semiconductors, the minimum of the conduction band and the maximum of the valence band occur for the same wave vector in the Brillouin zone, and the most likely transitions are between states close to the wave vector $\mathbf{k} = 0$. A theoretical calculation of the probability for these direct (allowed) transitions gives the following result for the absorption coefficient α_d as a function of the frequency ν [10,11]:

$$\alpha_d(\nu) = \alpha_0 \frac{(h\nu - E_g)^{1/2}}{h\nu} \tag{2.4}$$

E_g is the band gap energy, and α_0 is a constant that is obtained from the calculation, but is usually fitted to experimental data. The approximation is only valid for photon energies near the band edges and unpolarized light. A different energy dependence is obtained for direct *forbidden* transitions, but shall not be considered here (e.g., [11]).

In indirect semiconductors like silicon and germanium, where the minimum and maximum of the conduction and valence band, respectively, occur for different wave vectors, optical transitions between the states close to the band edges require a change in momentum of the electrons $\mathbf{p} = \hbar\mathbf{k}$ and are therefore only possible with the participation of phonons. Analogous to direct band-band transitions, allowed and forbidden indirect transitions have to be distinguished. In the case of the allowed transition, the calculations of the transition probabilities yield for the absorption coefficient

$$\alpha_{in}^{ij}(\nu) = \frac{(h\nu - E_{gi} + h\nu_{phj})^2}{\exp(h\nu_{phj}/KT) - 1} + \frac{(h\nu - E_{gi} - h\nu_{phj})^2}{1 - \exp(-h\nu_{phj}/KT)} \tag{2.5}$$

Since both phonon absorption and emission are possible, the absorption coefficient is the sum of both processes. Contrary to the direct transitions, the absorption coefficient here contains the temperature dependence explicitly. The absorption coefficients in both cases also depend on the temperature through the band gap energies, which usually decrease with increasing temperature.

The fundamental absorption processes for direct and indirect semiconductors described so far occur for photons with energies close to the band gap energy E_g. For larger photon energies, direct and phonon-assisted transitions become possible between other states of the band structure within the Brillouin zone. This may lead to additional features in the absorption spectra if the probability for a transition in a certain region of wave vectors of the band structure is high. Optical transitions within a band can also occur, but their probability is much lower, as might be expected, because the concentrations of electrons (in the conduction band) or unoccupied states (in the valence) are low.

An empirical expression for the absorption coefficient that takes into account indirect and direct transitions takes the form

$$\alpha(\nu) = \alpha_d + \sum_{i,j=1,2} C_{ij} \alpha_{in}^{ij} \tag{2.6}$$

The free parameters in this expression, C_{ij}, E_{gi}, and $h\nu_{phj}$, are usually determined from comparison with experimental measurements. For silicon, for instance, the values of the parameters are given in Table 2.2, which describe the experimental results over the photon range from 1.1 to 4.0 eV and for the temperature range

Table 2.2

Absorption Data for Silicon. (Parameters are defined in (2.5) and (2.6). E_{g1} and E_{g2} are the band gap values at $T = 0K$. The temperature dependence is given by $E_{gi}(T) = E_{gi} - \alpha_1 T^2/(T + \alpha_2)$, with $\alpha_1 = 7.021 \times 10^{-4}$ eV/K and $\alpha_2 = 1108K$.)

Parameter	Value
α_0	1.052×10^6 cm^{-1} eV$^{-1/2}$
C_{11}	1.777×10^3 cm^{-1} eV^{-2}
C_{12}	3.980×10^4 cm^{-1} eV^{-2}
C_{21}	1.292×10^3 cm^{-1} eV^{-2}
C_{22}	2.895×10^4 cm^{-1} eV^{-2}
E_{g1}	1.1557 eV
E_{g2}	2.5 eV
E_d	3.2 eV
$h\nu_{ph1}$	1.827×10^{-2} eV
$h\nu_{ph2}$	5.773×10^{-2} eV

20K to 500K. Some results for silicon and a few other semiconductors are depicted in Figure 2.2.

In many practical cases, the fundamental absorption edge even for a direct band gap semiconductor is not a sharp transition. Absorption can also occur below the band gap energy, and there are several effects that can contribute to electron-hole pair generation at lower energies. For the photovoltaic application, those effects are important that are related to the microstructure and chemistry of the material. Particularly when a new material is investigated, it can be difficult to

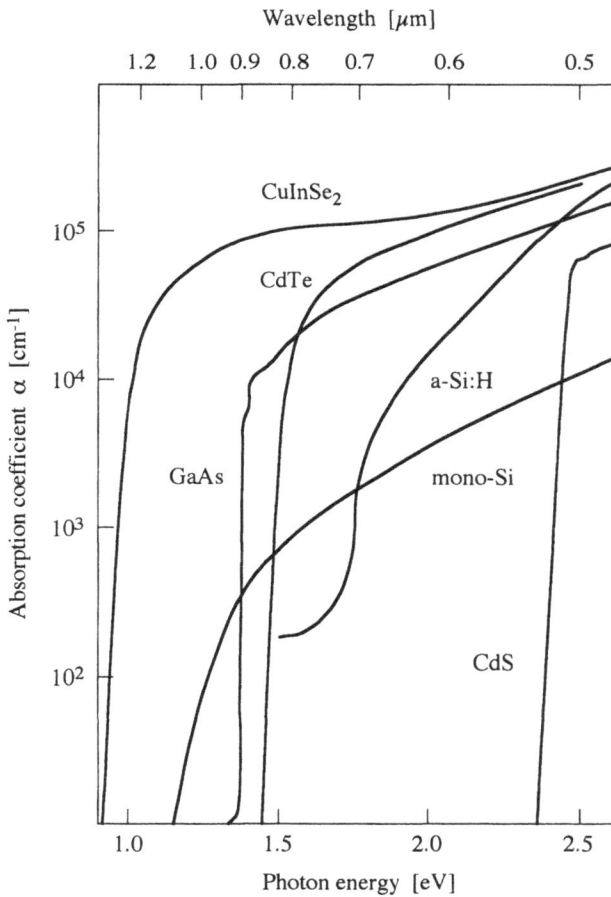

Figure 2.2 Absorption coefficient as a function of the photon energy for silicon and some direct band gap semiconductors.

determine the band gap energy and other details of the band structure precisely, without a simultaneous characterization of the microstructure and chemistry of the material.

Some processes shall be mentioned that change the absorption behavior of a semiconductor with a high defect concentration. Of particular importance are the effects of high dopant concentrations, which occur in the emitter region of *pn* junction solar cells. High defect concentrations also occur in polycrystalline thin films with a high density of grain boundaries (Section 5.5) or in nonstoichiometric ternary semiconductors where intrinsic point defects are incorporated (Chapter 8). The absorption behavior of these semiconductors can be considerably changed compared to the pure single crystal.

High dopant impurity concentrations affect the shape of the band edges of a semiconductor. The distributions of valence and conduction band states can be considered "smeared out" at the band edges (band tails), which effectively reduces the width of the band gap (Figure 2.3(a)). A calculation of the band gap narrowing ΔE_g as a function of the doping concentration N has been given by Lanyon and Tuft [12]:

$$\Delta E_g = \frac{3e^2}{16\pi\epsilon}\left\{\frac{e^2 N}{\epsilon KT}\right\}^{1/2} = \frac{3e^2}{16\pi\epsilon L_D} \tag{2.7}$$

and is in good agreement with experimental data in silicon, such as for $N \geq 10^{17}$ cm^{-3}. (ϵ is the permittivity of the semiconductor and L_D is the screening or Debye length). Band gap narrowing has to be taken into account for a realistic numerical calculation of the performance of a *pn* junction solar cell.

For very high dopant concentrations, the semiconductor finally becomes degenerate, and optical transitions in states below the Fermi energy (for instance, for an *n*-doped crystal) are almost impossible now because they are occupied. Therefore, the optical absorption edge is shifted by the energies $\Delta h\nu = E_F - E_C$ for *n*-type or $\Delta h\nu = E_V - E_F$ for *p*-type crystals, respectively, which is known as the Burstein Moss shift. The optical band gap thus appears larger in a degenerate semiconductor (Figure 2.3(b)).

If a heavily doped semiconductor is, however, partly compensated, the Fermi level lies within the forbidden gap and electron transitions between the band tail states become possible (Figure 2.3(a)). The absorption at photon energies below the band gap energy $h\nu < E_g$ is described then by the Urbach law [10,11]

$$\alpha_U(\nu) = C(\nu)\exp\left\{-\frac{E_g - h\nu}{E_0(T)}\right\} \tag{2.8}$$

where $C(\nu)$ is a slowly varying function of the frequency and $E_0(T)$ is a characteristic

(a)

(b)

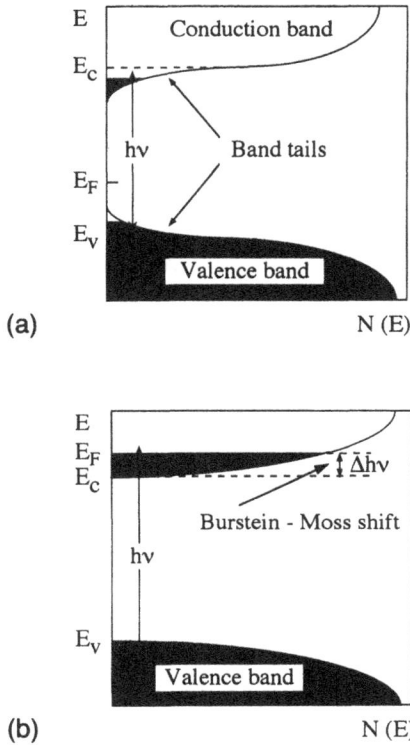

Figure 2.3 Schematic representation of optical transitions between valence and conduction band: (a) compensated semiconductor with band tail states showing possible transitions between tail states below the conduction and valence band edge; (b) highly doped, degenerate semiconductor showing transitions into empty states above the Fermi level and apparent increase of the optical band gap (Burstein-Moss shift).

energy that is proportional to the temperature for sufficiently high temperatures (e.g., in GaAs for $T > 100K$).

As in heavily doped semiconductors, the band structure of crystals with a high concentration of other lattice defects may also be characterized by band tails near the band edges. A typical example is an amorphous semiconductor (e.g., silicon) with a highly distorted lattice structure. Experimental data in fact show that the Urbach law can be used to describe the absorption behavior of some amorphous semiconductors below E_g, at least in the temperature range that is interesting for photovoltaic applications (above room temperature). Since the band gap energy E_g is not clearly defined in these materials, it has to be considered a free parameter which can be adjusted to fit experimental results. As will be discussed in Section 9.1, the optical band gap E_g is generally different from the single crystal value.

The absorption behavior can also be changed by high electrical fields which occur, for instance, in the space charge region of a *pn* junction. This effect becomes particularly important in thin-film solar cells with a *pin* structure (e.g., for amorphous silicon cells) where the major part of the absorbing volume is the intrinsic region. The electrical field in the space charge region and the corresponding distortion of the valence and conduction band allow band-band transitions where the electrons can tunnel through the forbidden gap (Figure 2.4). Optical transitions below the band gap energy become possible then and depend on the strength of the electrical field **E**, which is known as the Franz-Keldysch effect. For the simple case of a direct allowed transition without the participation of phonons, the absorption coefficient for energies $h\nu < E_g$ and $E_g - h\nu \gg E_r$ is given by

$$\alpha_{FK}(\nu) = \frac{AE_r^{3/2}}{\nu(E_g - h\nu)} \exp\left\{-\frac{4}{3}\left(\frac{E_g - h\nu}{E_r}\right)^{3/2}\right\} \tag{2.9}$$

with the critical parameter

$$E_r = \left(\frac{2\pi ehE}{m^{*1/2}}\right)^{2/3}$$

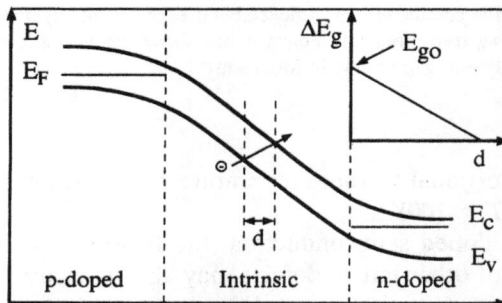

Figure 2.4 Optical transitions by tunneling in the presence of an electrical field (Franz-Keldysch effect) depicted for the electric field region of a *pin* diode. Possible transitions between valence and conduction band at neighboring positions are shown in the schematic band diagram. The insert shows the difference in band gap energies ΔE_g of adjacent positions as a function of the distance d. E_{g0} is the band gap of the semiconductor.

For energies $h\nu > E_g$, the expression for the absorption coefficient as derived in (2.4) is also changed. The complete equations and the constant A and mass m^* are given in [10,11].

2.3 SEMICONDUCTORS UNDER ILLUMINATION

The generation of electrons and holes by illumination and their recombination are nonequilibrium processes in a semiconductor. A consequence is that the Fermi-Dirac statistics, which describes the energy distribution of electrons and holes in thermodynamic equilibrium, cannot be applied anymore. In a doped semiconductor and for small light intensities, one can usually ignore the increase in the majority carrier concentration and mainly consider the change of the minority carriers. For higher intensities, such as in solar cells that operate under concentrated sunlight, the changes of the majority carriers have to be considered as well. In general, the light absorption and electron-hole pair generation is also inhomogeneous in space because of the decreasing intensity when the light penetrates the crystal.

A semiconductor under illumination shows an increased conductivity, which will be described first. The photoconductivity gives useful information about the fundamental optical properties of a semiconductor, such as the recombination processes and the quantum efficiency $\beta(\nu)$. For the photovoltaic conversion, it is necessary to separate the light-generated electrons and holes and collect them at external contacts. This requires an internal electrical field, which can be generated in semiconductors, for instance, by homojunctions and heterojunctions. Since pn junction solar cells are still the most widely used devices for solar energy conversion, the operation of an ideal device and the parameters that are important for the characterization of a solar cell will be discussed in the second part of this section.

2.3.1 Photoconductivity

We assume a homogeneous semiconductor under illumination first. If p_0 and n_0 are the carrier concentrations in thermodynamic equilibrium, which are determined by the doping concentrations of donors or acceptors, the changes in the electron δn and hole concentrations δp can be expressed by

$$\delta p = p - p_0 \qquad \delta n = n - n_0 \qquad (2.10)$$

The carrier concentrations for a non-degenerate semiconductor are given by

$$p = N_V \exp\!\left(\frac{E_V - e\phi - E_{Fp}}{KT}\right) \qquad (2.11a)$$

$$n = N_C \exp\left(-\frac{E_C - e\phi - E_{Fn}}{KT}\right) \tag{2.11b}$$

where the potential ϕ is related to the local electric field by $\mathbf{E} = -\text{grad } \phi$. N_C and N_V are the density of states of the conduction and valence band, respectively, E_{Fn} and E_{Fp} are the quasi-Fermi energies for electrons and holes, n and p are their concentrations, and e is the (positive) electron charge.

The basic equations that describe the flux of electrons and holes in a semiconductor under illumination are the current-density equations

$$\mathbf{J_p} = e(p\mu_p\mathbf{E} - D_p \text{ grad } p) \qquad \mathbf{J_n} = e(n\mu_n\mathbf{E} + D_n \text{ grad } n) \tag{2.12}$$

and the continuity equations

$$\frac{\partial p}{\partial t} = -\frac{1}{e} \text{ div } \mathbf{J_p} + G_p - \frac{\delta p}{\tau_p} \qquad \frac{\partial n}{\partial t} = \frac{1}{e} \text{ div } \mathbf{J_n} + G_n - \frac{\delta n}{\tau_n} \tag{2.13}$$

The diffusion coefficients D_n and D_p are related to the mobility of the carriers μ_n, μ_p by the general Einstein relationship $D = (kT/e)\mu$. For light generated carriers, the generation rates for electrons and holes are equal to $G_n = G_p = G$. If the excess concentrations are low, one can assume in addition that $\delta n = \delta p$ and combine (2.13) and (2.12). For instance, for holes, one obtains

$$\frac{\partial p}{\partial t} = -\mu\mathbf{E} \text{ grad } p + D\Delta p + G - \frac{\delta p}{\tau} \tag{2.14}$$

where the ambipolar diffusion coefficient D and ambipolar mobility μ are given by

$$D = \frac{(n\mu_n D_p + p\mu_p D_n)}{(n\mu_n + p\mu_p)} \qquad \mu = \frac{\mu_n\mu_p(n - p)}{(n\mu_n + p\mu_p)} \tag{2.15}$$

A corresponding equation is obtained for electrons if p is replaced by n. The lifetimes for electrons and holes are assumed equal in this case, $\tau_n = \tau_p = \tau$, and can be calculated, for instance, by the Shockley-Read-Hall statistics (see (4.29) in Section 4.2.2). These equations are required in the intrinsic case and can be simplified in a doped crystal.

Under steady-state conditions and a homogeneous generation of electrons and holes, the time derivative and gradients of p (or n) are zero, and (2.14) leads to the condition $\delta p = G\tau$. For negligible electric fields, the quasi-Fermi energies under illumination can then be calculated from (2.11) and yield

$$E_{Fp} = E_F + KT \ln\left(\frac{G\tau}{p_0} + 1\right) \qquad (2.16a)$$

$$E_{Fn} = E_F - KT \ln\left(\frac{G\tau}{n_0} + 1\right) \qquad (2.16b)$$

Under normal illumination conditions for solar cells (AM 1.5), the product $G\tau$ is usually smaller than the majority carrier concentration; therefore, only the quasi-Fermi energy of the minority carriers is essentially changed. When an external electrical field $\mathbf{E}$ is applied and a current flows, the total current density $\mathbf{J}$ derived from (2.12) is given by

$$\mathbf{J} = \mathbf{J_p} + \mathbf{J_n} = \sigma\mathbf{E} \qquad (2.17)$$

with the conductivity σ

$$\sigma = e(p\mu_p + n\mu_n) \qquad (2.18)$$

Inserting n and p from (2.10), one can separate the conductivity into two terms $\sigma = \sigma_0 + \sigma_{ph}$ with the following expressions for the dark conductivity σ_0 and the photoconductivity σ_{ph}:

$$\sigma_0 = e(p_0\mu_p + n_0\mu_n) \qquad (2.19)$$

$$\sigma_{ph} = e(\mu_p + \mu_n)G\tau \qquad (2.20)$$

In doped crystals, the dark conductivity depends on the concentration and mobility of the majority carriers, whereas the photoconductivity is mainly determined by the lifetime of the minority carriers.

2.3.2 The *pn* Junction

Photovoltaic energy conversion requires the separation of electrons and holes by an internal electrical field. For solar cell semiconductors, the most common device structures are the *pn* junctions, although other concepts, which will be discussed in Chapter 3, are also used. The basic characteristics of a *pn* junction have been presented in many textbooks and shall be discussed only briefly here. For a more detailed description, the reader is referred to the numerous references about the subject [2–5].

Solar cells require a particular *pn*-junction design, which is depicted in the schematic representation in Figure 2.5. It consists of a shallow junction formed near the front surface, a front ohmic contact in the form of stripes and fingers, and a back ohmic contact that covers the entire back surface. In most cases, the *pn*-junction solar cell is treated as a one-dimensional device, though for a realistic calculation of the current-voltage characteristics, the geometry of the particular device has to be taken into account. The internal electric field $\mathbf{E} = -\text{grad } \phi$ leads to an inhomogeneous distribution of electrons and holes, and the calculation of the currents requires the solution of the complete current-density equation (2.12) and continuity equation (2.13). The potential $\phi(\mathbf{r})$ is determined from Poisson's equation,

$$\Delta\phi = -\frac{e}{\epsilon\epsilon_0}(N_D - N_A - n + p) \qquad (2.21)$$

where $N_d(\mathbf{r})$ and $N_a(\mathbf{r})$ are, in general, functions of the position and are usually equal to the concentrations of completely ionized acceptors and donors on each side of the junction. ϵ is the dielectric constant of the material, and ϵ_0 is the permittivity of the vacuum.

For an abrupt *pn* junction with uniform doping concentrations on each side of the junction, the usual approximation is that within a certain width W the semiconductor is completely depleted from charge carriers. The depletion width W derived from (2.21) is then given by

$$W = \sqrt{2\epsilon\epsilon_0 V_B \frac{N_D + N_A}{N_D N_A}} \qquad (2.22)$$

where the internal potential barrier $V_B = -e\phi_B$ is determined by the doping concentrations N_D and N_A on either side of the junction

$$V_B = KT \ln\left(\frac{N_D N_A}{n_i^2}\right) \qquad (2.23)$$

The potential ϕ_B (or diffusion voltage) determines the maximum voltage that can be obtained from an ideal *pn* junction solar cell.

When light is incident on the front surface and penetrates the crystal, the number of electrons and holes generated at a distance x from the surface is given by the generation rate $G(x, \nu)$, which was determined in the previous section

Front surface metal grid

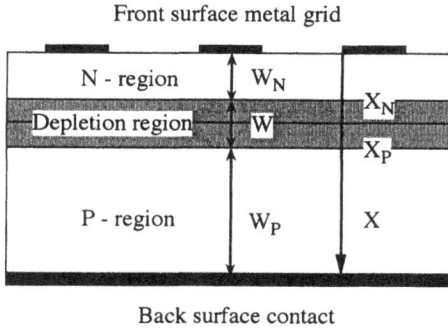

Back surface contact

Figure 2.5 Schematic diagram of a *pn*-junction solar cell, defining basic parameters.

(equation (2.3)). In thermodynamic equilibrium, when no current flows, minority carriers reaching the edges of the depletion region are immediately accelerated by the electric field to the opposite side of the junction. One can approximately assume in this case that the boundary conditions for the minority carrier concentrations at the depletion edges are $\delta p \approx 0$ at $x = x_n$ and $\delta n \approx 0$ at $x = x_p$.

A current flows under illumination when the two sides of the *pn* junction are connected externally. The corresponding voltage drop U across the junction in forward bias direction is determined by the external load resistance. If the current is small, the solution of Poisson's equation can easily be derived in the depletion approximation and is depicted schematically in Figure 2.6. The electric field outside the depletion region remains small compared to the field across the junction and can be ignored; thus, $\mathbf{E} = 0$.

The Fermi energies split into the quasi-Fermi energies for electrons and holes, E_{Fn} and E_{Fp}, respectively, according to (2.16). If the light intensities are low, which is the case under AM 1.5 conditions, for instance, only the quasi-Fermi energies of the minority carriers are substantially changed, whereas the quasi-Fermi energies of the majority carriers remain almost constant and equal to the Fermi energy in equilibrium E_F.

The quasi-Fermi energies vary only slowly as long as the carrier concentrations are large. Therefore, if the recombination in the depletion region is small, the quasi-Fermi energies of the majority carriers can be extended approximately horizontally into the depletion region. Considering the situation depicted in Figure 2.6, the quasi-Fermi energies of the minority carriers at the depletion edges are then given by

$$E_{Fp} = E_F - eU \quad \text{at} \quad x = x_n$$
$$E_{Fn} = E_F + eU \quad \text{at} \quad x = x_p$$

(2.24)

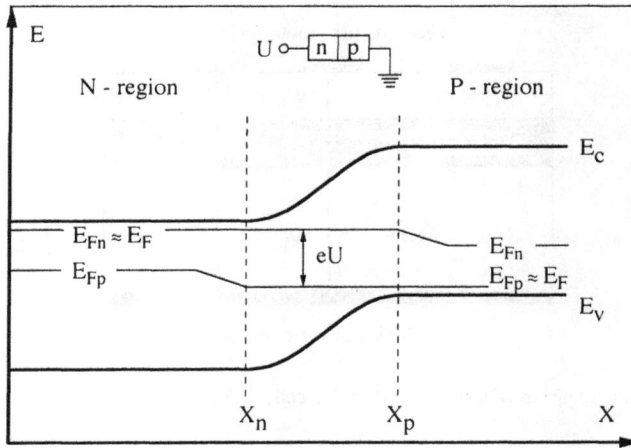

Figure 2.6 Schematic representation of the *pn*-junction band diagram. Under low illumination, the quasi-Fermi energies of majority carriers E_{Fn} and E_{Fnp} in the *n*- or *p*-doped region, respectively, remain almost constant and equal to the Fermi energy in thermal equilibrium E_F, and extend horizontally into the space charge region.

Inserting the expressions into (2.11) and (2.10) yields the excess carrier concentrations δp and δn at the depletion edges under applied bias and illumination:

$$\delta p = \frac{n_i^2}{N_D} \left(\exp \frac{eU}{KT} - 1 \right) \quad \text{at} \quad x = x_n \tag{2.25a}$$

$$\delta n = \frac{n_i^2}{N_A} \left(\exp \frac{eU}{KT} - 1 \right) \quad \text{at} \quad x = x_p \tag{2.25b}$$

The increased carrier concentrations at the edges of the depletion region lead to diffusion currents of the minority carriers (electrons and holes) into the *p*- and *n*-doped region, respectively. If recombination in the depletion region can be ignored, these currents flow unchanged through the depletion region. The total current can thus be determined by calculating the minority currents at the edges of the depletion region.

For *n*- or *p*-type semiconductors, the expressions in (2.12) for the minority currents can be simplified if the electrical field outside the depletion region is negligible. Under steady-state conditions, the time derivatives of *p* and *n* in the continuity equations are zero, and (2.14) outside the depletion region on the n- and p-side, respectively, reduces to

$$D_p \frac{\partial^2 p}{\partial x^2} = -G + \frac{\delta p}{\tau_p} \qquad D_n \frac{\partial^2 n}{\partial x^2} = -G + \frac{\delta n}{\tau_n} \qquad (2.26)$$

Inserting (2.9) for the generation rate G, one obtains, for instance, for the minority holes in the n-type front region the following differential equation:

$$D_p \frac{\partial^2 p}{\partial x^2} + \alpha \beta S_0 (1 - R) \exp(-\alpha x) - \frac{\delta p}{\tau_p} = 0 \qquad (2.27)$$

The corresponding equation for electrons in the p-base region is easily obtained from the second equation of (2.26). The general solution of the differential equation is

$$\delta p = c_1 \cosh \frac{x}{L_p} + c_2 \sinh \frac{x}{L_p} - \frac{\alpha \tau_p e \beta S_0 (1 - R)}{(\alpha L_p)^2 - 1} f(x) \qquad (2.28)$$

with

$$f(x) = e^{-\alpha x} + \frac{1}{2} \{ (\alpha L_p - 1) e^{-(c_3 - x)/L_p} - (\alpha L_p + 1) e^{(c_3 - x)/L_p} \}$$

c_1, c_2, and c_3 are constants which have to be determined from the boundary conditions. A corresponding equation is obtained for the electron concentrations in the p-doped base region.

In thermodynamical equilibrium without an applied bias ($U = 0$), the boundary conditions at the depletion edges are $\delta p \approx 0$ at $x = x_n$ and $\delta n \approx 0$ at $x = x_p$ 0, as discussed above. At the front and back surfaces, the excess concentration δp and δn, respectively, are determined by the surface recombination velocities S_p and S_n:

$$D_p \frac{\partial p}{\partial x} = S_p \delta p \quad \text{at} \quad x = 0 \qquad (2.29a)$$

$$-D_n \frac{\partial n}{\partial x} = S_n \delta n \quad \text{at} \quad x = W_p + x_p \qquad (2.29b)$$

W_p is the width of the p-base neutral region. These conditions are used to determine the constants c_1 and c_2 in the general solution (2.28), whereas c_3 is an arbitrary constant and determined, for instance, from the condition that $f(0) = 0$ for the hole current and $f(W_p + x_p) = 0$ for the electron current.

The minority currents from (2.12) in the field free region are given by

$$J_p = -eD_p \frac{\partial p}{\partial x} \qquad J_n = eD_n \frac{\partial n}{\partial x} \qquad (2.30)$$

and inserting the general solution (2.28), one obtains for the minority current density at $x = x_n$

$$J_p = e\beta S_0 (1 - R) \frac{\alpha L_p}{(\alpha L_p)^2 - 1} \{F_{p1} - F_{p2} \exp(-\alpha W_n)\} \qquad (2.31)$$

W_n is the width of the n-type front region, and the coefficients F_{p1} and F_{p2} are given by

$$F_{p1} = \left[\frac{S_p L_p}{D_p} + \alpha L_p \right] \Bigg/ \left[\frac{S_p L_p}{D_p} \sinh \frac{W_n}{L_p} + \cosh \frac{W_n}{L_p} \right] \qquad (2.32a)$$

$$F_{p2} = \alpha L_p + \left[\frac{S_p L_p}{D_p} \cosh \frac{W_n}{L_p} + \sinh \frac{W_n}{L_p} \right] \Bigg/ \left[\frac{S_p L_p}{D_p} \sinh \frac{W_n}{L_p} + \cosh \frac{W_n}{L_p} \right] \qquad (2.32b)$$

This photocurrent would be collected from the front side of an n-on-p-junction solar cell at a given wavelength, assuming this n-type region to be uniform in doping level, lifetime, and mobility. For the p-type base region, a similar set of equations can be derived. With the same boundary conditions given above for the back side surface at $x = W_p + x_p$ and for the depletion edge on the p-type side at $x = x_p$, the photocurrent collected from the base of the cell at $x = x_p$ is given by

$$J_n = e\beta S_0 (1 - R) \frac{\alpha L_n}{(\alpha L_n)^2 - 1} \exp(-\alpha x_p)\{F_{n1} - F_{n2} \exp(-\alpha W_p)\} \qquad (2.33)$$

with the coefficients given by

$$F_{n1} = \alpha L_n - \left[\frac{S_n L_n}{D_n} \cosh \frac{W_p}{L_n} + \sinh \frac{W_p}{L_n} \right] \Bigg/ \left[\frac{S_n L_n}{D_n} \sinh \frac{W_p}{L_n} + \cosh \frac{W_p}{L_n} \right] \qquad (2.34a)$$

$$F_{n2} = \left[\alpha L_n - \frac{S_n L_n}{D_n} \right] \Bigg/ \left[\frac{S_n L_n}{D_n} \sinh \frac{W_p}{L_n} + \cosh \frac{W_p}{L_n} \right] \qquad (2.34b)$$

Some photocurrent generation also takes place in the depletion region. Since the electric field in this region is high, the generated electrons and holes are accelerated out of the region. If recombination is again ignored, the photocurrent density in this case is equal to the number of photons absorbed:

$$J_{dp} = e\beta S_0(1 - R) \exp(-\alpha x_n)(1 - \exp(-\alpha W)) \tag{2.35}$$

where $W = x_p - x_n$ is the width of the depletion region. If the doping concentrations, lifetimes, and mobilities are a function of the position, analytical expressions cannot be derived anymore for the minority carrier currents and numerical methods must be used.

The derived expressions for the minority currents depend on the frequency of the absorbed light. Therefore, the total photocurrent I_L has to be calculated from the photocurrent density $J_L = J_p + J_n + J_{dp}$ by integrating over the entire solar spectrum, and it is given by

$$I_L = A \int_{\nu_m}^{\infty} (J_p + J_n + J_{dp})d\nu \tag{2.36}$$

where $h\nu_m$ is the smallest photon energy corresponding to the absorption edge of the semiconductor, and A is the active area of the solar cell. The depletion region and the usually much larger volume within the diffusion length are therefore the active collection regions of a pn-junction cell.

Each term in (2.36) is proprotional to the factor $eS_0(1 - R)$; therefore, the photocurrent I_L is proportional to the light intensity that enters the crystal. The optical performance of a solar cell is frequently characterized by the normalized photocurrent $J_L(h\nu)$ as a function of the photon energy $h\nu$ or wavelength λ. The *external spectral response* (or *quantum efficiency*) of the cell is the total photocurrent J_L divided by eS_0 and the *internal spectral response* $SR(h\nu)$ of the cell divided by $eS_0(1 - R)$, respectively:

$$SR(h\nu) = \frac{J_p + J_n + J_{dp}}{eS_0(1 - R)} \tag{2.37}$$

The individual contributions from each of the three regions are shown in Figure 2.7, where the internal spectral response of the cell $SR(h\nu)$ has been calculated for a silicon n-on-p solar cell [3]. In the ideal case, $SR = \beta(\nu)$, this is a step function that equals zero for energies below the band gap energy and unity for larger energies, if $\beta(\nu) = 1$. It can be seen that for small photon energies most of the current is generated in the base region, whereas for energies above 2.5 eV the front region takes over. The recombination velocity at the front surface has a

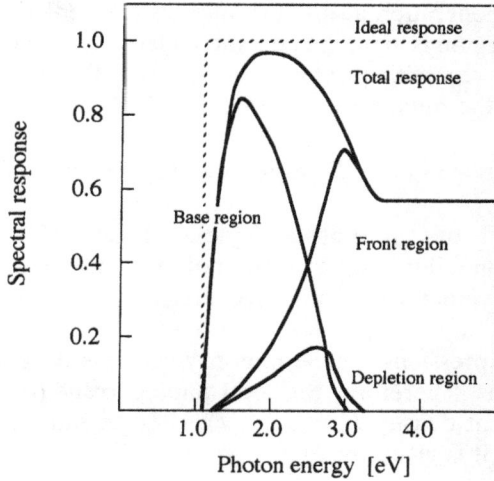

Figure 2.7 Total spectral response of a *pn*-junction solar cell as a function of the photon energy, together with the contributions of the front, base, and depletion regions. Dashed line indicates ideal behavior.

profound effect on the emitter photocurrent and this is the reason why the total current can depart substantially from the idealized step function for high photon energies. The surface recombination velocity is also particularly important in thin-film solar cells, where the light absorption in the bulk is always in proximity to a surface.

The maximum photocurrent that can be generated in a solar cell is given by I_L. A solar cell in an electrical circuit will produce a lower current which is determined by the external load resistance and the corresponding operation point on the current-voltage characteristics of the device. If a current flows, the minority carrier concentrations at the depletion edges are raised and the previous boundary conditions have to be replaced in (2.25) where U is the voltage drop across the junction. Determining again the constants c_1 and c_1 in the general solution (2.28) and solving for the total current I, one obtains the following well-known current-voltage characteristics for an ideal *pn*-junction solar cell [4]:

$$I = I_s \left\{ \exp \frac{eU}{KT} - 1 \right\} - I_L \tag{2.38}$$

where the saturation current is given by (A is the active solar cell area)

$$I_S = eAN_CN_V\left\{\frac{D_nF_p}{L_nN_A} + \frac{D_pF_n}{L_pN_D}\right\}\exp\left(-\frac{E_g}{KT}\right) \tag{2.39}$$

The factors F_n and F_p account for the finite recombination velocity of electrons and holes at the front and back surfaces S_n and S_p. F_p and F_n are given by

$$F_p = \frac{S_n\cosh(W_p/L_n) + (D_n/L_n)\sinh(W_p/L_n)}{S_n\sinh(W_p/L_n) + (D_n/L_n)\cosh(W_p/L_n)} \tag{2.40a}$$

$$F_n = \frac{S_p\cosh(W_n/L_p) + (D_p/L_p)\sinh(W_n/L_p)}{S_p\sinh(W_n/L_p) + (D_p/L_p)\cosh(W_n/L_p)} \tag{2.40b}$$

These equations which describe the current-voltage characteristics of an ideal *pn* junction under illumination have so far been derived under the assumption that recombination in the depletion region can be ignored. In many practical cases, however, recombination in the depletion region cannot be omitted. At forward bias, a recombination current has to be considered in addition to the diffusion current in (2.38).

If the recombination rate R_0 is constant in the depletion region, the recombination current I_{rec} is given by

$$I_{rec} = eA\int_{x_n}^{x_p} R_0(x)dx = eAWR_0 \tag{2.41}$$

where $W = x_p - x_n$ and A are the width and total area of the junction, respectively.

In the simple case that a single trap level in the band gap controls the recombination in the depletion region, one can apply the Shockley-Read-Hall statistics, which will be discussed in Section 4.2. Under low-level injection, the recombination rate R_0 in this case is given by

$$R_0 = \frac{np - n_i^2}{\tau_{p0}(n + n_{T1}) + \tau_{n0}(p + p_{T1})} \tag{2.42a}$$

with the individual lifetimes τ_{n0}, τ_{p0} and the parameters n_{T1} and p_{T1} given by (see also (4.22))

$$n_{T1} = N_C\exp\left(-\frac{E_C - E_T}{KT}\right) \qquad p_{T1} = N_V\exp\left(\frac{E_V - E_T}{KT}\right) \tag{2.42b}$$

Substituting (2.24) into (2.11) yields in the depletion region

$$np = N_C N_V \exp\left\{-\frac{E_g}{KT}\right\} \exp\frac{eU}{KT} = n_i^2 \exp\frac{eU}{KT} \qquad (2.43)$$

For a trap level near the middle of the band gap at $E_T - E_V = (KT \ln(N_V/N_C) + E_g)/2 \approx E_g/2$ one obtains $n_{T1} = p_{T1}$. The maximum of the recombination rate in the depletion region occurs for $n = p$; therefore, inserting (2.42) and (2.43) into (2.41) yields for the recombination current in the case of a near-midgap level:

$$I_{rec} = I_{SR}\left(\exp\frac{eU}{2KT} - 1\right) \qquad (2.44)$$

with

$$I_{SR} = \frac{eAWn_i}{2(\tau_{p0} + \tau_{n0})}$$

In general, recombination in the depletion region only partly contributes to the total current. A general expression which describes the current-voltage characteristics of a pn junction is obtained if one assumes the superposition of the two processes and combines the two expressions (2.37) and (2.44) linearly:

$$I = C_1 I_S\left\{\exp\frac{eU}{KT} - 1\right\} + C_2 I_{SR}\left\{\exp\frac{eU}{2KT} - 1\right\} - I_L \qquad (2.45)$$

where C_1 and C_2 are free parameters.

The current-voltage characteristics of the pn junction is further modified because of a parasitic series, R_s, and a shunt resistance, R_{sh}, associated with the solar cell. The bulk resistance of the semiconductor and the resistance of the contacts and interconnections are the origin of the series resistance. The shunt resistance can be caused by lattice defects in the depleted region, such as grain boundaries or large precipitates. Another reason may be leakage currents around the edges of the cell. To evaluate the effect of these parameters, one can describe the current-voltage behavior of the pn junction in the equivalent circuit shown in Figure 2.8 and obtain

$$I = C_1 I_S\left\{\exp\frac{e(U - IR_s)}{KT} - 1\right\} + C_2 I_{SR}\left\{\exp\frac{e(U - IR_s)}{2KT} - 1\right\}$$
$$+ \frac{U - IR_s}{R_{sh}} - I_L \qquad (2.46)$$

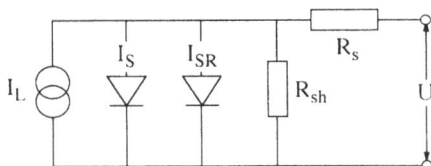

Figure 2.8 Equivalent electrical circuit of a *pn* junction for the two-diode model with diffusion I_S and recombination current I_{SR}, series R_S and shunt R_{Sh} resistance, and light-generated current I_L.

In many practical cases, this *two diode model* gives a good description of the current-voltage behavior of real *pn* junction solar cells. The parameters C_1 and C_2 are then fitted to experimental curves.

2.4 FUNDAMENTAL SOLAR CELL PARAMETERS

The performance of a solar cell under illumination can be completely described by the current-voltage dependence. For practical purposes, however, it is sufficient in many cases to characterize the current-voltage characteristics with a few parameters only. If we consider a typical current-voltage curve of a *pn*-junction diode in the dark and under illumination as depicted in Figure 2.9, we can define three parameters that give a rather complete description of the electrical behavior.

The first one is the *short-circuit current* I_{sc}, which is obtained for $U = 0$. Considering the analytical expression (2.46) for the current-voltage curve, it is evident that I_{sc} is equal to the light-generated current, $I_{sc} = I_L$, if the series resistance R_S is zero. A finite series resistance R_S reduces the short-circuit current.

The second parameter is the *open-circuit voltage* V_{oc}, which is obtained for $I = 0$. In the general two-diode model (equation (2.46)), this leads to a transcendental equation which can only be solved numerically. Only in the ideal case where $I_{SR} = R_S = 0$ and $R_{sh} = \infty$ an analytical expression can be derived:

$$V_{oc} = \frac{KT}{e} \ln \left\{ \frac{I_L}{I_S} + 1 \right\} \tag{2.47}$$

V_{oc} is determined by the ratio I_L/I_S and thus by the absorption and light-generation processes and the efficiency with which the charge carriers reach the depletion region.

The performance of the solar cell is eventually determined by the fraction of the total power of incident light that can be converted into electrical power. Under

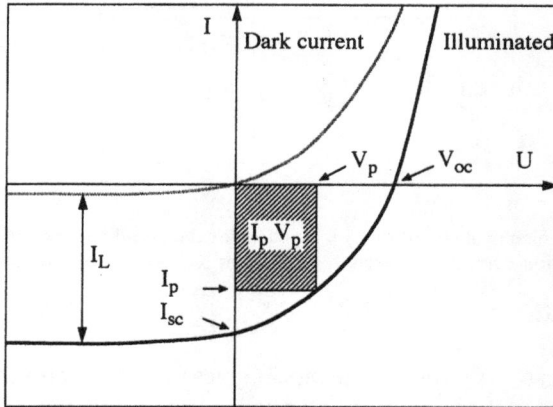

Figure 2.9 Current-voltage characteristics of a *pn*-junction diode in the dark and when illuminated, defining the basic parameters: light-generated current (I_L), short-circuit current (I_{sc}), and open-circuit voltage (V_{oc}). *Fill factor* (*FF*) is defined by the maximum rectangle given by V_{mp} and I_{mp}.

illumination, the junction is forward biased and the external load resistance determines an operating point on the current-voltage curve. The electrical power output $P = IV$ is equal to the area of the rectangle that is defined by the corresponding values V_p and I_p. For simplicity, the ideal diode behavior (equation (2.38)) shall be considered again, which yields for P

$$P = I_p \frac{KT}{e} \ln\left(1 + \frac{I_p + I_L}{I_s}\right) \tag{2.48}$$

In general, the solar cell will be operated under conditions that give the maximum power output. The maximum possible area $P_{max} = V_{mp}I_{mp}$ for a given current-voltage curve determines the fill factor *FF*, which is defined by

$$FF = \frac{V_{mp}I_{mp}}{V_{oc}I_{sc}} \tag{2.49}$$

Evidently, *FF* is larger the more "square-like" the current-voltage curve is. Typically, it has a value of 0.7 to 0.9 for cells with a reasonable efficiency. P_{max} can be calculated from (2.46) and taking the derivative with respect to U. In the ideal case, with $I_{SR} = R_S = 0$ and $R_{sh} = \infty$, the equation yields a relationship for FF which is only dependent on the open-circuit voltage and can be approximated by

$$FF = \frac{eV_{oc}/KT - \ln(0.72 + eV_{oc}/KT)}{1 + eV_{oc}/KT} \qquad (2.50)$$

In general, numerical methods have to be applied to determine the fill factor FF. The three parameters V_{oc}, I_{sc}, and FF are sufficient to calculate the *energy-conversion efficiency* η of the solar cell, which is defined by

$$\eta = \frac{V_{mp}I_{mp}}{P_{in}} = \frac{FFV_{oc}I_{sc}}{P_{in}} \qquad (2.51)$$

where P_{in} is the total power of the incident light. Considering the general expressions for V_{oc} and I_{sc}, the essential material parameters that determine the efficiency of the solar cell are the lifetime and mobility of the minority charge carriers, and the surface recombination velocities. These parameters are not independent from each other and are controlled by physical processes that will be discussed in the following chapters. The results will allow estimation of the efficiency of a solar cell from the perspective of the material properties. Since the technical design of a solar cell device is closely linked to the material parameters, practical solar cells will have to compromise between different aspects and will therefore have efficiencies lower than the ideal values. A discussion of the major limiting factors and basic requirements for a solar cell design will be given in the following section.

2.4.1 Efficiency of an Ideal *pn*-Junction Cell

In this part, the performance of an ideal *pn*-junction solar cell shall be analyzed. The results give an upper limit of the efficiencies that can be obtained for a particular semiconductor. Considering the standard *pn* junction depicted in Figure 2.5, the efficiency is given by (2.51). Three parameters describe the performance: the open-circuit voltage V_{oc}, the short-circuit current I_{sc}, and the fill factor FF. Since the fill factor FF can be expressed by the open-circuit voltage (equation (2.50)), only the ideal limits of V_{oc} and I_{sc} need to be examined.

It is relatively easy to calculate the upper limit of the short-circuit current $I_{sc} = I_L$ for any semiconductor from the light-generated current I_L given in (2.36). For simplicity, we assume a crystal of infinite thickness $W_p = \infty$, no current from the depletion region ($W = 0$), no surface recombination ($S_n = S_p = 0$), and a uniform diffusion length $L_n = L_p = L$. The total current density $J_L = J_n + J_p$ is then given by

$$J_L(\nu) = \frac{e\beta S_0(1 - R)}{\cosh(x_n/L)} \frac{\alpha L}{(\alpha L)^2 - 1} \{\alpha L - e^{-(\alpha L - 1)x_n/L}\} \qquad (2.52)$$

The current depends on the position of the *pn* junction below the surface and assumes a maximum value at $x_n^{\max}$, which can be determined from (2.52) by taking the derivative with respect to x_n. The corresponding current density is then given by

$$J_L^{\max}(\nu) = \frac{e\beta S_0(1 - R)}{\cosh(x_n^{\max}/L)} \quad \text{with} \quad x_n^{\max} \approx L \frac{\ln(\alpha L)}{\alpha L - 1} \qquad (2.53)$$

For a high-quality material, one can assume a large diffusion length L and $\alpha L \gg 1$ over most of the wavelength range of the sun spectrum. One obtains $x_n^{\max} < L$ and $\cosh(x_n^{\max}/L) \approx 1$ so that the current mainly depends on the photon spectrum for the sunlight $S_0(\nu)$. Integrating from the lowest possible photon energy that can generate an electron-hole pair, one obtains the maximum current as a function of

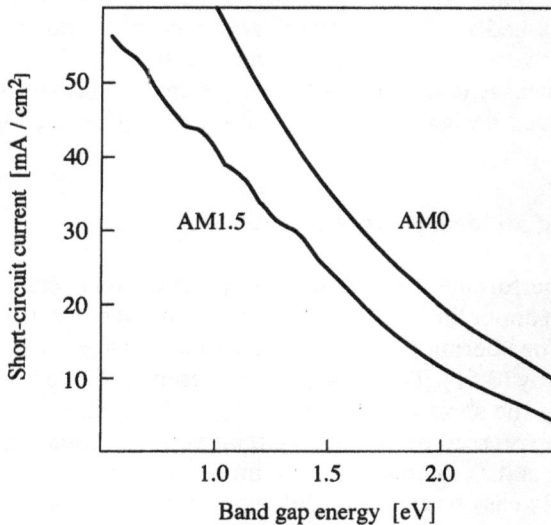

Figure 2.10 Calculated upper limits of the short-circuit current density as a function of the energy band gap of the solar cell semiconductor for a photon flux corresponding to AM0 and AM1.5 [2].

the band gap of the semiconductor. Figure 2.10 shows the result for the spectral distributions AM0 and AM1.5. It is quite evident that the current increases with decreasing band gap, since more photons have enough energy to generate charge carriers [2].

The limitations on the open-circuit voltage of the pn junction solar cell are less clearly defined. For the ideal cell, the expression for V_{oc} has been given in the previous section in (2.47). Inserting the saturation current I_S from (2.39),

$$I_S = I_{S0} \exp\left(-\frac{E_g}{KT}\right) \tag{2.54a}$$

V_{oc} can be approximated for $I_L > I_S$ by

$$V_{oc} = \frac{KT}{e} \ln\left(\frac{I_L}{I_S} + 1\right) \approx \frac{1}{e}\left(E_g + KT \ln \frac{I_L}{I_{S0}}\right) \tag{2.54b}$$

This expression is valid for low injection conditions where $I_L/I_{S0} < 1$ and $V_{oc} < E_g/e$. The prefactor of the saturation current I_{S0} depends on the mobilities and lifetimes of the charge carriers and can be expressed by

$$I_{S0} = eAN_CN_V\left(\frac{KT}{e}\right)^{1/2}\left(\frac{\mu_n^{1/2}}{\tau_n^{1/2}N_A} + \frac{\mu_p^{1/2}}{\tau_p^{1/2}N_D}\right) \tag{2.54c}$$

It is evident that the saturation current I_S needs to be as small as possible for a maximum V_{oc}. The parameter that depends most strongly on the choice of semiconductor material is the band gap energy, and the expression given above shows that with increasing E_g the saturation current decreases and the open-circuit voltage increases. This trend is thus opposite from that for I_{sc}; therefore, a maximum in the efficiency exists. One can assign favorable values to the mobilities and lifetimes and calculate the efficiency as a function of the band gap energy. For instance, the lifetimes of electrons and holes are intrinsically limited by the radiative recombination in the semiconductor (see Section 4.2). Calculations for two different sun spectra are given in Figure 2.11 [4,13] and show that the optimum band gap occurs between 1.4 and 1.6 eV. The near-optimal efficiency for AM1.5 (29%) occurs for GaAs (1.4 eV), whereas the peak efficiency for silicon (1.1 eV) of about 26% is lower than optimum but still relatively high. The corresponding open-circuit voltage is about 0.7 eV and the fill factor 0.84.

There are two fundamental reasons for the limited efficiency of a semiconductor solar cell based on an ideal pn junction device. First, losses occur because the energy of photons above E_g is wasted in the form of heat. Second, (2.54)

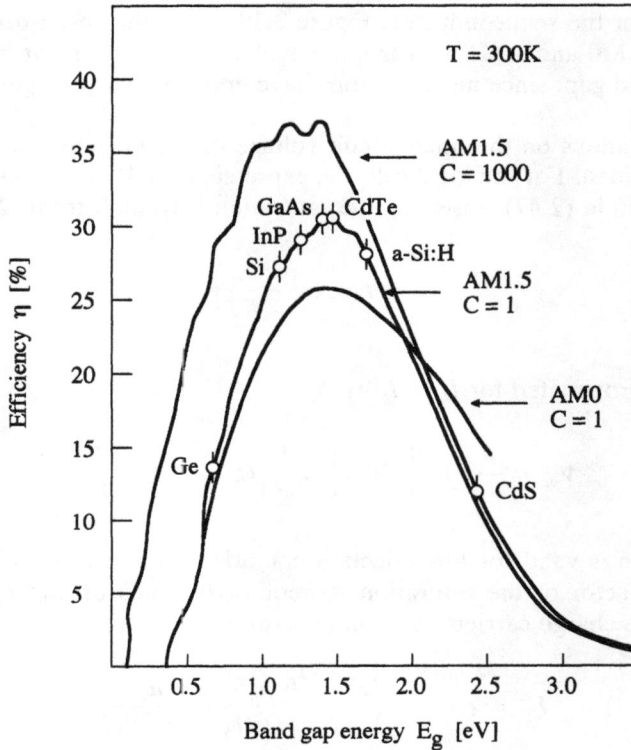

Figure 2.11 Ideal solar cell efficiency as a function of the band gap energy for the spectral distribution AM0 and AM1.5 with a power density of 1 sun, and for AM1.5 with 1000 sun (= 844 kW/m^2).

indicates that the output voltage is smaller than the maximum voltage which corresponds to the band gap energy E_g/e. In general, it can be shown that a *pn* junction is inherently incapable of fully utilizing the maximum voltage by which electron-hole pairs could be separated.

Equation (2.54) also suggests that the open-circuit voltage can be increased by enhancing the intensity of the incident sunlight. The numerical calculation shows that this effect is most effective near the maximum. For instance, for 1000 suns (i.e., 844 kW/m^2), the maximum efficiency increases to 37%. This increase is primarily caused by the open-circuit voltage V_{oc}, since the ratio I_{sc}/P_{in} in (2.51) is essentially independent of the incident power, as can be seen from (2.36). Technically, this concept is realized in concentrator solar cells, which need, however, direct sunlight and a tracking system to follow the path of the sun in the sky.

It is also obvious that the energy of the incident photons can be used more efficiently if the band gap of the semiconductor could be adjusted to the different wavelength ranges. This idea has lead to the development of multiple or tandem cells with different band gap energies, as will be discussed in the Chapter 3.

2.4.2 Design of a Solar Cell

The calculation in the previous chapter gives an upper limit for the efficiency of an ideal pn-junction solar cell. In practice, however, one has to take into account the geometry and finite dimensions of the cell, the surface and bulk recombination, and other factors. In general, the doping profile and the distance of the pn-junction from the front surface determine the open-circuit voltage and the current, which is generated by the absorbed light. Although numerical programs have to be used to optimize all parameters, it is possible to derive some guidelines for the design of the solar cell from the analysis of the basic equations (2.31) and (2.33), which describe the pn-junction characteristics. In the following discussion, the given parameters will refer to a silicon n-on-p-type solar cell, but the conclusions are valid for other semiconductors as well.

In considering the standard cell design depicted in Figure 2.5, the first requirement is that the major fraction of the incident light is absorbed into the crystal. Since the photons with the lowest possible energies also have the lowest absorption coefficients, the crystal thickness is determined by the absorption coefficients near the absorption edge. For silicon, one has $\alpha(\nu_{min}) \approx 10^{-2} - 10^{-3} \ \mu\text{m}^{-1}$, hence, a total thickness of the device of about 100–1000 μm is required to absorb the major fraction of the incident light. Typical silicon solar cells have a thickness of about 300 to 400 μm. The thickness can be reduced if light trapping techniques are applied so that the light is kept inside the crystal by multiple reflection between front and back surfaces. Semiconductors with a direct band gap have much higher absorption coefficients (see, for instance, Figure 2.2), and a thickness of the absorbing layer of about 1 μm is usually sufficient.

Once the total thickness $H = W_n + W + W_p$ is determined, the optimum conditions for the short-circuit current, which is given by (2.36), can be analyzed. The key parameters are the width of the n-type (emitter) and p-type (base) regions, W_n and W_p, the corresponding diffusion lengths, L_n and L_p, and the recombination velocities at the front and back surfaces, S_n and S_p. For simplicity, the width of the depletion region W will be ignored in the following. For a given frequency ν of the incident light and the corresponding absorption coefficient $\alpha(\nu)$, the total current collected from photons with the energy $h\nu$ is given by the expressions (2.31) and (2.33) for the diffusion currents of minority carriers at the edges of the depleted region. Both currents have a maximum value for a certain width of the emitter and base regions, which can be determined by differentiating the expressions with respect to these parameters. The optimum value for the base width depends on the surface recombination and can be approximated for high and low recombination velocities by

$$W_p^{\max} = \frac{L_n}{1 - \alpha L_n} \ln\left\{\frac{2}{1 + \alpha L_n}\right\} \qquad S_n L_n / D_n \gg 1 \qquad (2.55)$$

$$W_p^{\max} = \frac{L_n}{1 - \alpha L_n} \ln\left\{\frac{2}{\alpha L_n (1 + \alpha L_n)}\right\} \qquad S_n L_n / D_n \ll 1 \qquad (2.56)$$

Since the base width corresponds mainly to the total thickness of the cell, as will be shown later, these equations determine the diffusion length of the base region for maximum performance. Considering that light with longer wavelengths (with $\alpha L_n \leq 1$) penetrates deeper into the crystal and assuming that the recombination velocity at the back surface is high, the diffusion length of the base region becomes approximately $L_n \approx 1.4 \, W_p^{\max}$. The total base current in this approximation is given by

$$J_n = e S_0 (1 - R)\alpha L_n \exp(-\alpha(W + W_n)) \qquad S_n L_n / D_n \gg 1 \qquad (2.57)$$

and is proportional to the diffusion length in this region. For maximum performance, the diffusion length has to be slightly higher than the width of the base region, which yields values for the diffusion length in silicon; for instance, $L_n > 300 \, \mu$m. This is a large value and requires a high-quality semiconductor material.

If one takes the maximum base current, the best width for the emitter region can be determined from the optimization of the total current $J_L(\nu)$. In Figure 2.12, the normalized light generated current calculated from (2.37) (internal spectral response SR) is plotted as a function of the normalized emitter width W_n/L_p. The parameter αL_p, which depends on the wavelength, is varied. For low-absorption coefficients ($\alpha L_p < 1$), the optimum value for the emitter width is about twice the diffusion length of the n-type front region, whereas the maximum for high-absorption coefficients ($\alpha L_p > 1$) lies at the front surface. For silicon and many other semiconductors, the blue part of the spectrum is mainly absorbed in the front region. With typical values for the diffusion length in the front region, one obtains $\alpha L_p > 1$; hence, the current from that region decays very rapidly with increasing width of the emitter region. In order to maximize the total current (equation (2.36)), it is evident that the position of the junction should be as close as possible below the front surface so that the blue and violet parts of the sun spectrum can best be utilized. Typically, the junction depth is about 0.2–0.4 μm below the front surface.

Once the geometrical dimensions of the pn junction are specified, the doping levels of emitter and base regions have to be optimized. The design of the front

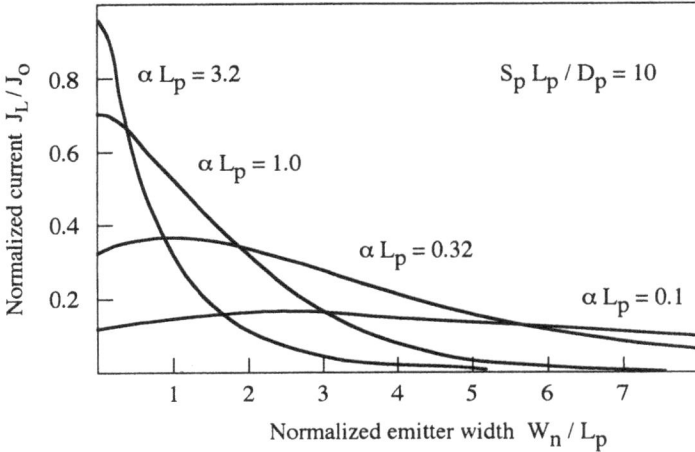

Figure 2.12 Internal spectral response $SR = J_L/J_0$ (normalized light-generated current) as a function of the normalized emitter width W_n/L_p for an n-on-p-base solar cell diode. The calculation is for an optimized base current and a negligible current from the depletion region ($W = 0$). Surface recombination $S_p = S_n$ and diffusion length $L_p = L_n$ are assumed to be equal in the n- and p-regions.

region of a solar cell is partly determined by the condition that the current that is extracted from the cell has to be collected at contacts which cover parts of the front surface of the cell. Whereas the current generally flows perpendicular to the surface, it flows laterally to the contact region in the top layer of the semiconductor (Figure 2.13). Since the series resistance of a solar cell is a critical parameter for the performance, the resistance of the top layer must be kept small. The inverse sheet resistivity of a nonuniformly doped layer is given by

$$\frac{1}{\rho_s} = e \int_0^t \mu_n(x) N_d(x) dx \qquad (2.58)$$

which yields for a uniform doping in the emitter region $\rho_s^{-1} = e\mu_n N_d W_n$. Considering the standard geometry for a contact grid in Figure 2.13, the current I which flows in a unit cell of the size $S \times D$ is zero at the midpoint between grating lines and increases linearly with the distance y; hence, $I = JDy$. The resistive power loss due to the lateral flow of the current is given by

$$P_{loss} = \int_0^S I^2 dR = J^2 D\rho_s \int_0^S y^2 dy = \frac{1}{3} J^2 D\rho_s S^3 \qquad (2.59)$$

Front surface metal grid

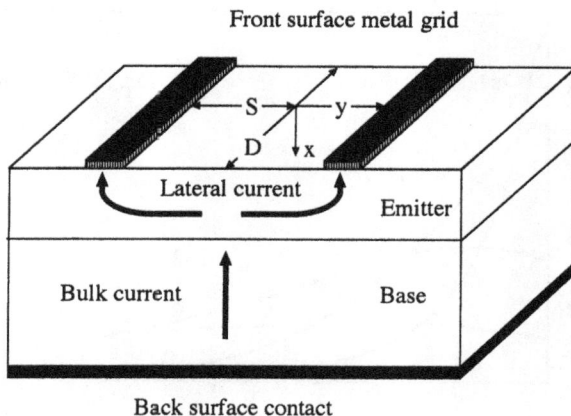

Figure 2.13 Current flow in a standard *pn*-junction solar cell. $2S$ distance between parallel fingers of the contact grid, and D distance between busbars connecting the finger structure.

where the lateral sheet resistance $R = \rho_s y/D$ is substituted. Compared with the generated power $P_{mp} = I_{mp}U_{mp}DS$ at the maximum power point, the fractional loss p_s is

$$p_s = \frac{P_{\text{loss}}}{P_{mp}} = \frac{\rho_s S^2 J_{mp}}{3U_{mp}} \qquad (2.60)$$

Since a smaller grid spacing S increases the losses due to shadowing of the incident light, an optimum value for S can be derived (see Section 2.3.3). The power loss due to lateral resistance effects is thus primarily determined by the sheet resistivity of the emitter layer. Typical values for the power loss due to lateral resistance effects for a commercial silicon cell are below 4% with an average spacing of the fingers of about 5 mm.

According to (2.58), the depth of the junction beneath the surface and the doping of the emitter region determine ρ_s. To achieve minimum sheet resistivity, the top layer is doped as high as possible. Typical values for the doping concentration are $N_d \approx 5 \times 10^{19}$ cm^{-3}, which yields for the diffusion length $L_p \approx 5$–10 μm in the case of an *n*-on-*p* silicon solar cell. The constraints of a low emitter resistivity have to be considered if one discusses the total light-generated current in the cell.

The doping level of the base region is determined from the condition of maximum efficiency. To estimate the efficiency of the solar cell, one has to consider the light-generated current I_L and the open-circuit voltage V_{oc}, which depends both on I_L and the saturation current I_s of the diode. Since the high doping level for the emitter region is required for a low series resistance, the key parameter for the determination of I_L is the diffusion length of the substrate material L_n, which is directly proportional to the base current, as discussed above. $L_n = (\mu_n \tau_n KT/e)^{1/2}$ depends in a complex way on the doping level. If one assumes that the lifetime of the minority carriers is mainly determined by recombination at point defects, it can be described by the SRH statistics, which will be discussed in Section 4.2 (equation (4.29)). The lifetime becomes essentially independent of the doping concentration above a certain level so that the doping dependence of the diffusion length is mainly determined by the mobility of majority carriers in the base region. The general trend is, according to (4.6) in Section 4.1, L_n and J_n , and therefore the short-circuit current $I_{sc} = I_L$ decreases with increasing doping level.

The doping dependence of the saturation current I_s can be seen from (2.54c). The parameters F_n and F_p which appear in the general expression (2.40) are also a function of the doping levels, but in general the dependence is weak and can be ignored. It is evident that the saturation current decreases with increasing doping in the emitter and base regions. Although the light-generated current also decreases with doping, the open-circuit voltage tends to increase with increasing (moderate) doping levels of the base region. Because of the opposite dependencies of I_{sc} and V_{oc} upon doping, there will be an optimum substrate doping for maximum efficiency of the cell. Because of the usually higher mobility of electrons in silicon, a p-doped substrate is preferred in most cases. Typical acceptor concentrations for a high-efficiency silicon cell are $N_a \approx 10^{16}$ cm^{-3} for the substrate material. The standard type of a silicon solar cell is therefore an n^+p junction for a p-type substrate with a highly doped front emitter region.

2.4.3 Efficiency Losses

The ideal limits of the efficiency have been calculated assuming optimal parameters for the material quality and the design of the solar cell. The important material parameters are the lifetime and mobility of the minority carriers in the bulk, and the recombination velocity at the front and back surfaces of the cell. Though for some semiconductors like silicon and GaAs a high quality material can be produced, it is generally difficult to retain the quality of the material during processing the cell, especially under the constraint of low production costs. Since the material parameters are closely linked to the technical design and the fabrication of the solar cell, the actual device characteristics of the pn junction will be lower than its ideal values. Some of the main sources of efficiency losses will be discussed next.

2.4.3.1 Optical Losses

The previous discussion has shown that losses in the light-generated current directly reduce the short-circuit current and the open-circuit voltage (equation (2.47)). It is evident from (2.36) that the incident light cannot be fully utilized because of the finite reflectivity R, which in the case of bare silicon is about 30%. Basically, two approaches are employed to decrease R. Most commonly used are *antireflection* (AR) coatings on the top surface of a material. The idea is to adjust the thickness of the layer so that reflected light at the top surface is out of phase with reflected light from the interface to the substrate. Extinction of the reflected light occurs for a phase difference of $\lambda/4$, and the minimum of reflection is then given by

$$R_{min} = \left(\frac{n_1^2 - n_0 n_2}{n_1^2 + n_0 n_2}\right)^{1/2} \tag{2.61}$$

where n_0, n_1, and n_2 are the refractive indices for air (or glass), coating, and substrate, respectively. The reflectivity is zero if $n_1^2 = n_0 n_2$ which can only be fulfilled for a small range of wavelengths (Figure 2.14). In practical cases, the condition is adjusted to the wavelength at the maximum intensity of the solar spectrum (600 nm), which reduces the total reflectivity for the coated silicon surface,

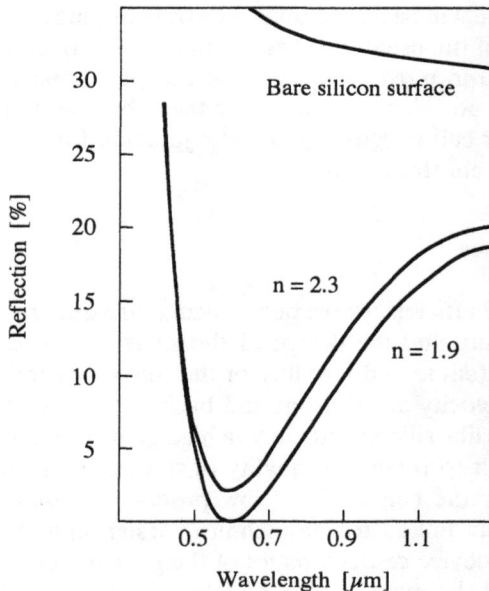

Figure 2.14 Percentage of normally incident light reflected from a silicon surface with an AR coating with different refractive indices. The reflectivity of a bare silicon surface is shown for comparison.

for instance, to about 10%. It is also evident that the AR coating has to be transparent, and some of the available coating materials are listed in Table 2.3. Usually AR coatings are deposited as amorphous layers to suppress the light scattering at grain boundaries.

A further improvement is possible by the use of multilayer coatings with different refractive indices. Another possibility for changing the reflectivity is by texturing the surface. This can be produced with particular etchants that preferentially attack inclined crystallographic planes so that pyramidal structures form (Figure 2.15) [14]. With the combination of both techniques, it is currently possible to keep the total reflectivity below 3%.

Optical losses also occur because of the finite thickness of the solar cell. In order to collect the major fraction of the sunlight inside the cell, a certain thickness of the material is required, which is particularly large for indirect semiconductors with a lower absorption coefficient. For instance, for silicon, a thickness of about 100 to 400 μm is needed to absorb most of the light. The optical thickness of a semiconductor can be reduced by *light trapping* the light inside the crystal so that it is reflected several times between front and back surfaces before it is finally absorbed. This requires a mirror at the back side and textured surfaces which reflect the light at oblique angles.

The incident sunlight is further reduced by the metal grid on the front side, which is necessary to make electrical contacts on the emitter side of the *pn* junction. Even under optimum conditions, the front contacts block about 5% to 10% of the incoming light unless more sophisticated cell structures are used (see Chapter 3). The design of the top contact grid is therefore an important area of cell design. Some guidelines can be developed if one considers the schematic representation of a simple metal grid in Figure 2.16, consisting of busbars and fingers. The busbars are heavier areas of metallization and are directly connected to the external leads of the cell. The fingers are finer contacts which collect the current from the cell and deliver it to the busbar. Ignoring the losses due to the busbars, the fractional

Table 2.3
AR Coating Materials for Solar
Cells and Refractive Indices.

Material	Refractive Index
SiO_2	1.4–1.5
Al_2O_3	1.8–1.9
Si_3N_4	1.9
MgF_2	1.3–1.4
TiO_2	2.3
ZnS	2.3–2.4

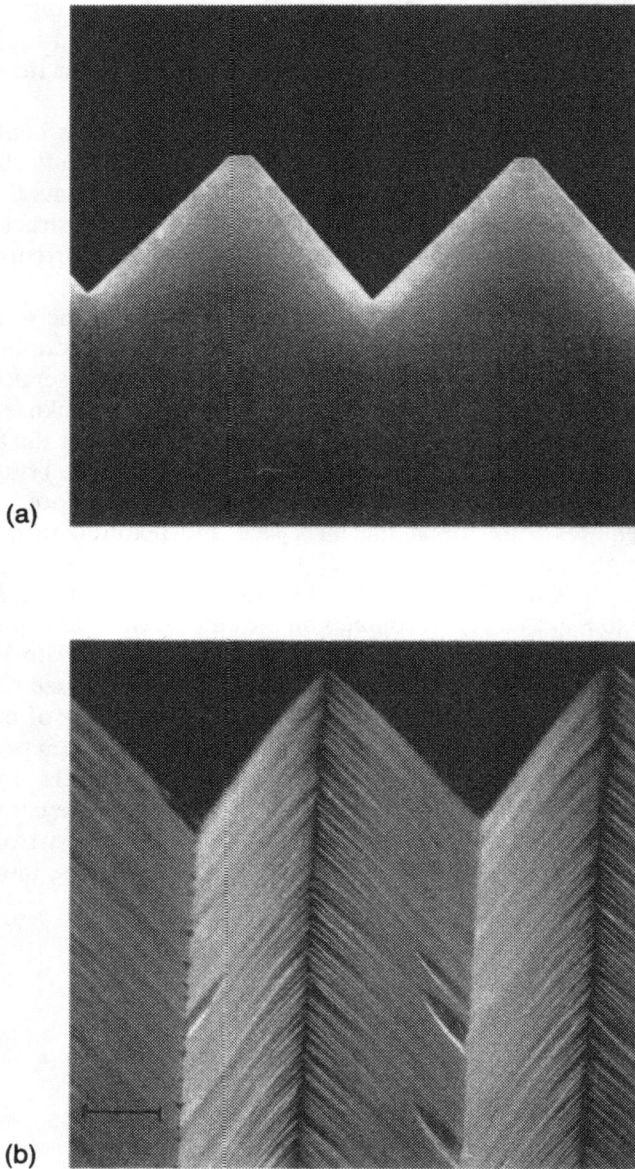

Figure 2.15 SEM image of a preferentially etched GaAs surface: (a) cross section of the textured surface; (b) plane view showing the corresponding micro-grooves (scale marker $\cong 1 \ \mu m$) (courtesy of S. Bailey).

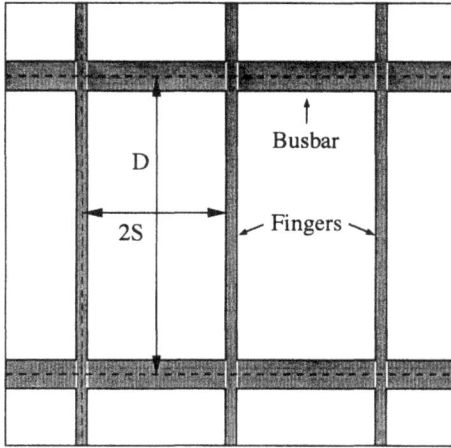

Figure 2.16 Schematic representation of a simplified design of a metal grid on a solar cell surface showing busbars and finger structure.

power loss in a single unit cell of size $S \times D$ consists of the resistive power losses in the metal fingers p_G and the emitter region p_S, the contact resistance loss p_C, and the power loss due to the shadowing of the fingers p_{SH}.

The resistive power losses in the metal fingers p_G can be calculated from (2.59), substituting the current J_{mp} at the maximum power point and the resistance of the metal finger $R = \rho_G D/W_G$:

$$p_G = \frac{\rho_G D^2 S J_{mp}}{W_G U_{mp}} \quad (2.62)$$

The fractional power loss due to the contact resistance is approximately given by

$$p_C = \frac{\rho_C S J_{mp}}{W_G U_{mp}} \quad (2.63)$$

and the losses due to the shadowing effect are

$$p_{SH} = \frac{W_G}{S} \quad (2.64)$$

The combined power loss $p = p_S + p_G + p_C + p_{SH}$ in a single cell can be minimized with respect to the finger distance $2S$. Summing (2.62) to (2.64), one obtains the optimum value after differentiating with respect to S from the equation:

$$S^3 + \frac{3\rho_G D^2 + \rho_C}{2\rho_S W_G} S^2 - \frac{3U_{mp} W_G}{2J_{mp}\rho_s} = 0 \tag{2.65}$$

This equation depends on the distance D of the busbars and the finger width W_G. In practice, the technology used for the top contacts will determine the gross features of the metal grid and place limits on the dimensions of these parameters. For instance, the application of the screen printing technique for the metallization does not allow finer structures than about 50 μm. If one assumes that the busbar and contact resistivities are negligible, one approximately obtains from (2.65) the minimum power loss

$$p_{\min} = \left(\frac{\rho_s J_{mp}}{U_{mp}}\right)^{1/3} \left(\frac{3W_G}{2}\right)^{2/3} \tag{2.66}$$

which occurs for

$$S = \left(\frac{3U_{mp} W_G}{2J_{mp}\rho_s}\right)^{1/3} \tag{2.67}$$

The optimum value for the finger distance $2S$ is thus primarily determined by the sheet resistivity of the emitter layer. Lower sheet resistivities reduce the power loss and allow larger finger distances. For example, for a commercial silicon cell, typical device parameters are $\rho_s = 40\ \Omega/\square$, $J_{mp} = 30$ mA/cm^2, and $V_{mp} = 0.45$ V which yields $S = 2.5$ mm for a finger width of 300 μm and a total power loss $p = 17\%$. Typical values for the power loss due to lateral resistance effects for a commercial silicon cell are below 4%, with an average spacing of the fingers of about 5 mm.

It should be noted that in practice more complex contact schemes are used, which require a more sophisticated analysis of the best parameters for the dimensions of the unit cell and the width of fingers and busbars.

2.4.3.2 Recombination Losses

A fraction of the charge carriers is always generated far away from the junction, and some losses occur because minority carriers recombine before they can diffuse to the device terminals. In Chapter 4, it will be shown that several recombination mechanisms can contribute to the lifetime. In a pure defect-free monocrystalline semiconductor, radiative recombination and Auger recombination determine the lifetime, which depends on the doping concentration and is approximately given for the *p*-type side, for instance, by (see also (4.17))

$$\tau_1 = \frac{1}{s_n N_a + s_{an} N_a^2} \tag{2.68}$$

s_n and s_{an} are the transition probabilities for the different recombination processes and, in practice, parameters that have to be determined experimentally. For silicon, the Auger recombination dominates for concentrations above $10^{18}\,\text{cm}^{-3}$, as can be seen in Figure 2.17. If deep trap impurities or other lattice defects are present, the lifetime for this process takes the approximate form for the *p*-type region (see also the general equation (4.29)),

$$\tau_2 = \frac{1}{\sigma_n v_T N_T} \left(1 + \frac{p_1}{N_a} \right) \tag{2.69}$$

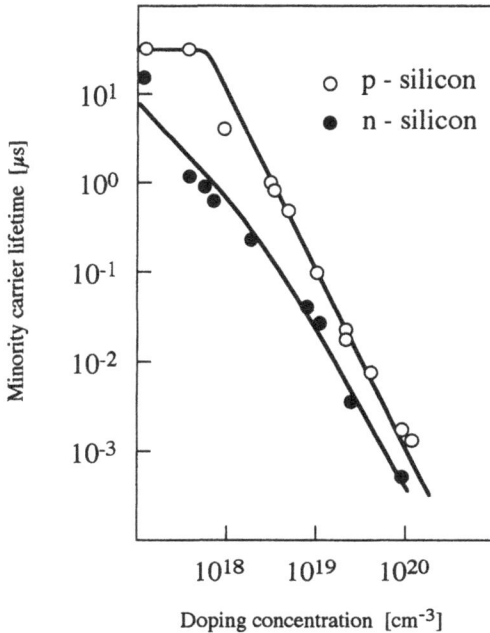

Figure 2.17 Experimental minority carrier lifetimes in high-quality *p*- and *n*-type silicon as a function of the doping concentration at different temperatures. Above a concentration of about $10^{18}\,\text{cm}^{-3}$, the lines represent the quadratic dependence predicted theoretically see (2.68).

where p_1 has been defined in (2.42b). N_T is the trap concentration, σ_n the capture cross section of the defect, and v_T the thermal velocity of electrons. Corresponding expressions can be derived for an n-type region. The total lifetime τ is given by

$$\tau = \frac{\tau_1 \tau_2}{\tau_1 + \tau_2} = \frac{1}{s_n N_a + s_{an} N_a^2 + \sigma_n v_T N_T N_a/(N_a + p_1)} \tag{2.70}$$

(see also (4.10)) and thus decreases with the doping level N_a. Since a shorter lifetime reduces the diffusion length of the minority carriers $L = (D\tau)^{1/2}$, higher doping levels also reduce the light-generated current I_L, as can be seen from (2.31) to (2.36).

Other important recombination centers are the surfaces, dislocations, grain boundaries in polycrystalline semiconductors, and interfaces in heterostructure solar cells. It will be discussed in Section 4.2 that the recombination behavior of extended lattice defects usually depends in a complicated way on the doping level and cannot easily be predicted.

The recombination at the surface and in the bulk are also the fundamental processes that determine V_{oc} according to (2.47). In general, the recombination processes in the entire cell should be minimized as much as possible. It can be seen from (2.42) for the bulk recombination at point defects that the recombination rate R_0 is especially high if the carrier concentrations n and p and the parameters n_{T1} and p_{T1} are small. This occurs for trap levels near the middle of the band gap and carrier concentrations near the intrinsic concentrations. If appropriate trap levels are present, these conditions are especially met in the depletion region of the pn junction. It was discussed in the previous section that in this case the recombination in the space charge region, which has been neglected in the calculation of the ideal current-voltage characteristics of the pn junction, can actually become very important. The addition of the recombination current in (2.45) yields an increased saturation current, which reduces the open-circuit voltage.

2.4.3.3 Series and Shunt Resistance

It was discussed in the previous section that the current-voltage characteristics of the pn junction are further modified because of a series R_s and a shunt resistance R_{sh} associated with the solar cell. The origin of the series resistance is the bulk resistance of the semiconductor and the resistance of the contacts and interconnections. The shunt resistance can be caused by extended lattice defects in the depleted region or leakage currents around the edges of the cell. Extended defects which can easily occur in low-cost semiconductors are dislocations, grain boundaries, and large precipitates.

To evaluate the ramifications of these parameters on the current-voltage characteristics, one can describe the electric behavior of the non-ideal pn junction according to (2.46). Plots of the equation for various combinations of the series and shunt resistance in Figure 2.18 show that essentially the shape of the current-voltage characteristics and hence the fill factor *FF* are changed. Whereas a shunt resistance as low as 100Ω does not appreciably change the power output of the device, it can be seen that a small series resistance of only 5Ω reduces the total efficiency by 30% already. For a typical *n*-on-*p* type silicon solar cell, the series resistance is about 0.7Ω.

2.4.3.4 Temperature Effect

A considerable fraction of the incident light (about 80% to 90%, depending on the efficiency) is transformed into heat and hence the operating temperature of a

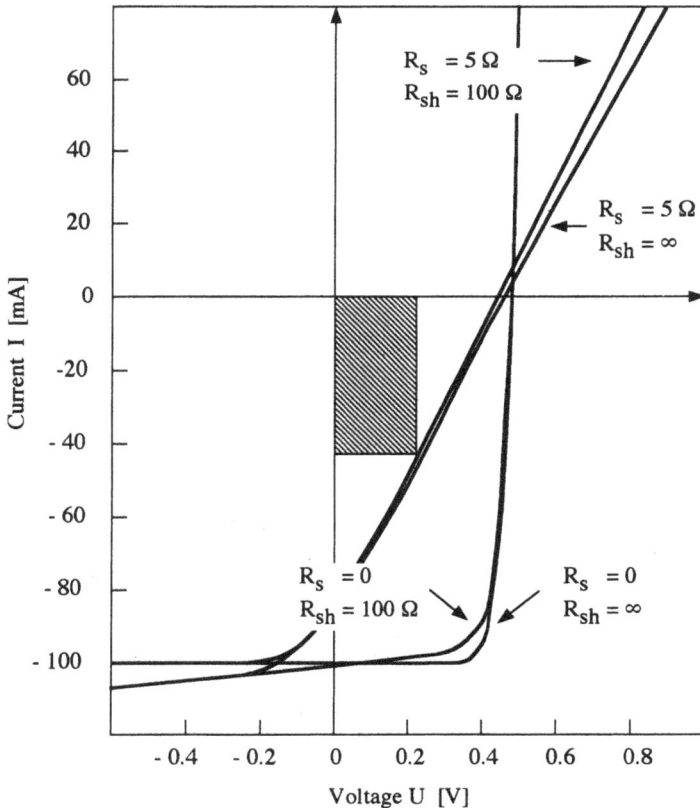

Figure 2.18 Calculated *I-V* characteristics for *pn*-junction solar cells under illumination with different series R_s and shunt resistance R_{sh} [4].

solar cell can vary over a wide range of temperatures, especially in the case of concentrator cells. The material parameters, which mainly change with the temperature, are the band gap energy, which usually decreases, and the minority lifetime, which generally increases, with increasing temperature. This will increase the light-generated current and thus I_{sc} slightly due to the increased light absorption and the increase in minority-carrier diffusion length. However, the open-circuit voltage will more rapidly decrease because of the exponential dependence of the saturation current on the temperature (equation (2.39)), and, correspondingly, the fill factor will degrade, too. Therefore, the overall effect causes a reduction of the efficiency as the temperature increases. One can approximately estimate the change ΔV_{oc} if one differentiates (2.47) with respect to the temperature and considers only the significant terms:

$$\Delta V_{oc} \approx \frac{eV_{oc} - \gamma KT - E_{g0}}{eT} \Delta T \tag{2.71}$$

E_{g0} is the band gap at $T = 0°K$, and γ is a parameter that summarizes the temperature dependence in the prefactors ($\sim T^{-\gamma}$) of I_L/I_S and varies approximately between 1 and 4. Inserting appropriate values for silicon, the open-circuit voltage decreases with temperature (≈ 2.3 mV/°C at room temperature) and the efficiency is reduced by about 0.5% per 1°C. This effect is reduced for larger band gap materials, such as GaAs, where the sensitivity to increasing temperature is about half as much compared to silicon.

REFERENCES

[1] Green, M. A., *High Efficiency Silicon Solar Cells*, Aedermannsdorf, Switzerland: Trans Tech Publ., 1987.
[2] Green, M. A., *Solar Cells*, Englewood Cliffs, NJ: Prentice Hall, 1982.
[3] Hovel, H. J., *Semiconductors and Semimetals*, Vol. 11, R. K. Willardson and A. C. Beer, eds., New York: Academic Press, 1975.
[4] Sze, S. M., *Physics of Semiconductor Devices*, New York: Wiley & Sons, 1981, p. 790.
[5] Thekackara, M. P., *NASA Technical Report*, No. R-351, 1970.
[6] Thekackara, M. P., *Suppl. Proc. 20th Annu. Meet. Inst. Environ. Sci.*, Vol. 21, 1974.
[7] Backus, C. E., ed., *Solar Cell*, New York: IEEE Press, 1980.
[8] Henry, C. H., *J. Appl. Phys.*, Vol. 51, 1980, p. 4494.
[9] *Terrestrial Photovoltaic Measurement Procedures*, ERDA/NASA Report 1022-77/16, 1977.
[10] Pankove, J. I., *Optical Processes in Semiconductors*, Englewood Cliffs, NJ: Prentice Hall, 1971.
[11] Bonc-Bruevic, V. L., and S. G. Kalasnikov, *Halbleiterphysik*, VEB Deutscher Verlag der Wissenschaften, Berlin, 1982.
[12] Lanyon, H. P. D., and R. A. Tuft, *IEEE Techn. Dig., Int. Electron Device Meet.*, 1978, p. 316.
[13] *Principal Conclusions of the American Physical Society Study Group on Solar Photovoltaic Energy Conversion*, New York: American Physical Society, 1979.
[14] Colemann, M. G., *Conf. Rec. 12th Photovol. Conf.*, New York: IEEE, 1976, p. 313.
[15] Dziewior, J., and W. Schmid, *Applied Physics Letters*, Vol. 31, 1977, p. 346.

Chapter 3
Technology of Solar Cell Devices

The standard type of n^+p-homojunction solar cell and the main parameters determined by the properties of the material were discussed in the previous chapters. In order to fully utilize the potential of a material, the device structures can be improved further. Important factors that can be controlled by the device configuration are, for instance, the position and doping concentrations of the pn junction, the geometry of the ohmic contacts which contribute to the series resistance and block part of the incident light on the front surface, and the surface reflectivity. Many variations of the basic solar cell concept have been proposed for achieving higher conversion efficiencies, and some modifications shall be considered next. There also exists a large variety of device concepts that are not based on the pn-homojunction cell. They are developed to use different and frequently novel materials and may offer the potential for higher efficiencies at lower costs in the future. Some of these concepts will be discussed in the second part of the chapter.

3.1 SINGLE-JUNCTION DEVICES

The discussion in the previous chapter showed how some sources of efficiency losses can be minimized by optimizing the geometry and doping profiles of the cell. Several device modifications of the standard cell have been developed to further improve the efficiency, and three different types will be discussed.

An important source of recombination losses is the recombination at the front and back surfaces. The recombination velocity at the back surface is usually very high because of the ohmic contacts covering the entire surface. The recombination at the front surface is generally lower because of the smaller contact area. Typical values for a silicon-metal interface are $S_n = 10^6$ cm/s. An analysis of the base current, which contributes to I_{sc}, and the saturation current I_S, which contributes to V_{oc}, shows that a reduced recombination at the back side can increase I_{sc} and reduce I_s, so that the overall effect is an increased efficiency. One way of creating an effective low recombination velocity is by using a back surface field (BSF), which can be formed by a junction between highly and lightly doped material of the same dopant (see Figure 3.1). Such a BSF causes a repulsion of minority carriers from this highly doped region near the back surface and can be described by an effective recombination velocity S_{eff}. A numerical calculation for the p^+nn^+ solar cell shows the dependence of S_{eff} on the BSF thickness in Figure 3.2 [1].

As with the cell with a BSF, the influence of the surface recombination is smaller if one forms a pn junction with nonuniform base doping. If the doping concentration in the base region increases towards the back surface, a drift field is produced that repels the minority carriers from the back contact. Thus, a further reduction of the saturation current and hence an increase of V_{oc} is possible. A typical drift-field $p^+n(x)$ silicon solar cell consists of a heavily doped p-type emitter and a nonuniform n-type base.

Usually, nonuniform doping profiles are also used for the emitter region. The dopant is diffused from the surface into the moderately doped substrate, which produces a concentration gradient in the emitter region. Very high dopant con-

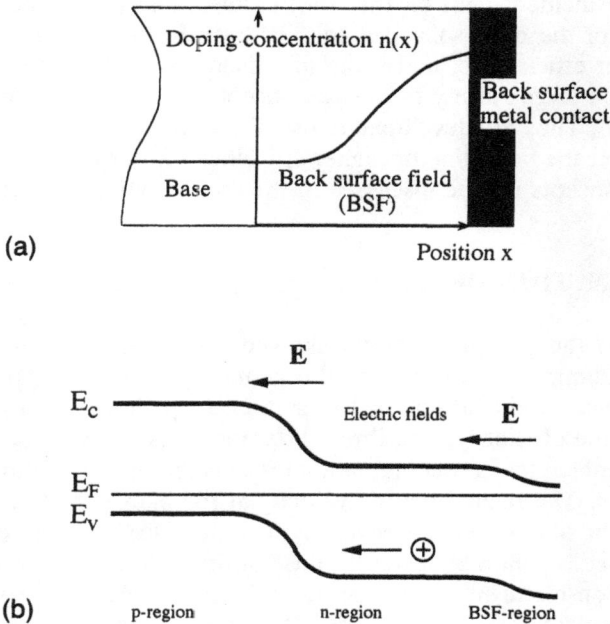

Figure 3.1 Schematic diagram of the doping profile at the back side contact of a *pn*-junction solar cell with a BSF (a) and corresponding band structure (b).

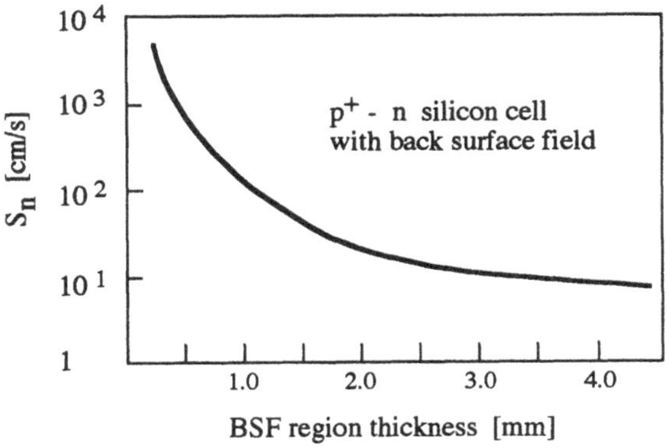

Figure 3.2 Calculated surface recombination velocity S_n as a function of the thickness of the BSF region for a p^+n silicon solar cell [1].

centrations can then occur near the surface, which may give rise to a variety of detrimental effects. For instance, highly doped phosphorus regions getter fast diffusing transition metal, which can lead to the formation of metal precipitates near the surface [2]. The accumulation of metal atoms or precipitates and the high doping concentrations enhance the recombination in the emitter region according to (2.42). The resulting lowering of the diffusion length may be so severe near the surface that no current can be obtained from photons with high energies that are preferentially absorbed near the surface. Such "dead layers" have been produced in earlier silicon solar cells when very high dopant concentrations were used to reduce the top-layer resistance. The problem has been solved by decreasing the doping level and reducing the width of the emitter region [3]. These much shallower solar cells are known as violet cells because of their better response to the shorter wavelengths ($hv > 2.75$ eV). The doping levels are typically about $N_d \approx 10^{18}$ cm^{-3} and the junction depth $W_n = 0.2$ μm. For a numerical analysis of the high doping region, one has to account for the narrowing of the band gap (Section 2.2) and the enhanced reduction of the lifetime due to Auger recombination.

The conventional pn-junction device concepts described so far yield efficiencies of less than 17% for monocrystalline silicon solar cells. With the subsequent availability of very high purity, float zone material, and the increased research in cell modeling and design, efficiencies up to 23% have evolved during the last few years [4–7]. These achievements in cell performance could be obtained by more complicated cell designs and structures. The main purpose is to collect the incident light better by minimizing the front contact area, improving the antireflection coating, and improving the junction properties. Some of the recent developments are illustrated in the device cross sections in Figure 3.3.

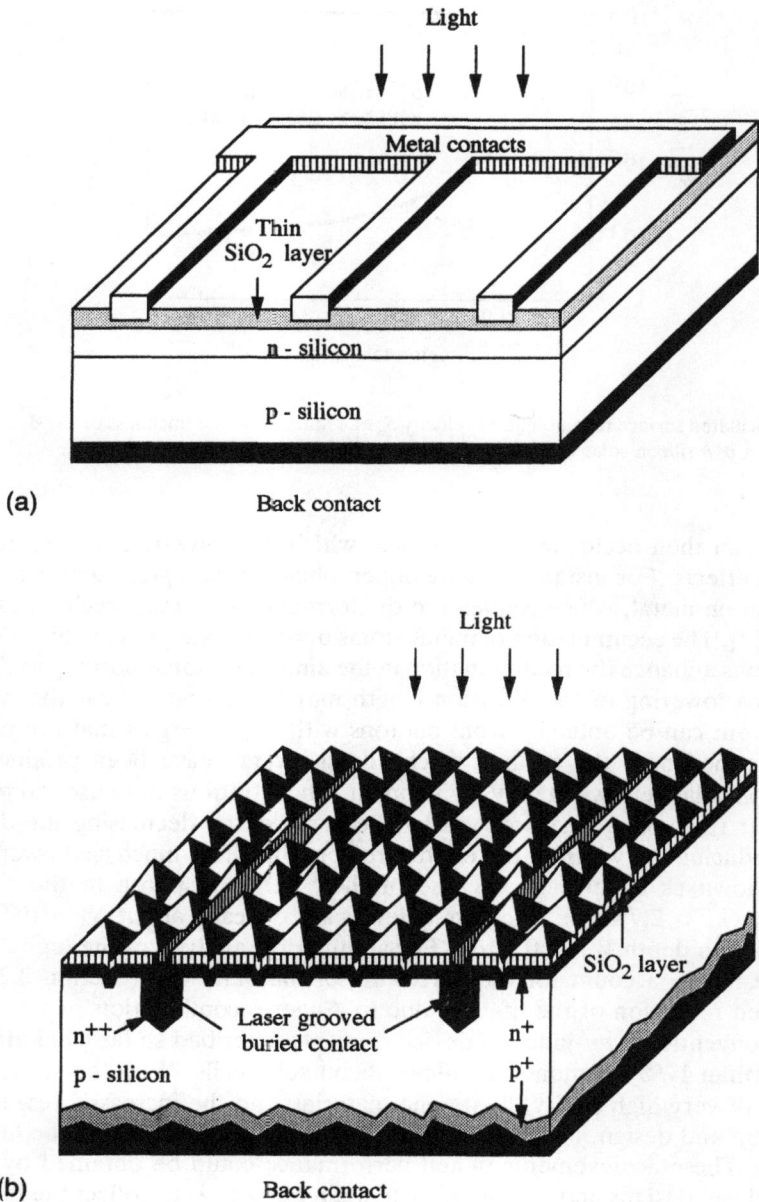

Figure 3.3 Schematic representations of high-efficiency designs for silicon solar cells: (a) PESC design [4]; (b) buried contact cell with laser grooved surface [5]; (c) back-side PCSC [6].

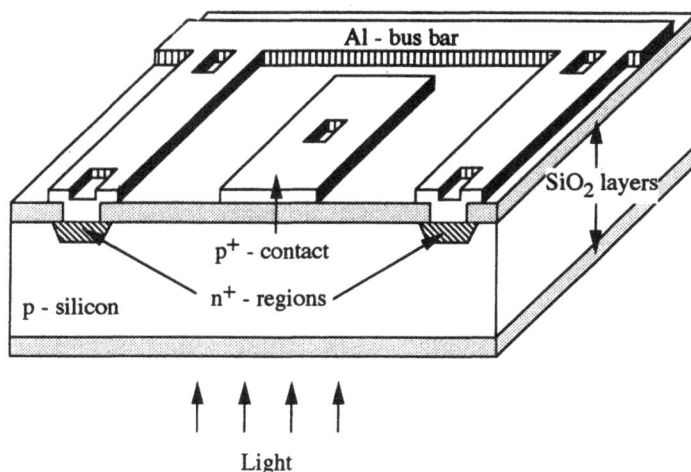

Figure 3.3 (Continued).

A more conventional design is used for the passivated emitter solar cell (PESC) (Figure 3.3(a)) [4], which has an especially high open-circuit voltage of about 0.66V and an efficiency of 20.9%. A low-resistivity, p-doped fz-silicon (0.2 Ωcm) with a high lifetime is used for this cell. The front surface is textured with microgrooves, which together with a double antireflection coating yields a high light collection efficiency. A thin SiO_2 layer with only small openings for the contacts is used to reduce the front surface recombination. A BSF provides a very low recombination at the back side, and a very thin emitter region reduces the bulk recombination. A further evolution of this cell concept is the laser grooved, buried contact silicon cell shown in Figure 3.3(b) [5]. This approach uses junctions perpendicular to the incident light and has thus only a minimal front contact area. In addition, the cell has front and back surface texturing and a BSF.

Another possibility to eliminate the shadowing losses of the front contacts is to bring both contacts out to the rear of the cell. Figure 3.3(c) presents a configuration of a point contact solar cell (PCSC), which is mainly developed as a concentrator cell [6]. The volume of the highly p- or n-doped regions is kept small so that the recombination currents from the space charge regions are minimized. The surface recombination at the front and back surfaces is reduced by a thin SiO_2 layer which also partly covers the doped region. Thus, only small point-like contact areas remain where the surface recombination is high. The metal contacts have to be very fine, but thick enough to obtain a low series resistance. The cell is only about 100 μm thick, which requires light trapping techniques for complete absorption of the radiation. The problems of the cell are the high short-circuit current densities in combination with the difficulty in contacting. In particular, the operation under concentrated light requires an effective cooling of the cell. Efficiencies of about 22% for a radiation of 1 sun (AM1.5) and 27.5% for a radiation of 100 sun have been reported.

The major drawbacks of these sophisticated cell designs are that very clean fabrication processes and microelectronic device techniques have to be employed, which raise the production costs. However, the excellent performance demonstrations have provided the knowledge and the basis for further development of less expensive high-efficiency cells.

Though most pn-homojunction solar cells use monocrystalline or polycrystalline silicon, they are not restricted to this material. In fact, compound semiconductors, such as GaAs, InP, CdTe, and various ternary/quaternary compositions of these compounds, such as AlGaAs or GaAsInP, have been receiving considerable attention for photovoltaic applications [8,9] because of their excellent photovoltaic characteristics. GaAs, for instance, has an almost perfect band gap energy for optimal efficiencies (see Figure 2.11). The cost of the high-quality single crystals is a major factor in these technologies, and solar cells using compound semiconductors are usually considered for space applications or terrestrial concentrator cells (see Section 3.6), in which the cost of the cell is less significant compared to other requirements, such as higher performance and long-term stability under the intense irradiation in space (irradiation resistance). In particular, much of the work on InP has recently been directed towards space applications, primarily because the cells offer a better irradiation resistance compared to silicon and GaAs solar cells [10].

Most compound semiconductors are direct band gap materials and have therefore higher absorption coefficients. The devices require less material compared to silicon, and problems with high surface recombination velocities are even more serious for the conventional homojunction structures. The technique to reduce the recombination influence of the top surface was to make the top layer as thin as possible. For a highly absorbing semiconductor like GaAs, the emitter region in this case has to be an order of magnitude thinner ($\approx 500\text{Å}$) compared to silicon. Therefore, for fabricating cells on the basis of GaAs and other compound semiconductors, epitaxial growth techniques such as metalorganic chemical vapor deposition (MOCVD), liquid phase epitaxy (LPE), and MBE are usually preferred. A common pn-homojunction structure for a GaAs-based solar cell is shown in Figure 3.4. Efficiencies of about 24% for GaAs and about 19% for InP [11] single-crystal cells have been reported. A summary of the status of current (1989) cell structures and parameters for a variety of technologies has recently been given by Kazmerski [12].

Material-related problems for some semiconductors may require other device structures which are more suitable for the particular material. A device type that has recently been used for CdTe [13] and amorphous silicon (a-Si) is the pin-diode structure (Figure 3.5). This design differs from the conventional *pn* homojunction because of the extended depletion region where the semiconductor is basically intrinsic. Most of the photocurrent for this structure is generated in the depletion region, as given by (2.35), and the design of the intrinsic region offers great flexibility for cell optimization. The thin films are either deposited on a glass substrate covering the front surface of the cell (Figure 3.5(a)) or on a metal (steel) substrate, which can be used as a back-side contact (Figure 3.5(b)).

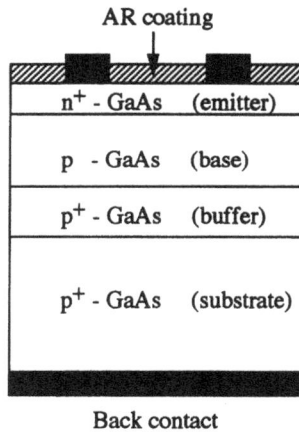

Figure 3.4 Cross-sectional representations of a high-efficiency structure for a GaAs-based *n*-on-*p*-homojunction solar cell with a thin ($\approx$500Å) n^+-GaAs emitter region. The thick substrate (100 to 200 μm) requires a high doping level for a low series resistance.

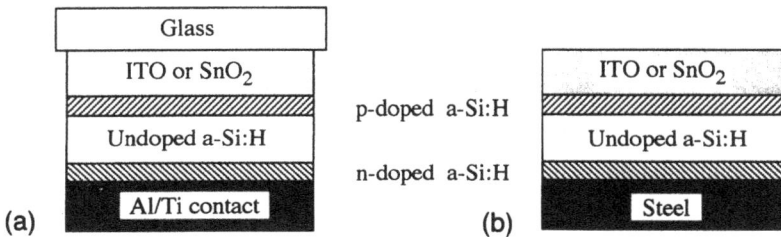

Figure 3.5 Cross-sectional representation of two *pin* cell designs for an amorphous Si:H thin-film cell: (a) films on glass substrate; (b) on steel substrate, which also serves as a back-side contact.

3.2 HETEROJUNCTION DEVICES

The previous discussion showed that the energy of the incident light is only partly utilized for a particular semiconductor. Theoretically, one can expect an improvement of the efficiency of a solar cell device if it consists of materials with different band gap energies which match different parts of the solar spectrum. This concept is realized in a heterojunction device formed between semiconductors with different band gap energies. A typical n-on-p-heterojunction band diagram is shown in Figure 3.6. The material with the larger band gap E_{g1} is on the top. Light with energy less than the band gap energy E_{g1} but greater than E_{g2} passes through the first semiconductor, which acts as a window, and will be absorbed by the second semiconductor. Carriers generated in the depletion region and within a diffusion length of the junction are collected similarly to a pn-homojunction solar cell. However, light with photon energies larger than E_{g1} will be more efficiently utilized by the first semiconductor. The advantages of heterojunction solar cells over conventional cells include enhanced short wavelength response, lower series resistance, if the first semiconductor can be heavily doped, and higher irradiation resistance.

The band diagram of the heterojunction and the depletion region is determined by the work function (the difference between the vacuum level and the Fermi energy) of the two semiconductors ϕ_i, the electron affinity χ_i (the difference between the vacuum level and the conduction band edge), and the band gap energies E_{gi}. The built-in potential φ_{np} is simply the difference between the work functions $\varphi_{np} = \phi_2 - \phi_1$, and the spatial distribution of the space charge can be calculated similarly to the pn homojunction. The main difference for a pn heterojunction is the discontinuity in the conduction band energies, which is equal to the difference in the electron affinities

$$\Delta E_C = \chi_2 - \chi_1 \tag{3.1a}$$

and the band offset

$$\Delta E_V = E_{g2} - E_{g1} + \chi_2 - \chi_1 \tag{3.1b}$$

A negative ΔE_C (or ΔE_V for a p-on-n heterojunction) produces a spike in the conduction (or valence band, respectively) which is undesirable for photovoltaic applications. The spike impedes the flow of minority carriers across the junction from the p-type to the n-type regions, and the photocurrent will be reduced. Such spikes can, however, be avoided by a suitable combination of electron affinities and band gap energies [14]. For the ideal case with negligible spikes, the expressions for the photocurrents in heterojunctions are essentially the same as in homojunctions. For an n-on-p heterojunction, the minority electron and hole photocurrent densities are given by the expressions (2.31) and (2.33), except that the absorption coefficient α and the diffusion lengths L_p and L_n are replaced by the corresponding

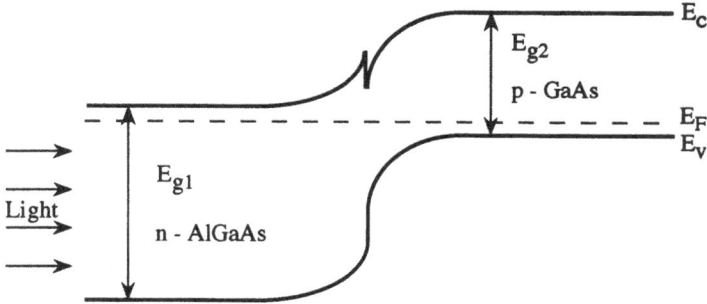

Figure 3.6 Schematic band structure of an *n*-on-*p* AlGaAs-GaAs solar cell showing a spike in the conduction band at the interface due to a difference in the electron affinities $\Delta\chi_i$ of the two semiconductors. The internal potential barrier is determined by the difference in the work functions $\Delta\phi_i$. The light enters through the semiconductor with the larger band gap first.

quantities for each semiconductor. In addition, the photocurrent density from the depletion region is now given by

$$J_{dp} = e\beta S_0(1 - R)\{\exp(-\alpha_n x_n)(1 - \exp(-\alpha_n W_1)) \\ + \exp(-\alpha_p x_p)(1 - \exp(-\alpha_p W_2))\} \tag{3.2}$$

where W_1 and W_2 are the widths of the depleted regions on either side of the heterojunction and the total width $W = W_1 + W_2$. In the latter expression, it has been assumed that no recombination takes place in the depletion region. For a normal homojunction in a high-quality single crystal, the recombination losses in the depletion region can in fact be ignored if the concentration of recombination centers is small. For heterojunctions, there is the inherent problem that the crystal structure changes across the junction and an interface is formed between the two semiconductors. Interfaces can be efficient recombination centers because they introduce deep trap levels in the band gap (Chapter 5). They can also provide sites for quantum mechanical tunneling processes, which is important for current loss mechanisms across the junction. In both cases, the interface traps degrade the performance of the solar cell, and it becomes essential to produce heterojunctions with a low density of interface traps. Though the origin of the interface states is not completely clear, it appears that the density of interface traps is to some extent related to the degree of mismatch between the crystal lattices of the two semiconductors.

Therefore, the requirements for a good *n*-on-*p*-heterojunction solar cell are a small ΔE_C (for a *p*-on-*n* heterojunction, ΔE_V has to be small) and a good lattice match. A system that fulfills these requirements is $n\text{-Al}_x\text{Ga}_{1-x}\text{As} - p\text{-GaAs}$ in which the AlGaAs layer is the semiconductor with the wider band gap. With increasing composition x of the AlGaAs layer, the band gap energy E_{g1} increases and the spectral response extends to higher photon energies. The AlGaAs layers can be deposited on single-crystal GaAs substrates by MOCVD or MBE. These are standard techniques used for microelectronic devices and produce high-quality, single-crystal epitaxial layers.

A derivative of the heterojunction cell is the heteroface cell, which is a *pn* homojunction to which a semiconductor with a larger energy gap has been added. The most common cell structure for the AlGaAs-GaAs system is represented in Figure 3.7. Because of the good lattice match of the AlGaAs layer, it can be grown epitaxially with a low density of interface states. For compositions of $x > 0.4$, the $Al_xGa_{1-x}As$ becomes an indirect semiconductor and the absorption is therefore low. In essence, this top layer will absorb only little light and act as a window for the underlying *pn* homojunction. The main advantage is that the AlGaAs layer passivates the surface of the underlying GaAs and effectively reduces the surface recombination velocity. In fact, this device structure has resulted in the best efficiencies reported for single-junction cells so far, with values of about 24% [12].

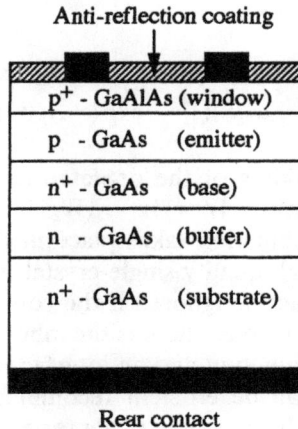

Figure 3.7 Schematic representation of a GaAs-based heteroface *p*-on-*n* cell with a thin p-$Al_xGa_{1-x}As$ window region (≈ 500Å). $Al_xGa_{1-x}As$ has an indirect band gap for $x > 0.4$, and thus a low absorption in the window.

Particular wide band gap semiconductors are a class of materials frequently used in heterojunction devices. They serve not only as part of the heterostructure, but also as antireflection coating and contact material. Typical semiconductors (see Table 3.1) include oxides such as ZnO, SnO_2, InO, and ITO, a mixture of about

Table 3.1
Large Band Gap Window Materials for Solar Cells

Material	Band Gap E_g [eV]	Electron Affinity χ [eV]
ZnO	3.35	4.8
SnO$_2$	3.50	4.8
InO	3.50	4.45
ITO (In$_2$O$_3$—9% SnO$_2$)	3.70	4.2–4.5
CdS	2.43	
ZnS	3.60	

9 mol% SnO$_2$ and In$_2$O$_3$, which have good electrical conductance and optical transparency in thin-film form.

3.3 SCHOTTKY BARRIER AND MIS SOLAR CELLS

A Schottky barrier is formed when a metal and a semiconductor are brought into contact. In a Schottky barrier solar cell, the light-generated charge carriers are separated by the electric field created by the space charge near the interfacial region. A schematic band diagram is shown in Figure 3.8. The difference of the work functions of the metal and the semiconductor gives the contact potential $\varphi_{SC} = \phi_M - \phi_S$. Since the work function of the semiconductor depends on the position of the Fermi energy, the contact potential varies with the doping level. More important is the barrier height φ_{BS}, which determines the current transport over the Schottky barrier and is given for an ideal diode by the difference of the metal work function and the electron affinity of the semiconductor $\varphi_{BS} = \phi_M - \chi_S$. In practice, however, it is frequently found that the barrier height is independent of the work function of the metal. This can be explained by the presence of electronic states in the band gap, which are introduced by the metal-semiconductor interface. Though their origin can generally be attributed to the atomic defect structure of the interface and the effect of impurity atoms, many details are still not clear and are primarily determined experimentally (see Table 3.2). The barrier height in the presence of interface states can be approximately [15] expressed by

$$\varphi_{BS} = c(\phi_M - \chi_S) + (1 - c)\frac{1}{e}(E_g - E_T) - \Delta\varphi_{BS} \qquad (3.3)$$

with $c = \dfrac{\epsilon_i}{\epsilon_i + e^2 \delta N_{ss}}$ and $\Delta\varphi_{BS} = \left(\dfrac{eE}{4\pi\epsilon_s\epsilon_0}\right)^{1/2}$

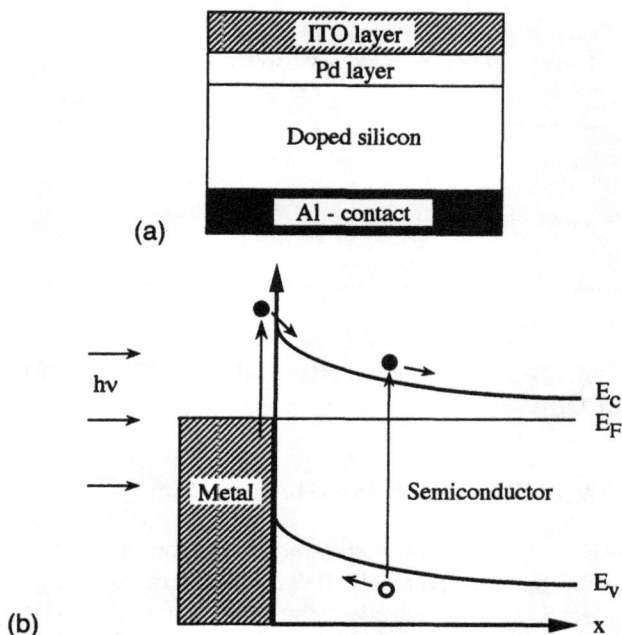

Figure 3.8 Cross section of a solar cell having a metal-semiconductor structure (a). Corresponding band diagram of the Schottky barrier under illumination (b).

Table 3.2

Schottky Barrier Heights for Some Transition Metals on Different Semiconductors (E_g is the band gap energy at room temperature)

Semiconductor		Au	Al	Pt	Pd	Ti	Mo	W	E_g [eV]
Si	n	0.80	0.72	0.90	0.81	0.50	0.68	0.67	1.12
Si	p	0.34	0.58			0.61	0.42	0.45	
SiC (β)	n	1.95	2.00						3.0
GaAs	n	0.9	0.8	0.84				0.80	1.42
GaAs	p	0.42							
CdTe	n	0.71	0.76	0.76					1.50
CdS	n	0.78		1.1	0.62	0.84			2.43
CuInSe$_2$	n	0.5							1.04
CuInSe$_2$	p		0.4						

where ϵ_i and ϵ_s are the permittivities of the interface layer and the semiconductor, respectively, N_{SS} is the density of states of the interface levels, E_T is the occupation level of the neutral interface, and δ is the thickness of the interface layer. Under applied bias, the barrier height changes slightly because of the image force-induced lowering of the potential energy $\Delta\varphi_{BS}$ for charge carrier emission into the semiconductor (Schottky effect); but this effect is small for low electric fields ($E < 10^5$ V/cm) applied in solar cells. For many semiconductors, experimental results have shown that the metal-semiconductor interface introduces a distribution of trap levels centered in the lower half of the band gap approximately at $E_T \approx E_g/3$. The density of states distribution is in the range of $N_{SS} = 10^{13}$ eV^{-1}cm^{-3}.

For photovoltaic applications, the metal must be thin enough to allow a substantial amount of light to penetrate into the semiconductor. Typically, the metal layers are about 100Å thick. For instance, for a gold film with antireflection coating, one obtains for the transmission coefficient $D = 1 - R \approx 0.9$. The advantages of Schottky barriers include (1) low-temperature processing, so that the starting properties of the substrate material can be maintained, (2) adaptation to polycrystalline and thin-film solar cells, (3) high current output and good spectral response because of the presence of the depletion layer right at the surface, and (4) high radiation resistance due to a high electric field near the surface.

The short-wavelength light entering the semiconductor is mainly absorbed in the depletion region, whereas long-wavelength light is absorbed in the neutral base region. The major contribution to the photocurrent I_L is therefore from the depletion and the neutral base regions, and can be calculated in a manner similar to that of the *pn* junction. The basic current-voltage characteristics of an ideal Schottky barrier solar cell under illumination is given by

$$I = I_S \left(\exp\frac{eU}{nKT} - 1 \right) - I_L$$

$$I_S = AA^*T^2 \exp\frac{e\varphi_{BS}}{KT}$$

$$(3.4)$$

where U is the applied voltage, n the ideality factor, A the total illuminated area, and A^* the effective Richardson constant, which varies between about 30 to 120 A/cm^{2}K^2. The conversion efficiency given by (2.51) increases with the barrier height. Taking $\varphi_{BS} = E_g$ as the limiting case, the maximum efficiency is about 25%. However, for most metal-semiconductor systems made on uniformly doped substrates, the maximum barrier height that can be achieved is only about $2E_g/3$, which reduces the efficiency considerably: about 10% for Si and 15% for GaAs for an ideal solar cell. The performance of Schottky barrier solar cells is therefore limited.

These practical limitations can be partly removed by inserting a thin insulating layer between the metal and the semiconductor. The current-voltage characteristics of such an MIS solar cell device is similar to the Schottky diode (Figure 3.9). The saturation current contains an additional term due to tunneling currents through the insulating layer and is given by

$$I_S = AA^*T^2 \exp\left(\frac{e\varphi_{BS}}{KT}\right) \exp[-(e\varphi_T)^{1/2}w_i] \tag{3.5}$$

where φ_T is the barrier height presented by the insulating layer and w_i is the insulator thickness.

The open-circuit voltage can be easily determined from (3.4) and is given by [16]

$$V_{OC} = \frac{nKT}{e}\left\{\ln\frac{I_L}{A^*T^2} + \frac{e\varphi_{BS}}{KT} + (e\varphi_T)^{1/2}w_i\right\} \tag{3.6}$$

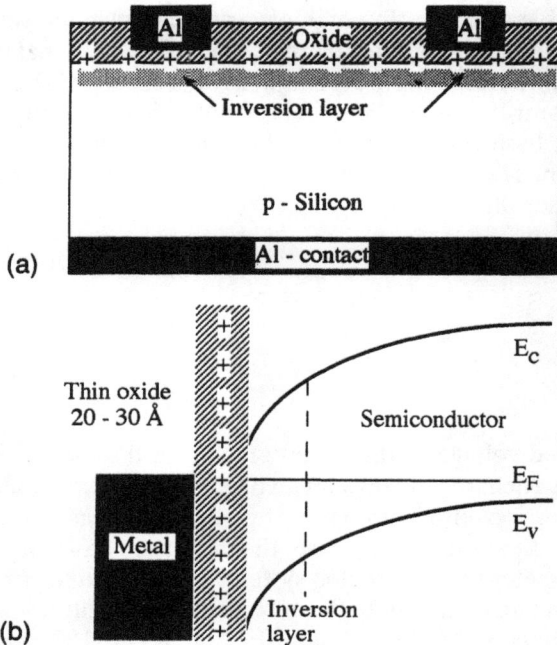

Figure 3.9 Cross section of a solar cell having an MIS structure (a). Corresponding band structure of the MIS cell (b). The thickness of the oxide layer below the contacts is about 20Å to 30Å. Positively charged ionic impurities in the oxide layer enhance the inversion.

This equation shows that V_{OC} increases with the insulator thickness. On the other hand, the short-circuit I_{sc} will decrease, causing a degradation of the conversion efficiency. An optimum oxide thickness for a metal-SiO$_2$-Si system is found to be about 20Å. A recently developed MIS solar cell has shown efficiencies of about 15%.

3.4 THIN-FILM SOLAR CELLS

In thin-film solar cells the active semiconductor is a polycrystalline or amorphous thin film that has been deposited on a supporting substrate made from glass, ceramic, metal, plastic, or another semiconductor. Various deposition techniques such as evaporation, CVD, or sputtering are available today and offer a great flexibility in forming semiconductor films of various compositions. The basic requirement is that the thickness of the film is larger than the inverse of the absorption coefficient for the longer wavelengths of the spectrum so that most of the light can be absorbed, and secondly that the diffusion length is larger than the film thickness so that most light-generated carriers can be collected. Thin-film solar cells are therefore mainly made from compound semiconductors with direct band gaps and high absorption coefficients.

The main advantage of thin-film solar cells is their promise of lower costs, since less energy for processing and relatively lower costs for the materials are required, and large-scale production is feasible. The flexibility in the deposition techniques also allows the development and utilization of novel semiconductors which otherwise might be difficult to produce. The deposition of semiconductors on foreign substrates usually results in polycrystalline or amorphous films with optical and electrical properties that can be substantially different from the single-crystal behavior. This is mainly due to the large number of grain boundaries and other lattice defects, as will be discussed in Chapter 5. However, one of the major problems with thin-film cells is that in many cases the higher defect density also reduces the efficiency and stability of the semiconductor compared to the single-crystal cells. Great efforts have therefore been made to understand the influence of lattice defects on the photovoltaic properties of the semiconductor. Though remarkable improvements have been obtained in particular cases, many fundamental problems, which will be addressed in the following chapters, still need to be solved.

The first polycrystalline thin-film solar cells were fabricated with an active layer of a polycrystalline cuprous sulfide (Cu$_2$S) film of about 0.1 to 0.3 μm thickness on a 20 μm thick polycrystalline layer of CdS. A schematic diagram in Figure 3.10, which is also typical for other thin-film cells, shows basically a heterojunction structure for the cell design. The Cu$_2$S material is p-type with a band gap of 1.2 eV, whereas the CdS is n-type with a gap of 2.3 eV. Most of the light-generated

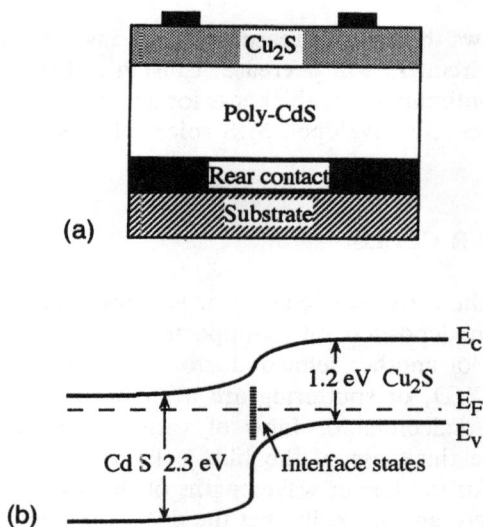

Figure 3.10 Schematic diagram of a polycrystalline Cu₂S/CdS thin-film solar cell (a) and corresponding band structure (b).

current comes from the thin Cu_2S layer and relatively little from the CdS layer with its larger band gap.

Despite the advantage of low processing costs of the Cu_2S/CdS solar cell and reported efficiencies of about 10%, reliability problems virtually eliminated all work on this particular cell. The solar cells degraded under high humidity and at higher temperatures ($>60°C$), or when the load voltage exceeded 0.33V [17]. These problems could be overcome with the further development of copper-based ternary semiconductors (see Chapter 8). Among these, $CuInSe_2$ has received the most attention because of its extremely high absorption coefficient and stable electro-optical properties. Other advantages include a band gap that is suitable for either homojunction or heterojunction device types and a lattice structure and electron affinity that matches with common n-type window materials.

Generally, the device structures suitable for maximum performance vary with the semiconductor materials used in the solar cell. They are determined by the specific properties of the materials and the fabrication techniques that can be applied. A typical device concept and the corresponding band structure are given in Figure 3.11 for a $CuInSe_2$ solar cell, but devices have also been fabricated as homojunctions, *pin* structures, or Schottky diodes. A major source of efficiency losses in the depicted structure is the recombination at the $CuInSe_2$/CdS-window interface, and special precautions have to be taken to improve the cell performance

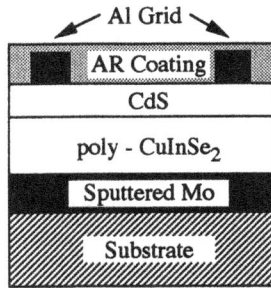

Figure 3.11 Device cross section for a CdS/CuInSe$_2$ solar cell. The substrate can be a low-cost material such as glass or alumina.

(see Chapter 8). The best devices now reach efficiencies up to 12.5% and have the potential for further improvements.

Many thin-film deposition techniques produce an amorphous semiconductor film if the processing temperatures are kept low. Despite this possibility, only amorphous silicon has attracted wide interest as a photovoltaic material. The demonstration of the ability to dope controllable amorphous silicon n- and p-types and to fabricate a Schottky barrier was an important first step for electronic and photovoltaic applications. Since the first discoveries, amorphous silicon has been thoroughly investigated and has been used in solar cells that power watches and calculators. Amorphous silicon cells now account for about one-third of the world's photovoltaic market.

Amorphous silicon usually contains significant concentrations of hydrogen (10% to 40%), which is mostly bonded to silicon atoms and necessary to saturate free covalent bonds in the disordered crystal structure. In contrast to crystalline silicon as an indirect semiconductor, hydrogenated a-Si:H has an optical absorption characteristic that resembles the behavior of a crystal with a direct band gap of 1.7 eV, which is somewhat larger than the band gap width for optimal performance. A film thickness between 1 and 3 μm is sufficient to absorb the light, and a-Si:H films can be easily grown by plasma deposition techniques onto metal or glass substrates. The minority carrier diffusion length is below 1 μm; therefore, the depletion region forms most of the active carrier collecting region. Another consequence is a rather high bulk series resistance, which is also undesirable for a solar cell, as has been discussed before. Most current cells are based on the heteroface pin structure depicted in Figure 3.7, but cells have also been designed as Schottky barriers. A very undesirable property of the a-Si:H solar cells is their instability: the efficiency degrades by about 15% to 20% (relative change) when exposed to light, usually within the first 48 hours of operation. This light-induced degradation problem has become one of the most investigated problems in recent years because of its impact on this promising technology. Although a generally accepted explanation has not been found for this material-related (Staebler-Wronski) effect, it could be shown that improvements in the efficiency and stability can be obtained for special cell designs. Recent measurements suggest that multijunction *pin* cells are more stable compared to single-junction cells (Figure 3.12).

Materials and device studies have recently focused on the development of a-SiGeH and a-SiC compounds for high-performance multijunction cells [18]. The alloying of amorphous silicon with germanium and carbon offers the possibility of varying the optical properties over a wider range so that the device designer has greater flexibility in constructing devices. The band gaps of a-SiGe (≈ 1.4 eV) and a-SiC (1.85 eV) make these compounds suitable for triple-junction devices, and recent results have shown efficiencies exceeding 10%. Stacked junction devices degrade more slowly primarily because thinner films are introduced and the overall degradation rate is approximately an average over the single cells.

Polycrystalline and amorphous thin-film solar cells, especially in connection with tandem device structures, have great potential for further development because of the flexibility in the utilization of novel materials and processing techniques; but they also present a great challenge for the fundamental understanding of the physical properties of the semiconductors and their relationship to crystal defects.

3.5 MULTIGAP STRUCTURES

In connection with the heterostructure cells, it has already been discussed that the spectrum of the incident sunlight can be used more effectively if semiconductors

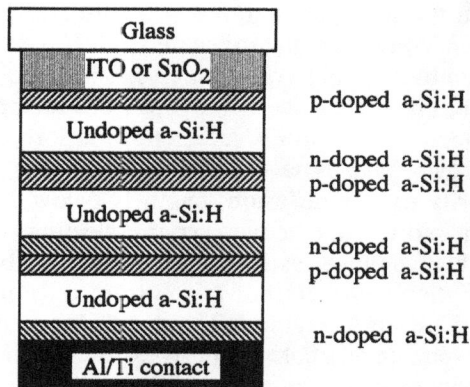

Figure 3.12 Cross-sectional representation of a tandem *pin* structure for an a-Si:H solar cell.

with different band gaps are employed. Practical systems differ in the method by which the light is directed towards the semiconductor with the appropriate band gap. One concept separates the light by using spectrally sensitive mirrors before it is directed to the appropriate cell, whereas in the second case, the tandem cell approach, a series of cells is stacked upon each other. The cells are arranged from the top side with decreasing band gaps, so that light with longer wavelengths and lower absorption coefficients passes through the upper cells and is absorbed into the bottom cells. The latter concept seems to be more compatible with today's semiconductor thin-film technologies and has been the one mainly pursued. Since the design of a cascade system adds extra complexity to the solar cell, it is better suited for concentrator concepts.

Each band gap cell has a different current and voltage output. Although it is possible to have an individual circuit for each cell, the device structures become more complex. Therefore, in most cases the cells are connected in series. Tandem configurations can be completely integrated with two terminals or connected only mechanically, with four terminals, as shown in Figure 3.13. The advantage of the

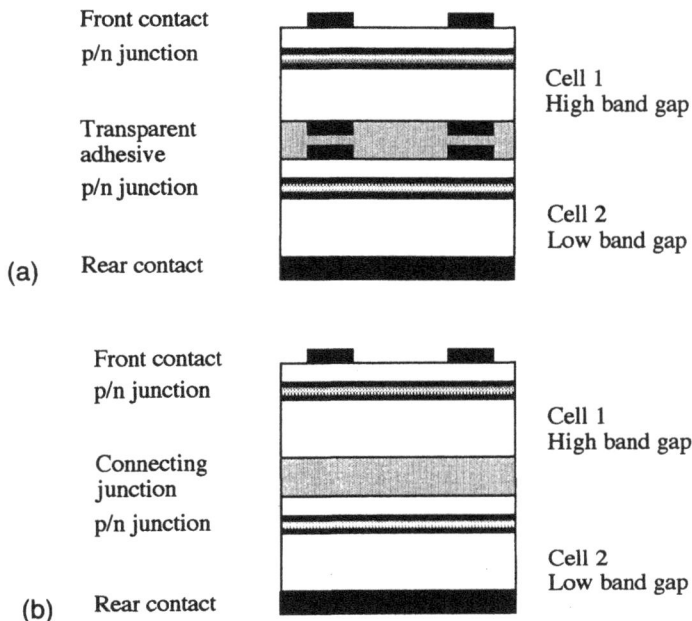

Figure 3.13 Cross-sectional representations of tandem cell structures: (a) four-terminal, mechanically stacked, the connecting material is a transparent adhesive; (b) two-terminal, monolithic cell, the connecting junction is transparent and of low resistance.

four-terminal design is that the number of available semiconductors is significantly larger for this structure and thus allows greater flexibility. In the first case, there is the problem of fabricating a conductive and optically transparent connection between the top and bottom cell.

A difficulty of series-connected cells is that the performance is determined by the worst cell in the system. This can be seen by considering a tandem system of two cells where each diode can be described by an ideal current-voltage equation:

$$I = I_{S1}\left\{\exp \frac{eU_{D1}}{KT} - 1\right\} - I_{L1} \qquad I = I_{S2}\left\{\exp \frac{eU_{D2}}{KT} - 1\right\} - I_{L2} \qquad (3.7)$$

In a series connection, the external voltage U_a is the sum of those of the individual cells $U_a = U_{D1} + U_{D2}$. Since the current is the same in the circuit, one can eliminate the voltages U_{D1} and U_{D2} and derive the current-voltage equation for the tandem cell, which is

$$I = \left\{I_{S1}I_{S2} \exp \frac{eU_a}{KT} + \frac{1}{4}(I_{L1} + I_{S1} - I_{L2} - I_{S2})^2\right\}^{1/2}$$
$$- \frac{1}{2}(I_{L1} + I_{S1} - I_{L2} - I_{S2}) \qquad (3.8)$$

A schematic representation is depicted in Figure 3.14. It is easy to show that the open-circuit voltage is the sum of the open-circuit voltages of each individual cell. Another property becomes apparent if one considers a tandem system where one diode has a much lower short-circuit current; for instance, $I_{L1} \ll I_{L2}$. In this case, one can approximate equation 3.8 and obtains

$$I = I_{S1}\left\{\frac{I_{S2}}{I_{L2} - I_{L1}} \exp \frac{eU_a}{KT} - 1\right\} - I_{L1} \qquad (3.9)$$

The short-circuit current $I_{sc} \approx -(I_{L1} + I_{S1})$ is thus mainly determined by the short-circuit current of the bad cell. To maintain a good efficiency of the multicell, it is thus necessary to develop a design where each cell has approximately the same short-circuit current.

Numerical programs that incorporate individual cell performance parameters (e.g., current-voltage characteristics and spectral quantum efficiencies) are currently applied to calculate optimal configurations for tandem solar cells. Some calculations of the efficiencies of a tandem structure as a function of the band gaps for the top and bottom cells are shown in Figure 3.15. The band gap of the top cell can vary between 1.5 and 1.9 eV, and for the bottom cell between 0.95 and

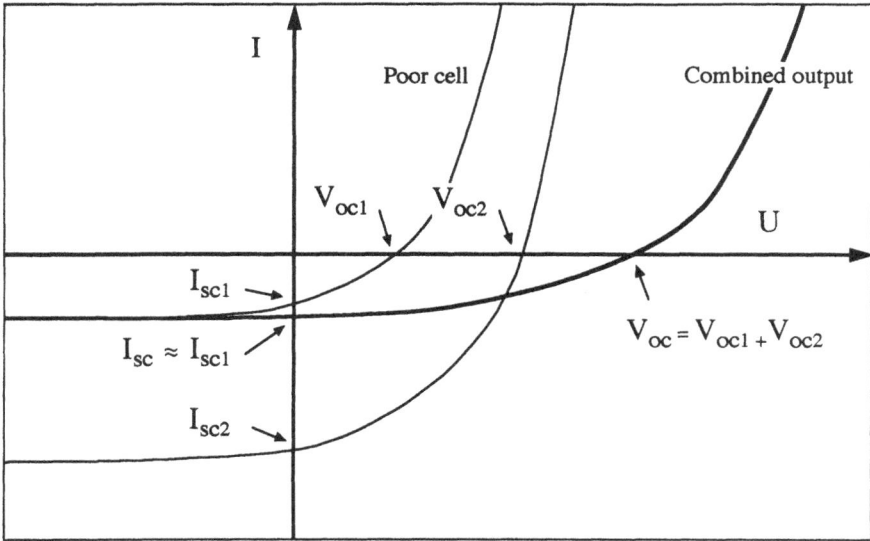

Figure 3.14 Current-voltage curve for two series connected cells. The short-circuit current of the combined output is determined by the poor cell.

1.2 eV. Recently, a tandem cell using a silicon bottom cell and a GaAs top cell was reported with a terrestrial concentrator efficiency of 35% at $C = 350$, which is the highest efficiency reported so far for a photovoltaic device [19,20].

The tandem concept seems especially suitable for thin-film solar cells, and currently a number of systems such as a-Si:H/a-Si-Ge:H (efficiency 12.7%), a-Si:H/ CuInSe$_2$ (efficiency 14.6%) or CuGaSe$_2$/CuInSe$_2$ are being considered [12].

Calculations of the efficiencies of cascade systems with more than two semiconductors show that for an ideal system the efficiency saturates for about $n > 7$ [21]. Ideally, this approach doubles the efficiency compared to a single-cell system. In reality, however, one has to consider additional unavoidable optical losses when connecting the different cells, which will eventually decrease the total efficiency of a multigap system (Figure 3.16). An optimized system will therefore most likely consist of about two to four cells stacked upon each other.

3.6 CONCENTRATOR CELLS

The concentration of sunlight by lenses and mirrors offers an attractive and flexible approach to reducing cell costs. In general, the higher the concentration ratio C

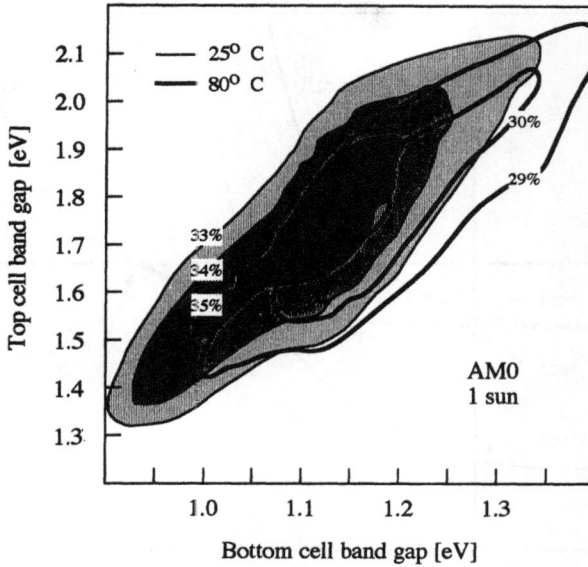

Figure 3.15 Calculated isoefficiency contours for a voltage-matched two-junction tandem structure as a function of the band gaps (AM0 spectrum at 1 sun, and for 25°C and 80°C). The vertical axis is the band gap of the upper cell and the horizontal axis is the band gap of the lower cell [20].

the smaller is the range of angles of the incident light rays that can be accepted by the system. Once the concentration ratio exceeds about 10, the system can use direct sunlight only and a tracking system is required to follow the sun across the sky. The diffuse part of the sunlight, which amounts to about 20% on a clear sky, is wasted in such a system. For low concentration ratios (<20), it may be sufficient to adjust the concentrator periodically.

A novel approach to concentrating the sunlight uses luminescent concentrators. These are glass or plastic sheets doped with luminescent dyes that absorb sunlight and re-emit it in a narrow frequency band in all directions. The light is partly trapped in the sheet by total reflection and can be collected at the edges by the attached solar cell. Such a concentrator can accept light from all directions and therefore does not require a tracking system. Another advantage is the emission of light in a narrow range of wavelengths which can be used to adjust the semiconductor band gap of the cell material to the maximum intensity of the emitted spectrum. The concentration of the light is determined by the absorption in the sheet and the efficiency losses of the luminescence of the material.

Concentrating the sunlight on a cell will also raise the operating temperature, which will decrease the efficiency, as was discussed in Section 2.4.3. Active cooling

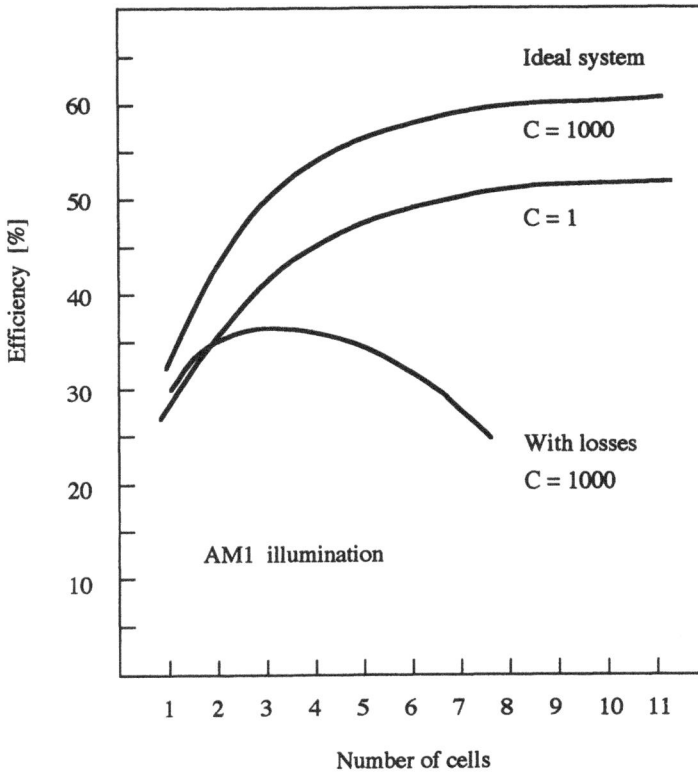

Figure 3.16 Calculated maximum efficiency for an ideal multigap system for normal (1 sun) and concentrated light (1000 sun) and an AM 1 spectrum. The effect of optical losses for concentrated light is also shown [21].

is therefore required for concentration ratios above 50. It is entirely feasible that a photovoltaic system in the future uses both the electric power and the thermal energy collected by the cooling system.

The geometrical concentration ratio C of an ideal lens or mirror system as a function of the range of angles, θ, is different for a linear- and point-concentrating system, and is given by

$$C_{2D} = \frac{1}{\sin(\theta/2)} \qquad C_{3D} = \frac{1}{\sin^2(\theta/2)} \tag{3.10}$$

The three-dimensional concentrators give a higher concentration ratio for the same range of angles and allow a maximum concentration of 45,000 for an angle of about $\theta = 0.5$ deg, which is determined by the finite size of the sun's disk. Almost ideal concentration ratios are obtained by the *compound parabolic concentrator* (CPC). This system consists of two parabolic reflectors that are separately aligned with their focal points, as shown in Figure 3.17. Other systems use standard planoconvex or Fresnel lenses.

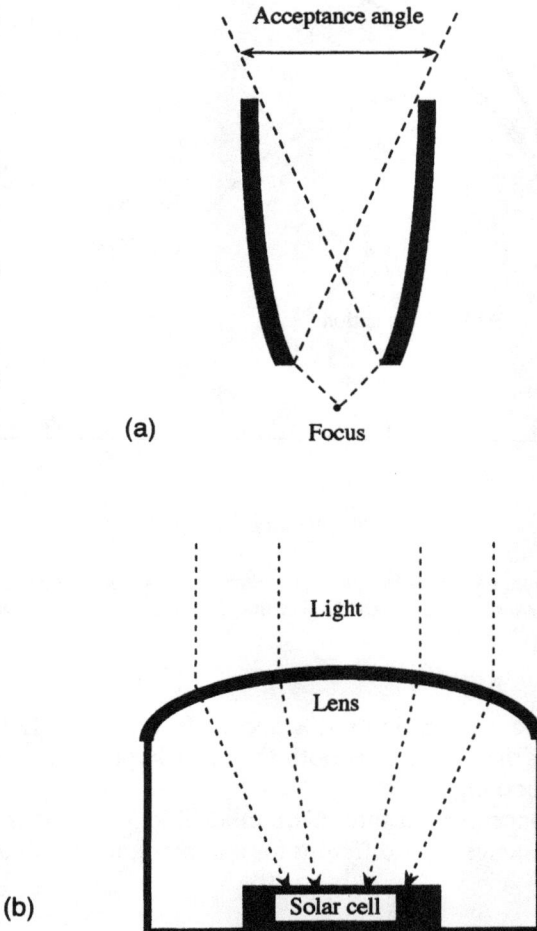

Figure 3.17 Schematic diagrams: (a) nonimaging compound concentrator with parabolic reflecting surfaces; (b) solar cell with an integrated concentrator lens.

The higher light intensity primarily changes the light-generated current I_L, which increases linearly with the intensity according to (2.36). The total current I under these conditions is still approximately given by (2.38), with a saturation current given by (2.39):

$$I_S = I_1 T^3 \exp\left(-\frac{E_g}{KT}\right) \qquad (3.11)$$

If the minority carrier density approaches that of the substrate doping, a high-injection condition prevails and the saturation current is given by [22]

$$I_S = I_2 T^{3/2} \exp\left(-\frac{E_g}{2KT}\right) \qquad (3.12)$$

where I_1 and I_2 are weakly dependent on the temperature and combine all transport and geometrical parameters and the doping concentrations of the device. They are usually obtained empirically. The open-circuit voltage becomes

$$V_{oc} = \frac{2KT}{e} \ln\left(\frac{I_L}{I_S} + 1\right) \qquad (3.13)$$

and increases logarithmically with the light intensity. Thus, for a fixed temperature the efficiency under light concentration increases, since both open-circuit voltage and short-circuit current increase. Since the open-circuit voltage decreases as a function of the temperature, the efficiency is reduced if the solar cell operates at elevated temperatures, as was discussed in the previous chapter.

With increasing current densities, the series resistance becomes increasingly important and eventually decreases the efficiency again due to a reduced fill factor, as was shown in Section 2.4. Solar cells that are intended to operate in a concentrator system have to be designed to minimize the series resistance. This requires the use of low-resistivity substrates, minimization of the sheet resistance of the top emitter layer, and reduction of the contact resistance. The general behavior of a solar cell is therefore an increase in the efficiency at lower light intensities and a decrease for higher values. In practical systems, the peak efficiencies occur at concentration ratios between twenty and several hundred. Efficiencies of about 20% have been reported for concentrator cells on the basis of silicon [23] and about 25% for GaAs [24], which is still well below the extreme efficiency limit of about 37% (see Figure 2.11).

REFERENCES

[1] Verhoef, L. A., *Ph.D. Thesis*, University of Utrecht, Holland, 1990.

[2] Ourmazd, A., and W. Schröter, *Appl. Phys. Lett.*, Vol. 45, 1984, p. 781.

[3] Lindmayer, J., and J. F. Allison, *Conf. Rec. 11th Photovolt. Conf.*, New York: IEEE, 1972, p. 40.

[4] Green, M. A., A. W. Blakers, S. R. Wenham, S. S. Narayanan, M. R. Willison, M. Taouk, and T. Szpitalak, *Proc. 18th Photovolt. Specialists Conf.*, New York: IEEE, 1985, p. 39.

[5] Green, M. A., J. Zhao, A. Wang, C. M. Chong, S. S. Narayanan, A. W. Blakers, ad S. R. Wenham, *Proc. 8th E.C. Photovolt. Solar Energy Conf.*, Dordrecht: Kluwer Academic Publ., 1988, p. 167.

[6] Swanson, R. M., *Proc. 8th E.C. Photovolt. Solar Energy Conf.*, Dordrecht: Kluwer Academic Publ., 1988, p. 156.

[7] Mitchel, K. W., C. Ebersbacher, J. Ermer, D. Pier, and A. Milla, *Proc. 8th E.C. Photovolt. Solar Energy Conf.*, Dordrecht: Kluwer Academic Publ., 1988, p. 167.

[8] MacMillan, H. F., N. R. Kaminar, M. S. Kuryla, M. J. Ladle, D. D. Liu, and G. G. Virshup, *Proc. 20th IEEE Photovolt. Specialists Conf.*, New York: IEEE, 1988, p. 462.

[9] Fraas, L. M., *Current Topics in Photovoltaics*, T. J. Coutts and J. D. Meakin, eds., New York: Academic Press, 1985, p. 169.

[10] Weinberg, I., *Solar Cells*, Vol. 29, 1990, p. 225.

[11] Keavney, C. J., V. E. Haven, and S. M. Vernon, *Proc. 2nd Int. Conf. on InP and Related Materials*, Piscataway, NJ: IEEE-LEOS, 1990.

[12] Kazmerski, L. L., *Intern. Mat. Rev.*, Vol. 34, 1989, p. 185.

[13] Meyers, P. V., *Solar Cells*, Vol. 27, 1988, p. 91.

[14] Johnston, W. D., and W. M. Callahan, *Appl. Physics Lett.*, Vol. 28, 1976, p. 150.

[15] Sze, S. M., *Physics of Semiconductor Devices*, New York: Wiley & Sons, 1981, p. 273.

[16] Sze, S. M., *Physics of Semiconductor Devices*, New York: Wiley & Sons, 1981, p. 825.

[17] Hovel, H. J., *Semiconductors and Semimetals*, Vol. 11, R. K. Willardson and A. C. Beer, eds., New York: Academic Press, 1975, p. 195.

[18] Catalano, A., R. R. Arya, M. Bennett, L. Yang, J. Morris, B. Goldstein, B. Fieselmann, J. Newton, and S., Wiedemann, *Solar Cells*, Vol. 27, 1988, p. 25.

[19] Gee, J. M., and G. F. Virshup, Proc. *20th IEEE Photovolt. Specialists Conf.*, New York: IEEE, 1989, p. 754.

[20] Wanlass, M. W., K. A. Emery, T. A. Gessert, G. S. Horner, C. R. Osterwald, and T. J. Coutts, *Solar Cells*, Vol. 27, 1989, p. 191.

[21] Bennett, A., and L. C. Olsen, *Conf. Rec. 13th IEEE Photovolt. Specialists Conf.*, New York: IEEE, 1988, p. 868.

[22] Sze, S. M., *Physics of Semiconductor Devices*, New York: Wiley & Sons, 1981, p. 94.

[23] Boes, E. C., *Proc. 14th IEEE Photovolt. Specialists Conf.*, New York: IEEE, 1980, p. 994.

[24] Sahai, R., D. D. Edwall, and J. S. Harris, *Proc. 13th IEEE Photovolt. Specialists Conf.*, New York: IEEE, 1978, p. 946.

Chapter 4
Fundamental Material Parameters

The band gap, the lifetime of minority carriers, and the mobility of these carriers have been identified as the main material parameters controlling the performance of solar cells. The perfect crystal determines the band structure of a semiconductor and the ideal values for the lifetime and the mobility. Real crystals, however, contain imperfections in the crystal structure that change these parameters and are primarily responsible for the degradation of the efficiency of a solar cell.

Impurities, grain boundaries and interfaces, and dislocations influence the electrical properties of a semiconductor in two ways: the trapping states introduced by the lattice defects enhance the recombination velocity, and the scattering of the charge carriers at the lattice defect reduces the mobility. Consequently, the elimination of lattice defects and the use of a high-quality material is one possibility for enhancing the performance of the solar cells. However, under the constraint of low production costs, expensive material purification and processing steps may be prohibitive in the future. In silicon, for instance, dislocation densities $N_D < 10^6$ cm^{-2} and grain sizes larger than about 1 mm are considered tolerable for a solar cell with an efficiency of about 10%. Whereas in many semiconductors the dislocation densities can be kept below this value, as will be discussed in later chapters, polycrystalline ribbons and thin-film materials have grain sizes below 1 mm or in the micron range, respectively. In this case, only semiconductors with high absorption coefficients are suitable so that electron-hole pairs can be generated close to the *pn* junction.

The major goal in improving the performance of a solar cell must be to limit the number of electrically active lattice defects and reduce the recombination activity of these defects, for instance, by passivation. This objective requires a thorough understanding of the origin and physical behavior of these defects in each material.

4.1 MOBILITY OF CHARGE CARRIERS

Generally, the mobility μ of electrons and holes is determined by a variety of scattering processes. If several mechanisms operate at the same time, the contribution of the individual values μ_i to the total mobility is usually given by the summation rule

$$\frac{1}{\mu} = \sum \frac{1}{\mu_i} \qquad (4.1)$$

A particular situation occurs if the charge transport involves several mechanisms with different scattering processes. This can be the case, for instance, for amorphous semiconductors where electrons (or holes) occupy both extended and localized states and the conduction mechanisms in these states are different (see Chapter 9). If the number of electrons in the different states is N_i, one instead obtains for the mobility [1]

$$\frac{1}{\mu} = \sum \frac{a_i}{\mu_i} \qquad a_i = \frac{N_i \mu_i^2}{\sum N_j \mu_j^2} \qquad (4.2)$$

The drift mobility of charge carriers in extended states (conduction and valence band) is determined by scattering mechanisms involving both lattice vibrations (phonons) and lattice imperfections. In single-crystal semiconductors, dopant impurities are usually the crystal defects with the highest concentrations, whereas in polycrystalline thin films, the grain boundaries play the dominant role as scattering centers. It should be noted that in non-stoichiometric compound semiconductors doping can also occur through intrinsic point defects such as vacancies or antisite defects. These point defects scatter electrons and holes in a way similar to foreign atom impurities. Since their concentrations can become quite high, these point defects also contribute significantly to a decrease of the mobility. Theoretical calculations for various scattering mechanisms in single crystals give the following general relationship for the temperature dependence of the mobility:

$$\mu_i = CT^s \qquad (4.3)$$

where C is a constant or only weakly (logarithmically) dependent on the temperature. A fundamental scattering mechanism in any crystal is the interaction with phonons. Calculations for non-polar crystals such as silicon show that scattering with *acoustic* and *non-polar optical* phonons is characterized by the exponents $s = -3/2$ and the constants C given by [2–4]

$$C_{AP} = \frac{eMh^4c_s^2}{3\pi^{7/2}(2m^*)^{5/2}K^{3/2}\Omega_0 E_1^2} \tag{4.4a}$$

$$C_{ONP} = \frac{eMh^4\omega_0^2}{3\pi^{7/2}(2m^*)^{5/2}K^{3/2}\Omega_0 E_0^2} \tag{4.4b}$$

with the effective mass of the scattered charge carrier, m^*, the volume of the unit cell of the crystal, Ω_0, the atomic mass, M, Planck's constant, h, the electronic charge, e, and the sound velocity, c_s. E_1 and E_0 (per unit length) are the deformation potentials for acoustic and optical phonons. In polar crystals with two different kinds of atoms of mass M_1 and M_2 (e.g., in compound semiconductors), the interaction with *polar optical* phonons is an essential contribution and characterized by the exponent $s = -1/2$ and

$$C_{OP} = \frac{M_1M_2h^2\Omega_0\omega_0^2}{3\,(M_1 + M_2)Z^2e^3\pi^{7/2}(2m^*)^{3/2}K^{1/2}} \tag{4.5}$$

where ω_0 is the characteristic frequency of optical phonons. Phonon scattering is different for electrons and holes because of their different effective masses, and it is characterized by a decrease of their mobility with increasing temperature.

The theory for scattering at ionized (usually doping) point defects yields $s = 3/2$, which yields a positive slope for the temperature dependence. For neutral impurities, which dominate at low temperatures, one obtains $s = 0$, and the mobility becomes temperature independent. In the first case, C depends on the concentration N_T of the defects and its charge Ze, and the mobility is given by the Brooks-Herring equation [5]:

$$C = \left(\frac{2K}{\pi}\right)^{3/2} \frac{4e^2}{Z^2e^3m^{*1/2}N_T} \left\{\ln\frac{24\pi m^* \epsilon\epsilon_0(KT)^2}{e^2hN_T}\right\}^{-1} \sim \frac{1}{(N_T)^r} \tag{4.6}$$

If one ignores the weak dependence on N_T in the logarithm, one obtains $r = 1$. In addition to the scattering mechanism discussed above, other mechanisms also affect the mobility. Therefore, in real crystals, the exponents and prefactors differ from the theoretical values and are better determined empirically. For instance, the temperature dependence of the mobilities of electrons and holes in pure silicon at room temperature is characterized by the exponents $s = 2.42$ and $s = 2.20$, respectively, and in GaAs $s = 1.0$ and $s = 2.1$, respectively.

In crystals containing charged point defects, one measures a combined mobility according to Matthiessen's rule (equation (4.1)). Because of the positive and negative temperature coefficients for the point defect and phonon scattering, the mobility has a maximum in a certain temperature range which depends on the impurity concentration. Measurements on n-type silicon, for instance, show a maximum of $\mu \approx 5000$ cm^2/Vs at $T = 50$K for a doping concentration $N \approx 10^{17}$ cm^{-3}. At room temperature, the mobility is mainly determined by phonon scattering, but the additional impurity contribution decreases the mobility with increasing impurity concentrations $N \geq 10^{15}$ cm^{-3}. Exponents of $r = 0.38$ for holes and $r = 0.42$ for electrons have been determined, which is smaller compared to the theoretical estimation $r \approx 1$ [6,7]. Such high doping concentrations occur in various types of silicon solar cells in the emitter region (see Chapter 3); hence, the doping dependence of the mobility has to be taken into account in a realistic calculation of a solar cell device [8,9].

Dislocations and grain boundaries also scatter charge carriers and reduce the mobility in deformed and polycrystalline semiconductors, respectively. A major factor that contributes to the scattering mechanism is the Coulomb interaction with the electrical charge that accumulates at the lattice defects and the corresponding screening charge in the vicinity. The origin of the defect charge is a consequence of the electronic levels of these defects, which are introduced in the forbidden band gap, as will be discussed in detail in the next section for surfaces, grain boundaries, and dislocations. Experimental data on the mobility in the presence of grain boundaries and dislocations can be obtained from conductivity and Hall effect measurements in polycrystalline or deformed semiconductors, respectively.

For *dislocations*, a comprehensive study of the mobility μ_D in deformed low-doped silicon and germanium has shown that at low temperatures ($T < 150$K) scattering at dislocations dominates, and the mobility decreases with decreasing temperature [10]. A theoretical calculation [11] of the mobility yields two terms which are the contributions of the scattering of charge carriers in the long-range strain field and at the line and screening charge:

$$\frac{1}{\mu_D} = A\,\frac{N_D}{T} + B\,\frac{\lambda N_D Q_D^2}{T^{3/2}} \tag{4.7}$$

where Q_D is the line charge (per unit length) and N_D is the dislocation density. The line and compensating screening charge produce an electric field and a potential barrier V_D at the dislocation, which can be calculated from Poisson's equation (Figure 4.1). The results for V_D show that the decay of the potential with the distance from the dislocation r is characterized by a screening length λ. Several authors have derived approximations for λ that are valid for different doping and temperature regimes. If the temperatures are not too low, λ is given by the Debye

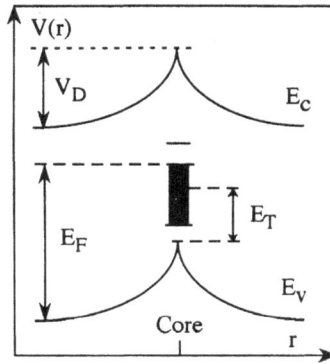

Figure 4.1 Schematic representation of the potential barrier $V(r)$ at a charged dislocation as a function of the distance r from the dislocation line. E_T is the occupation limit of the energy states of neutral dislocation.

screening length, and for a p-type crystal, the following expressions for the dislocation charges Q_D and λ are obtained:

$$\lambda = \left\{ \frac{\epsilon\epsilon_0 KT}{e^2 p} \right\}^{1/2} \qquad Q_D = e\,\frac{N_a - p}{N_D} \tag{4.8}$$

where N_a is the acceptor and p is the average hole concentration. Because of the trapping of charge carriers at the dislocations, the average free carrier concentration p in the crystal differs from the acceptor concentration, particularly in cases where the dislocation density is high. In general, the line charges Q_D and p are complex functions of the temperature and doping level. For low doping levels and high dislocation densities, one can approximate $Q_D \approx eN_a/N_D$, whereas for high doping levels, all dislocation traps are filled so that Q_D remains constant and $p \approx N_a$. The constants A and B in (4.7) are parameters that are calculated in the model, but can also be determined empirically. A good agreement with mobility measurements as a function of the temperature has been obtained with adjusted parameters in silicon and germanium. The results show that dislocation scattering becomes important for dislocation densities $N_D > 10^7$ cm^{-2}. In solar cell semiconductors, the dislocation densities usually can be kept below this value so that their contribution to the mobility can be ignored, but they can still be effective recombination centers at lower densities, as will be discussed in Section 4.2.4.

Mobility measurements on polycrystalline thin-film semiconductors yield information about the influence of *grain boundaries*. Because of the technological importance of polycrystalline materials for applications other than solar cell applications, the conductivity and mobility of charge carriers has been studied extensively in a number of semiconductors and will be discussed separately in Section 5.5. The main result is that, in the case of small grain sizes, the crystal is almost completely depleted from carriers that are trapped in the grain boundary states. The potential barriers V_B and screening charges of adjacent grain boundaries overlap in this case. For this situation, the mobility μ_B of electrons (for an n-type crystal, for instance) can be obtained from the zero bias conductivity $\sigma = en\mu_B$, which has been calculated by several authors for polycrystalline thin films and is discussed in Section 5.5 (equation (5.106))

$$\mu_B = A(T) \exp\left\{-\frac{V_B}{KT}\right\} \tag{4.9}$$

$A(T)$ is a prefactor that depends on the model for the calculation of the conductivity. In all models, $A(T)$ is weakly dependent on the temperature. The grain boundary potential (as will be discussed in Section 5.3) is a complex function of the doping concentration, temperature, and grain size. Numerical calculations and experimental results have shown that the potential as a function of the doping concentration N_d (for n-type crystals, for instance) has a maximum at $N_d \approx N_B/L$, where N_B is the number of grain boundary states per unit area and L is the grain size. Correspondingly, the mobility has a minimum, whereas for low doping levels μ_B remains constant [12,13]. For high doping levels and larger grain sizes, the mobility increases with the doping concentration and approaches the single-crystal values. The mobility in fine-grained polycrystalline semiconductors is usually reduced compared to single crystals and can thus be a major limitation for the efficiency of a polycrystalline material.

The mobility is frequently determined by Hall and conductance measurements. If only one scattering mechanism is dominant, the Hall mobility μ_H differs from the drift mobility μ by a constant factor only. The proportionality constant depends on the scattering mechanism, and for acoustic and non-polar phonon scattering, it is $\mu_H/\mu = 1.18$, for polar phonon scattering 1.11, and for ionized impurity scattering 1.93, respectively. In heavily deformed and polycrystalline semiconductors, the crystals become rather inhomogeneous, and the interpretation of Hall effect measurements is less straightforward, as is discussed in [10]. Therefore, the relationship between Hall and drift mobility is more complicated, so that in many practical cases the difference is ignored completely. This may be justified if the electrical properties of the semiconductor or the efficiency of a solar cell are determined by other, more important factors.

Hall mobility measurements can be very useful in the characterization of new semiconductor materials, where information about the defect structure and impurity concentrations are very limited. If it can be established experimentally that one scattering mechanism is dominant in a certain temperature range, the doping and temperature dependence can be used to determine some of the properties of the scattering defect, such as the concentration of ionized impurities or dislocations.

4.2 RECOMBINATION OF CHARGE CARRIERS

Whenever the thermal equilibrium of a semiconductor is disturbed by the generation of electrons and holes, recombination processes exist to restore the system to equilibrium. In the non-equilibrium situation, the electron and hole concentrations deviate from their equilibrium values and are determined by the quasi-Fermi energies (equation (2.11)). The charge carrier product np differs from the equilibrium value $np \neq n_i^2$, where n_i is the intrinsic carrier concentration. The recombination processes are characterized by the lifetimes of electrons and holes, which can vary over several orders of magnitude—from several hours to 10^{-8} sec or less—depending on the physical process. Recombination also occurs in thermal equilibrium and balances the thermal generation of electron-hole pairs. The thermal equilibrium is characterized by the condition that the rate of thermally stimulated transitions between any two energy levels is equal to the rate of the recombination process (principle of detailed balance). It is useful to distinguish between two classes of recombination mechanisms: the band-band recombination and recombination through the participation of crystal defects. Generally, all processes occur simultaneously and the total recombination rate is the sum of all individual processes. If the individual lifetimes are τ_i, their contribution to the total lifetime is given by the summation rule

$$\frac{1}{\tau} = \sum_i \frac{1}{\tau_i} \qquad (4.10)$$

The shortest lifetime thus usually dominates the total recombination in the crystal. The lifetime for a particular recombination process can be calculated by considering the generation and recombination rates G_n and G_p, and R_n and R_p, respectively, between two energy levels. For bulk recombination processes at a single trap level, the recombination rates are defined as the number of transitions of electrons or holes per unit time and volume. For recombination at extended defects, such as surfaces, interfaces, or dislocations, R_n and R_p are defined per unit time and area.

If δn and δp are the excess concentrations of electrons and holes as defined in (2.10), their time dependence in a non-equilibrium situation is given by the continuity equation. For the case that no drift or diffusion currents flow in the semiconductor, one can write

$$\frac{\partial \delta p}{\partial t} = G_p - R_p \qquad \frac{\partial \delta n}{\partial t} = G_n - R_n \qquad (4.11)$$

It is usual to separate the thermal generation of charge carriers and include it in the recombination rates R_n and R_p. If g_n and g_p are the thermal generation rates of electrons and holes and r_n and r_p are the corresponding capture rates, one can write

$$R_n = r_n - g_n \qquad R_p = r_p - g_p \qquad (4.12)$$

In thermal equilibrium, the recombination rates are zero and $g_n = r_{n0}$ and $g_p = r_{p0}$. This condition can be used to determine the thermal generation rates from the equilibrium recombination rates r_{n0} and r_{p0}.

It will be shown in the following that in many cases R_n and R_p are proportional to the excess carrier concentrations δn and δp:

$$R_n = \frac{\delta n}{\tau_n} \qquad R_p = \frac{\delta p}{\tau_p} \qquad (4.13)$$

where τ_n and τ_p are defined as the lifetimes for electrons and holes, respectively. (This approximation is frequently used in the general expression for the continuity equation. See (2.13).) For extended defects, the proportionality factors have the dimensions of an inverse velocity; therefore, the lifetimes have to be replaced by the recombination velocities S_n and S_p. Under illumination, the generation rates for electrons and holes are equal: $G_p = G_n = G$. Therefore, one obtains in the steady state from (4.11) and (4.13) the conditions that

$$R_n = R_p \quad \text{and} \quad \delta n\, \tau_p = \delta p\, \tau_n \qquad (4.14)$$

If the concentrations of recombination centers are small, one can assume that $\delta n = \delta p$ and obtain a unique lifetime $\tau = \tau_p = \tau_n$. The calculation of the lifetimes or of the recombination velocities requires a detailed analysis of the particular recombination mechanism.

4.2.1 Band-Band Recombination

The direct recombination of an electron-hole pair when an electron descends from the conduction to the valence band is a process that cannot be suppressed in a crystal, and the corresponding lifetime thus gives an upper limit for the total lifetime in a semiconductor. Analogous to the band-band transition of electrons during the absorption of a photon, the total energy and momentum of the system have to be conserved. Several processes have to be distinguished depending on how the excess energy is dissipated and the momentum exchanged. The energy can be released by the emission of a photon (radiative recombination) or by the excitation of a second free electron or hole (Auger recombination). In the latter case, the excited electron (or hole) finally transfers its energy to the lattice by the interaction with phonons. In indirect semiconductors, the electron transition from the conduction to the valence band requires a change in the momentum through the participation of phonons so that part of the energy is exchanged with the lattice directly.

For the calculation of the lifetime, one can begin with the general expression for the number of transitions r (per unit time and volume) of electrons or holes between two groups of energy levels, which may be characterized by their density of state distributions $N_1(E)$ and $N_2(E)$. The transition rate is proportional to the number of electrons in group 1 and proportional to the number of unoccupied states in group 2 and is thus given by

$$r = \int s(E, E')N_1(E)f(E)N_2(E')(1 - f(E'))dEdE' \qquad (4.15)$$

where $s(E, E')$ is the probability (per unit time) for the transition between the energy levels at E and E'. In general, the function $s(E, E')$ has to be calculated by quantum mechanical methods; however, in many cases it will be considered a constant parameter which can be determined from experimental measurements. In non-equilibrium, the occupation probability $f(E)$ has to be calculated separately, whereas in thermodynamic equilibrium it is equal to the Fermi-Dirac distribution $f_0(E)$.

In general, one assumes that the transition probability $s(E, E')$ for the radiative recombination is a constant, whereas for the Auger process it depends on the number of collisions with free electrons and holes in the conduction and valence bands and is proportional to their concentrations. Since all processes occur simultaneously, one can write for electrons and holes

$$s_n(E, E') = s_n + s_{an1}n + s_{an2}p \quad \text{(electrons)}$$
$$s_p(E, E') = s_p + s_{ap1}p + s_{ap2}n \quad \text{(holes)}$$

Substituting the expressions in (4.15), one can evaluate the integral directly and obtain the following recombination rates for electrons R_n and holes R_p (equation (4.12)):

$$R_n = s_n np + s_{an1}n^2p + s_{an2}p^2n - g_n \qquad (4.16a)$$
$$R_p = s_p np + s_{ap1}p^2n + s_{ap2}n^2p - g_p \qquad (4.16b)$$

The thermal generation rates can be derived from the condition that in thermal equilibrium $R_n = R_p = 0$ and $n = n_0$ and $p = p_0$. Substituting the non-equilibrium electron and hole concentrations $n = n_0 + \delta n$ and $p = p_0 + \delta p$ and assuming

$\delta n = \delta p \ll n_0$ (for n-type material) or $\delta n = \delta p \ll p_0$ (for p-type material), one can calculate the lifetimes from (4.13) and obtain

$$\frac{1}{\tau} = s_n p_0 + s_{cn1} p_0^2 + s_{an2} n_0 p_0 = \frac{1}{\tau_n} + \frac{1}{\tau_{an}} \quad (p\text{-type}) \quad (4.17)$$

$$\frac{1}{\tau} = s_p n_0 + s_{ap1} n_0^2 + s_{ap2} n_0 p_0 = \frac{1}{\tau_p} + \frac{1}{\tau_{ap}} \quad (n\text{-type}) \quad (4.18)$$

The equations show that the total lifetime is determined by the transition probabilities of the minority charge carriers and that the lifetime decreases with the doping concentration. For low doping levels, the probability of the Auger recombination (τ_{an}, τ_{ap}) is usually small compared to the radiative recombination, but because of the quadratic term it increases rapidly with the majority carrier concentration. In fact, in silicon it becomes the dominant recombination mechanism in heavily doped material [14], as can be seen in Figure 2.17, where the theoretically predicted quadratic dependence can be observed for doping concentrations $N \geq 10^{18}$ cm^{-3}. These are typical doping levels for the emitter region of a silicon pn junction solar cell.

The transition probabilities s_n, s_p, s_{an1}, s_{an2}, s_{ap1} and s_{ap2} are usually determined empirically from measurements of the doping dependence. Radiative recombination lifetimes are much larger in indirect than in direct semiconductors, which include most of the compound semiconductors. In pure intrinsic silicon it can be as long as 3 hr compared, for instance, to 6×10^{-7} sec for InSb. The band-band recombination, however, is not the only possible recombination process, and in semiconductors with a large band gap the lifetime is usually dominated by recombination at impurities and other lattice defects. For example, the recombination through impurity traps often reduces the lifetimes in doped silicon to 100 μs or less.

4.2.2 Shockley-Read-Hall Statistics

Point defects and extended lattice defects which introduce energy (trap) levels in the forbidden band gap reduce the lifetime of minority carriers and are thus important for the performance of photovoltaic semiconductors. The investigation of the electronic properties of crystal defects in semiconductors is a major research area, and the following chapters will review the electronic properties of crystal defects. For the following considerations, it suffices to describe the defects by the position of the trap levels they introduce.

Point defects are characterized by single energy levels, and basically two different types of defects can be distinguished: defects that introduce either one trap level E_T or defects with several levels E_{Ti}, which can be charged successively with electrons or holes. Typical impurities in the first group are mainly doping elements, and examples of multi-level defects are transition metal impurities in silicon.

The recombination through a single-level trap E_T, which can either capture or release an electron, will be derived first. In this case, $N_1(E)$ in (4.15) reduces to a sharp peak at the trap level energy E_T and can be described by a delta-function $N_1(E) = N_T\delta(E - E_T)$ if N_T is the total concentration of traps. In thermodynamic equilibrium, the occupation probability is given by the Fermi-Dirac distribution f_0, and there are $N_T f_0(E_T)$ occupied trap levels and $N_T[1 - f_0(E_T)]$ unoccupied levels. If the thermodynamic equilibrium is perturbed, by illumination for instance, the occupation function $f(E)$ is changed and has to be determined from the balance of the transition rates of the electrons and holes at the level. The occupation of the trap is determined by four different processes (Figure 4.2): the capture of electrons in the trap level and the thermal generation into the conduction band, and the capture of holes and the thermal generation into the valence band. The starting point for the calculation of the capture rates for electrons and holes is (4.15), which reduces for a single level to

$$r_n = \sigma_n v_T n N_T [1 - f(E_T)] \qquad r_p = \sigma_p v_T p N_T f(E_T) \qquad (4.19)$$

Instead of the transition probabilities $s(E, E')$, the constant capture cross sections σ_n and σ_p of electrons and holes, respectively, are introduced here. The thermal velocity of electrons is $v_T = (8KT/\pi m_0)^{1/2}$, and m_0 is the electron mass. The transition probabilities per unit time δt and volume are compared here with the number of

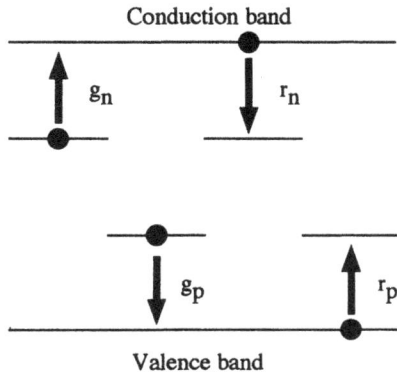

Figure 4.2 Schematic diagram of the electron and hole transitions for a recombination process at a trap level, defining basic parameters.

collisions of electrons and holes in a cylinder with the cross section σ_n or σ_p and a length $v_T \delta t$. For instance, for electrons, one obtains $s_n = \sigma_n v_T \delta t / \delta t = \sigma_n v_T$. The thermal generation rates are proportional to the number of occupied (for electrons) or unoccupied states (for holes) and can be expressed by

$$g_n = g_{n0} N_T f(E_T) \qquad g_p = g_{p0} N_T [1 - f(E_T)] \qquad (4.20)$$

In thermal equilibrium, the principle of detailed balance requires $r_n = g_n$ and $r_p = g_p$, which allows the determination of the coefficients g_{n0} and g_{p0}. One obtains for the probability (per unit time) that an electron (g_n) or hole (g_p) is thermally emitted from the trap level

$$g_n = \sigma_n v_T N_T n_1(E_T) f(E_T) \qquad g_p = \sigma_p v_T N_T p_1(E_T)[1 - f(E_T)] \qquad (4.21)$$

where $n_1(E)$, $p_1(E)$ are defined by

$$n_1(E) = N_C \exp\left(-\frac{E_c - E}{KT}\right) \qquad p_1(E) = N_V \exp\left(\frac{E_v - E}{KT}\right) \qquad (4.22)$$

Inserting the expressions (4.19) and (4.21) into the continuity equation (4.11) gives a set of differential equations, which can be solved for δn and δp. In the steady state and for $G_n = G_p = G$, one obtains a unique recombination rate R_0 and from (4.14) the condition

$$\sigma_n \{n(1 - f) - n_1 f\} = \sigma_p \{pf - p_1(1 - f)\} \qquad (4.23)$$

where f, n_1 and p_1 are taken at $E = E_T$. The equation can be solved for $f(E_T)$ and yields

$$f(E_T) = \frac{\sigma_n n + \sigma_p p_1}{\sigma_n(n + n_1) + \sigma_p(p + p_1)} \qquad (4.24)$$

Substituting (4.24) into the expression for R_0, one obtains, with $n_1 p_1 = n_i^2$,

$$R_0 = \frac{np - n_i^2}{\tau_{p0}(n + n_1) + \tau_{n0}(p + p_1)} \qquad (4.25)$$

The constants τ_{n0} and τ_{p0} are related to the concentration and capture coefficients of the trap center and are given by

$$\tau_{n0} = \frac{1}{\sigma_n v_T N_T} \qquad \tau_{p0} = \frac{1}{\sigma_p v_T N_T} \qquad (4.26)$$

The excess carrier concentrations $\delta n = n - n_0$ and $\delta_p = p - p_0$ (equation (2.10)) are related by a quasi-neutrality equation:

$$\delta p = \delta n + N_T(f - f_0) \qquad (4.27)$$

The Fermi-Dirac distribution can be expressed, for instance, by $f_0 = n_0/(n_0 + n_1)$. Combining (4.25) through (4.27), the recombination rate R_0 becomes proportional to δp or δn, respectively. Accordingly, one can determine the lifetimes for electrons and holes, which are approximately given for low excitation rates by

$$\tau_n = \frac{\tau_{p0}(n_0 + n_1) + \tau_{n0}(p_0 + p_1) + \tau_{n0}N_T f_0}{n_0 + p_0 + N_T f_0(1 - f_0)} \qquad (4.28a)$$

$$\tau_p = \frac{\tau_{p0}(n_0 + n_1) + \tau_{n0}(p_0 + p_1) + \tau_{p0}N_T(1 - f_0)}{n_0 + p_0 + N_T f_0(1 - f_0)} \qquad (4.28b)$$

For low trap concentrations N_T the lifetimes become equal and (4.28a) and (4.28b) reduce to the usual Shockley-Read-Hall (SRH) relationship:

$$\tau = \frac{\tau_{p0}(n_0 + n_1) + \tau_{n0}(p_0 + p_1)}{n_0 + p_0} \qquad (4.29)$$

The lifetime τ depends on the doping level, the excitation rate, the temperature, and the nature of the impurity (through the capture cross sections and the position of the trap level). The cross sections reported in the literature can vary over several orders of magnitude from 10^{-12} cm^2 to 10^{-22} cm^2. The large cross sections 10^{-12} cm^2 to 10^{-15} cm^2 are associated with attractive centers (e.g., positive charge states for electron trapping). The cross sections 10^{-15} cm^2 to 10^{-17} cm^2 are usually found for neutral centers, and the small cross sections 10^{-22} cm^2 for repulsive (negatively charged for electron trapping) centers.

The analysis of the SRH expression shows that the lifetime is mostly affected by (deep) trap centers in the middle of the band gap (n_1 and p_1 are large), whereas shallow level impurities reduce the lifetime less efficiently unless the capture cross sections are large. The lifetime τ decreases with increasing doping concentration first and remains constant for the higher doping level. In n-type and p-type material, one obtains $\tau \approx \tau_{p0}$ or τ_{n0}, respectively. The total lifetime is thus determined by the lifetime of minority carriers and depends inversely on the concentrations of trap level impurities.

Deep trap levels are frequently introduced by impurities such as transition elements, which can be occupied by several electrons or holes. Their level structure is characterized by a separate energy level E_{Ti} for each charge state ($i = 1, \ldots, \alpha$; see Section 5.1.3). The occupation of these multilevel impurities is not described by the usual Fermi-Dirac statistics anymore and requires a more general statistical approach. Generally, for recombination, the participation of all energy levels of the impurity has to be taken into account, which leads to a more complicated expression for the lifetime. However, if the different energy levels are separated by more than a few KT, which is usually the case, mainly one level (close to the Fermi level) participates in the recombination process, and one can apply approximately the SHR statistics for a single level. The occupation of the single level (at E_{Tm}, for instance) can be expressed then by a simplified expression. If N_T^m is the number of defects (per unit volume) occupied with m electrons and N_T^{m-1} with $m - 1$ electrons, they are related by

$$\frac{N_T^m}{N_T^{m-1}} = \exp \frac{E_F - E_{Tm}}{KT} \tag{4.30}$$

The other trap levels above E_{Tm} are mainly empty then, whereas levels below the Fermi energy are predominantly occupied. The total concentration N_{tot} is given by the sum over the concentrations of the defect in its different charge states:

$$N_{tot} = N_T^0 + \sum_{m=1}^{M1} N_T^{m-} + \sum_{m=1}^{M2} N_T^{m+} \tag{4.31}$$

$M1$ and $M2$ are the numbers of acceptor and donor states, respectively. The main difference to a single-level trap is that the active recombination level of a multilevel impurity changes as a function of the Fermi energy, which depends on the doping level and temperature. Hence, the influence of the impurity on the lifetime in the semiconductor may change considerably as a function of these parameters.

4.2.3 Recombination at Grain Boundaries

Point defects influence the bulk properties of a semiconductor, the lifetime and mobility of charge carriers, whereas extended lattice defects, such as surfaces, dislocations, grain boundaries, or interfaces, change the lifetime and mobility only locally. In the following, we will consider the electrical properties of a single grain boundary first. The main results, however, can also be applied to the recombination at interfaces, surfaces, and dislocation, which will be discussed in the following chapter. If the density of the extended defects is so high that their space charges interact with the adjacent defect, they cannot be considered independently from each other and a separate discussion is required. This situation can occur in crystals with high densities of grain boundaries or dislocations, such as polycrystalline semiconductors, which are of particular importance for solar cell applications. The electrical properties of this class of semiconductors will be discussed in Section 5.5 from the perspective of the properties of the individual defects.

The enhanced recombination at extended defects is due to additional electronic states in the band gap, which are introduced by the disorder of the crystal lattice and the occurrence of broken ("*dangling*") *bonds*. It will be discussed in more detail in Section 5.3 that the position and distribution of these energy states is related to the distortion and the reconstruction of covalent bonds, and also to the presence of impurities, which preferentially segregate or precipitate at the defect. Because of the two-dimensional nature of planar defects, their electronic states form a two-dimensional band in the band gap, which in most cases can be described by a continuous density of states distribution (per unit area), such as for a grain boundary $N_B(E)$. The band may be partly filled with electrons (Figure 4.3) up to a level E_{B0}, which is characteristic for the neutral defect and determined by the atomic structure and chemistry of the surface. If E_{B0} differs from the Fermi

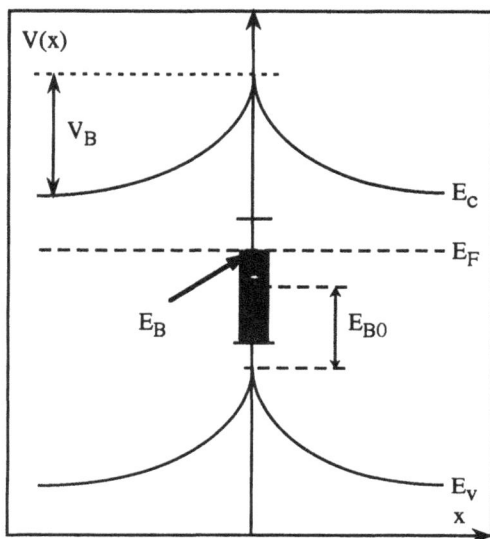

Figure 4.3 Schematic band diagram showing the symmetric potential barrier at a charged grain boundary. E_B and E_{B0} are the occupation limits of the energy states of the charged and neutral grain boundaries, respectively.

energy of the crystal E_F, the surface states will be filled with either electrons or holes. Since these charge carriers are localized at the defect, a charge Q_B builds up which is compensated by an equivalent space charge in the semiconductor to maintain neutrality. The corresponding electrical field and potential $\varphi_B(x)$ leads to a band bending that shifts the defect states until the occupation limit of the charged defect E_B equals the Fermi energy (or Quasi-Fermi level) of the crystal, $E_B = E_F$.

For the grain boundary, the compensating space charge lies symmetric to the boundary plane for a homogeneously doped crystal. The resulting potential barrier (sometimes described as a *double Schottky barrier*) and the corresponding depleted region in the vicinity of the boundary plane are regions of enhanced resistance. In polycrystalline and heterostructure solar cells, grain boundaries and interfaces are thus important for the electrical transport properties. In general, one has to consider that, under an applied voltage U, the occupation of the grain boundary states changes, and therefore the recombination and electrical transport across the grain boundary changes as well. The general problem is thus to calculate the potential barrier under bias U and illumination.

The potential $\varphi_B(x)$ and the grain boundary charge Q_B are calculated from Poisson's equation. Assuming an n-doped crystal (N_d doping concentration), for instance, and total depletion between the "screening lengths" x_L and x_R, one obtains

$$\frac{\partial^2 \varphi_B}{\partial x^2} = -\frac{1}{\epsilon\epsilon_0} \{Q_B \delta(x) + eN_d[\Theta(x + x_L) - \Theta(x - x_R)]\} \qquad (4.32)$$

with the boundary conditions $\varphi_B(-x_L) = 0$ and $\varphi_B(x_R) = U$. The charge distribution perpendicular to the grain boundary is approximated here by a step-like function. $\delta(x)$ and $\Theta(x)$ denote the delta and Heaviside functions, respectively. The general case that the space charge is additionally screened by deep trap centers in the vicinity of the boundary is be treated in Section 5.3. The problem's solution gives the following expression for the grain boundary barrier: $V_B = -e\varphi_L = -e\varphi_B(0)$ as a function of the applied voltage [15]:

$$V_B = \frac{V_c}{4} \left(1 - \frac{eU}{V_c}\right)^2 \qquad (4.33)$$

where the potential V_c is given by

$$V_c = \frac{Q_B^2}{8\epsilon\epsilon_0 N_d} \qquad (4.34)$$

The boundary charge $Q_B = -eN_{Be}$ is related to the number of excess or deficiency electrons at the grain boundary N_{Be} (see Figure 4.4)

$$N_{Be} = \int_{E_v + V_B}^{\infty} N_B(E) f_B(E + V_B) dE - N_{Bn} \qquad (4.35)$$

where N_{Bn} is the number of occupied states of the neutral grain boundary, and $f_B(E)$ is the occupation statistics for the interface states. The equation shows that the grain boundary charge depends on the height of the potential barrier, since the position of the surface states relative to the band edges is assumed to be fixed. An additional charge at the grain boundary changes the potential barrier and shifts all levels until the new equilibrium is reached for $E_B = E_F$. Thus, the barrier height and the boundary charge have to be determined self-consistently from (4.33) through (4.35). The potential barrier in this approximation is positive for an n-doped crystal and negative for p-type material. In general, however, the situation can be more complicated. For instance, in an n-type crystal, if the potential barrier is very large, an inversion layer could form near the grain boundary where the hole concentration exceeds the electron concentration. The barrier can also become

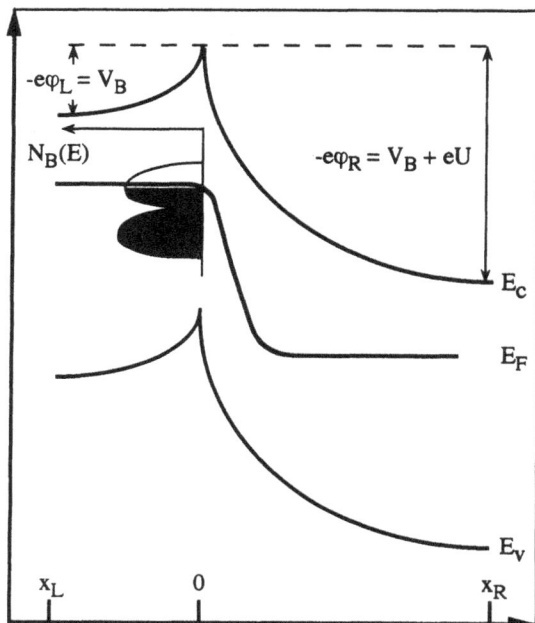

Figure 4.4 Band structure and potential barrier V_B of a grain boundary under bias voltage U. Density of states distribution $N_B(E)$ of grain boundary states in the insert.

negative in n-type material depending on the position of the Quasi-Femi level of the neutral boundary; thus, an accumulation layer would form. In these general cases, the depletion approximation for the charge distribution is not correct anymore and the potential has to be calculated using the complete expression for the charge density in Poisson's equation, which usually requires numerical methods.

It is evident that the calculation of Q_B requires the density of states distribution $N_B(E)$, which contains all the information about the electrical properties of the grain boundary. $N_B(E)$ is usually not known and has to be determined experimentally; however, for the following discussion these details are not necessary. The calculation of Q_B also requires the determination of the occupation function $f_B(E)$. In thermodynamic equilibrium, $f_B(E)$ is equal to the Fermi-Dirac distribution $f_0(E)$ but in a non-equilibrium situation, such as under bias and illumination, $f_B(E)$ has to be determined first.

Analogous to the case of a single defect level, $f_B(E)$ can be calculated from the recombination statistics for the grain boundary states. The starting point is the general expression (4.15) for the recombination probability between two different groups of energy states. Since a planar defect is considered here, the recombination rates for electrons R_{Bn} and holes R_{Bp}, which include the thermal generation of electrons and holes, are defined per unit time and area. If one assumes that the principle of detailed balance can be applied to each trap level separately, one can integrate over the conduction and valence band states and obtain

$$R_{Bn} = \int \{s_n(E)n_B N_B(E)(1 - f_B(E + V_B))$$
$$- g_n(E)N_B(E)f_B(E + V_B)\}dE \qquad (4.36a)$$

$$R_{Bp} = \int \{s_p(E)p_B N_B(E) f_B(E + V_B)$$
$$- g_p(E)N_B(E)f_B(E + V_B)\}dE \qquad (4.36b)$$

where n_B and p_B are the concentrations of electrons and holes in the conduction and valence band, respectively, at the grain boundary. The dimension of the capture coefficients $s_n(E)$ and $s_p(E)$ is that of a velocity in this case; therefore, they are usually denoted as the recombination velocities of grain boundary. The thermal emission of electrons and holes from the trap level into the conduction and valence bands, respectively, is characterized by the thermal emission rates g_n and g_p. They can be determined from the thermal equilibrium condition $R_{Bn} = R_{Bp} = 0$ and $f_B = f_0$, where f_0 is the Fermi-Dirac statistics.

A simplified treatment of the recombination process at a grain boundary under illumination can be obtained under the following assumptions (for a comprehensive discussion, see [16–18]). We assume a uniform generation of electron-hole pairs in the crystal under illumination and steady-state conditions. The recombination rates at the grain boundary states are therefore equal for electrons and holes: $R_{Bn} = R_{Bp} = R_0$ and determine the occupation function $f_B(E)$. Assuming again independence for each energy level, analogous to (4.24), one obtains

$$f_B(E) = \frac{s_n n_B + s_p p_1}{s_n(n_B + n_1) + s_p(p_B + p_1)} \quad (4.37)$$

where $n_1(E)$ and $p_1(E)$ are functions of the energy E and defined in (4.22). Inserting the expression into (4.36) yields

$$R_0 = \int \left\{ s_n(E) s_p(E) N_B(E) \frac{n_B p_B - n_1 p_1}{s_n(n_B + n_1) + s_p(p_B + p_1)} \right\} dE \quad (4.38)$$

The recombination rate R_0 of electrons and holes is completely determined, if one knows their concentrations n_B and p_B at the grain boundary position. Their calculation generally requires the solution of the continuity equation in the vicinity of the grain boundary. For simplicity, we assume in the following an n-type semiconductor and $n_B + n_1 \gg (p_B + p_1) s_p / s_n$. In this case, (4.38) with $n_1 p_1 = n_i^2$ reduces to

$$R_0 = \frac{n_B p_B - n_i^2}{n_B} \int s_p(E) N_B(E) f_B(E) dE \quad (4.39)$$

The occupation probability $f_B(E)$ in this approximation is given by

$$f_B(E) = \frac{n_B}{n_B + n_1} \quad (4.40)$$

In the approximation of the thermionic emission model, the concentration of the majority carrier n_B is related to the electron concentration at the edges of the depletion region, which is equal to the bulk concentration n far away from the barrier (see Figure 4.4) and given by

$$n_B = n \frac{1}{2} \left\{ \exp\left(-\frac{e\varphi_L}{KT} \right) + \exp\left(-\frac{e\varphi_R}{KT} \right) \right\} = n B_n \quad (4.41)$$

with

$$B_n = \frac{1}{2} \exp\left(-\frac{V_B}{KT} \right) \left\{ 1 + \exp\left(-\frac{eU}{KT} \right) \right\}$$

The applied voltage is determined by $U = \varphi_L - \varphi_R$. If n in the quasi-neutral bulk region and n_B at the grain boundary are expressed by (2.11), one obtains from (4.40) that f_B has the form of the Fermi-Dirac statistics:

$$f_B(E + V_B) = \frac{1}{1 + \exp\{(E + V_B - E_{FB})/KT\}} \tag{4.42}$$

with a quasi-Fermi level E_{FB} for the grain boundary states, which is related to the quasi-Fermi level in the bulk region F_{Fn} by

$$E_{FB} = E_{Fn} - KT \ln \frac{2}{1 + \exp(-eU/KT)} \tag{4.43}$$

The calculation of the minority concentration p_B is less straight-forward. Since holes, which reach the depletion zone edge by diffusion in the quasi-neutral bulk region, are immediately driven by the electrical field to the interface where they can recombine, the concentrations p_L and p_R at the edges are determined by the generation of holes in the bulk, the recombination at the grain boundary, and the diffusion current from the bulk to the interface. Solving the continuity equation (2.13) in the bulk region with the boundary condition that the holes in the depletion region are driven to the interface by the electrical field, one obtains the following conditions for the minority diffusion currents

$$J_L = \frac{eD_p}{L_p}(p - p_L) = ev_T\left\{p_L - p_B \exp\left(-\frac{e\varphi_L}{KT}\right)\right\} \tag{4.44a}$$

$$J_R = \frac{eD_p}{L_p}(p - p_R) = ev_T\left\{p_R - p_B \exp\left(-\frac{e\varphi_R}{KT}\right)\right\} \tag{4.44b}$$

D_p and L_p are the diffusion coefficient and the diffusion length, respectively, of the minority holes. The recombination current at the interface $J_B = eR_0$ is equal to the total diffusion current from the left and right hand sides, $J_B = J_L + J_R$. Eliminating p_L and p_R from (4.44) and substituting the expression (4.39) for R_0, for p_B one obtains

$$p_B = \frac{1}{n_B} \frac{B_n np + Bn_i^2}{B_n + B} \tag{4.45}$$

The parameter B is given by

$$B = \frac{2v_T D_p}{v_T L_p + D_p} \int s_p(E) N_B(E) f_B(E) dE \qquad (4.46)$$

Substituting the expressions for n_B and p_B in (4.39) and using the expansion $n = n_0 + \delta n$ and $p = p_0 + \delta p$, for low generations rates one obtains

$$R_0 = \frac{\delta_p}{B_n + B} \int s_p(E) N_B(E) f_B(E) dE \qquad (4.47)$$

For higher illumination levels, the hole concentration at the grain boundary p_B cannot be ignored anymore, and one has to use the general expression for the occupation function given by (4.37). The result yields a rather complex expression, which was derived in [19,20]. If the energy dependence of the capture coefficient is ignored and $s(E)$ is expressed by a constant capture cross section $s(E) = \sigma_p v_T$, the remaining integral in (4.46) and (4.47) yields the number of occupied grain boundary states N_{B0}:

$$\sigma_p v_T N_{B0} = \int s_p(E) N_B(E) f_B(E) dE \qquad (4.48)$$

In order to examine the recombination at the grain boundary under illumination, the recombination velocity defined by $R_0 = S_p \delta p$ shall be considered. S_p can be simplified in the two limiting cases of a high or low recombination rate of holes:

$$S_{p1} = \frac{2\sigma_p v_T N_{B0}}{1 + \exp\left(-\frac{eU}{KT}\right)} \exp \frac{V_B}{KT} \qquad (B_n > B, \text{ low } \sigma_p) \qquad (4.49a)$$

$$S_{p2} = \frac{2v_T D_p}{v_T L_p + D_p} \qquad (B_n < B, \text{ high } \sigma_p) \qquad (4.49b)$$

The corresponding equations for a p-type crystal are

$$S_{n1} = \frac{2\sigma_n v_T N_{Bu}}{1 + \exp\left(-\frac{eU}{KT}\right)} \exp\left(-\frac{V_B}{KT}\right) \qquad S_{n2} = \frac{2v_T D_n}{v_T L_n + D_n} \qquad (4.50)$$

where N_{Bu} is the number of empty grain boundary states and the potential V_B is negative. In general, the recombination velocities are given by $S_n = S_{n1} S_{n2}/(S_{n1} + S_{n2})$ or $S_p = S_{p1} S_{p2}/(S_{p1} + S_{p2})$, respectively. The recombination velocities are thus determined by the capture cross sections, diffusion coefficients, and diffusion lengths of minority carriers. In the first case, the recombination velocity depends in a complex way on U, since the quasi-Fermi level for the majority carriers at the grain boundary is a function of the applied voltage. Since the potential barrier $V_B(U)$ and the number of occupied states $N_{B0}(U)$ change as well, V_B has to be determined self-consistently from (4.33) through (4.35) and (4.42), as mentioned above.

A calculation of the grain boundary potential without illumination as a function of the applied voltage will be discussed in Section 5.3. The general result is that the potential remains almost constant for small voltages and decreases rapidly with increasing voltage when all trap states are filled (see Figure 5.32). The details of the voltage dependence are determined by the density of states distribution. It is evident from (4.49) that the recombination velocity of individual grain boundaries may thus vary considerably, depending on their charge state, and that even a neutral grain boundary ($V_B = 0$) remains a recombination-active defect. In a polycrystalline solar cell, the voltages applied across the boundaries are usually small so that, in this case, $E_{FB} \approx E_{Fn}$ is a good approximation. Therefore, both quantities $V_B(U)$ and $N_{B0}(U)$ remain almost constant and can be replaced by their equilibrium values.

A numerical calculation of the grain boundary potential under illumination is given in Figure 4.5. In this calculation, it is assumed that $U = 0$ and that no deep trap levels are present in the vicinity of the grain boundary. The results show that under illumination levels typical for photovoltaic applications the grain boundary potential decreases considerably. The recombination velocity will decrease as well and approach a constant value according to (4.49). Although the general expression for the dependence on the illumination level is more complex, the result derived here is essentially the same as in [18].

If the capture cross section at the interface is high, the recombination velocity is constant and depends only on the supply of minority carriers from the bulk (holes in the present case), which is determined by the diffusion coefficient and the diffusion length in the vicinity of the interface. Since for high illumination levels a constant recombination velocity is also obtained in the first case, one cannot easily distinguish experimentally between both cases. A powerful experimental technique to study recombination at extended defects is the EBIC technique, which will be discussed in Section 4.2.5. In this technique, it is frequently assumed that the second case is valid and that the diffusion length in the bulk can be determined from the measurements.

Finally, it should be mentioned that without illumination or other generation processes ($\delta n = \delta p = 0$), the recombination rate for electron-hole pairs at the grain boundary becomes zero ($R_0 = 0$). This result is important in view of a description of the electrical transport over the grain boundary barrier, which will be discussed in Section 5.3. Without electron-hole recombination, the current that

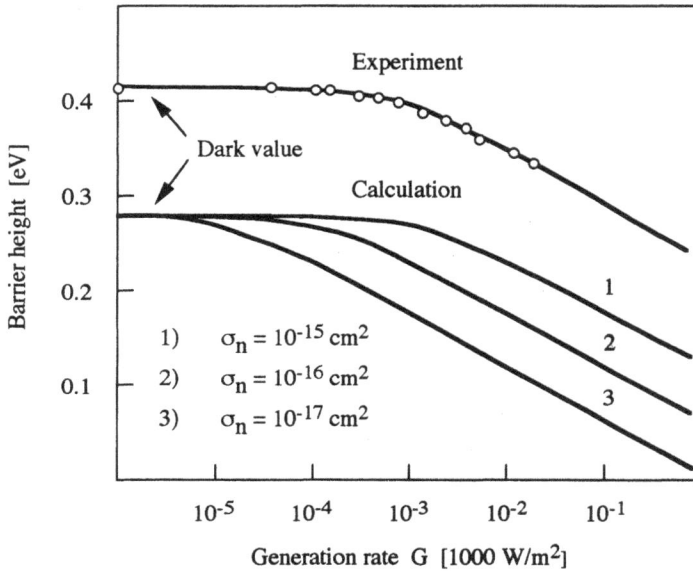

Figure 4.5 Measurement of the grain boundary potential upon illumination for a silicon bicrystal (open circles). Computed variation of the barrier height as a function of the illumination level for different capture cross sections σ_n of the grain boundary states. Parameters are the doping concentration $N_d = 1.3 \times 10^{16}$ cm^{-3}, $E_T = 0.1$ meV, $\sigma_p = 2 \times 10^{-14}$ cm^2, $N_B = 2.0 \times 10^{12}$ cm^{-2}, $L_p = 50$ μm, and $T = 300$K, (1 sun $= 1000$ W/m^2) [18].

flows through the grain boundary under an applied bias is totally dominated by majority carrier transport, and minority currents can then be completely ignored.

4.2.4 Recombination at Surfaces and Dislocations

4.2.4.1 Surfaces

The discussion of the electrical characteristics of a *pn*-junction solar cell has already shown that recombination at the surfaces of the device is an important parameter for the performance and efficiency of the cell. Like grain boundaries, surfaces introduce electronic states in the band gap with a continuous density of states distribution (per unit area) $N_S(E)$. For a clean surface, the position and distribution of these energy states is related to the atomic structure and the reconstruction of bonds. In practice, however, surfaces are usually covered with an oxide layer or with adsorbed foreign atoms. Depending on whether it is energetically more favorable to supply or attract electrons, these atoms can exchange electrons with the crystal and become positively or negatively charged. A similar role is played by structural defects, such as vacancies, in the semiconductor-oxide interface. It is therefore not surprising to observe that the surface properties can be greatly modified by surface preparation techniques like polishing, etching, and ion beam bombardment or by exposure to different ambients. Therefore, the control of the surface processing techniques and protection of the surface to maintain the desired properties is crucial for any semiconductor devices particularly for thin-film semiconductors, where the surface-to-volume ratio is greatly enhanced.

The previous calculation of the recombination velocity at grain boundaries can easily be extended to the recombination at surfaces if one takes into account that the surface charge is only compensated by a space charge on one side (Figure 4.6). If we consider the recombination at a free surface, no voltage is applied, and, analogous to the case of a charged grain boundary, the surface potential V_s can be calculated from Poisson's equation (equation (4. 32)) and is given by

$$V_s = \frac{Q_s^2}{8\epsilon\epsilon_0 N_d} \tag{4.51}$$

where N_d is the doping concentration. The surface charge Q_s can be calculated from the density of states distribution of the surface $N_s(E)$ using (4.35), where the electron concentration (per unit area) of the neutral surface N_{sn} and the occupation statistics for the interface states $f_s(E)$ are introduced. The calculation of the surface

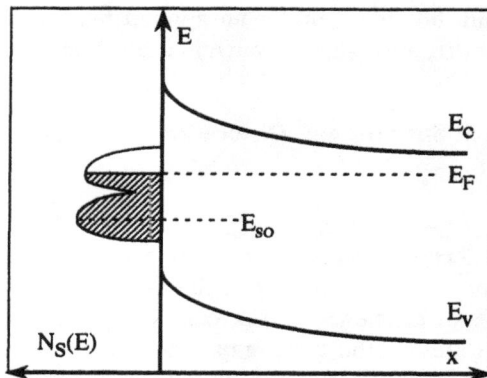

Figure 4.6 Schematic diagram of the band structure of a charged surface with a continuous density of states distribution $N_s(E)$ of the surface states. E_{so} is the occupation limit of the uncharged surface states.

recombination velcocity follows the same procedure as for grain boundaries, and for n-type and p-type crystals corresponding to (4.49) and (4.50) for low-capture cross sections, one obtains

$$S_{p1} = \sigma_p v_T N_{so} \exp \frac{V_s}{KT} \quad S_{n1} = \sigma_n v_T N_{Su} \exp\left(-\frac{V_s}{KT}\right) \tag{4.52}$$

The corresponding equations for high-capture cross sections remain the same as for the grain boundary case (equation (4.49)). The recombination velocities in this case depend on the capture rates of minority carriers, the densities of occupied (N_{so}) or unoccupied surface states (N_{su}), respectively, and the surface potential, which is positive for an n-type crystal and negative for a p-type crystal. The latter two parameters are both related to the density of states distribution of the surface levels, as discussed above for the grain boundary (equation (4.35)). The most effective surfaces are those with a quasi-Fermi level E_{so} near the middle of the band gap, which is frequently the case for surfaces and interface states.

4.2.4.2 Dislocations

The previous calculation of the recombination velocity at grain boundaries can also be used to describe the recombination at dislocations. Since dislocations are one-dimensional defects, the recombination velocities S_n or S_p have to be defined for the surface of a cylinder with radius r_0, corresponding to the radius of the dislocation core around the dislocation line (Figure 4.1). The particular features of the dislocations, the line character, the strain field, and the dislocation density, enter the final equations only in the calculation of the potential barrier V_D and the dislocation charge Q_D. V_D has been calculated from Poisson's equation by various authors [21,22]. An approximate solution for small potential barriers in a p-doped crystal is

$$V_D = \frac{Q_D}{2\pi\epsilon\epsilon_0} \ln \frac{r_0}{\lambda} \tag{4.53}$$

with the Debye screening length λ:

$$\lambda = \left\{ \frac{\epsilon\epsilon_0 KT}{e^2 p} \right\}^{1/2}$$

The line charge Q_D can be calculated from the occupation of the dislocation states by $Q_D = -eN_{De}$, where N_{De} is given by (4.35) with $N_{DL}(E)$, the density of states distribution (per unit length) for a dislocation, N_{Dn}, the density of states of the neutral dislocation, and $f_{DL}(E)$, the corresponding occupation function given in (4.42). In general, the average electron and hole concentrations in the bulk, and thus Q_D and V_D, are complex functions of the doping concentration and temperature. For a low occupation of the dislocation states, the dependence on the doping concentration and dislocation density can be approximated by $Q_D \approx eN_a/N_D$, and the barrier height becomes proportional to the doping concentration $V_D \approx N_a/N_D$ if one ignores the weak logarithmic dependence on $\lambda(p)$. For high doping concentrations, when all trap states are filled, the hole concentration becomes $p \approx N_a$ and the dislocation charge remains constant.

The calculation of the recombination velocity yields expressions similar to the recombination velocity for the surface. Since an applied voltage does not change the dislocation barrier height, the case $U = 0$ has to be considered. For low capture rates s_n and s_p, one obtains

$$S_{p1} = s_p N_{D0} \exp \frac{V_D}{KT} \qquad S_{n1} = s_n N_{Du} \exp\left(-\frac{V_D}{KT}\right) \tag{4.54}$$

where N_{D0} and N_{Du} are the density of occupied and unoccupied dislocation states, respectively. Since, according to (4.53), the dislocation potential is negative for a p-type crystal, in both cases an increase of the potential barrier increases the recombination velocity. The equations for high capture rates s_n and s_p are again the same as for grain boundaries and surfaces (equations (4.49) and (4.52)), so that in this case there is no difference between the recombination behavior of these extended defects.

The present discussion and the previous one in Section 4.1 have demonstrated that the presence of dislocations alters both the carrier mobility μ and the lifetime τ in a crystal. This has direct ramifications on the diffusion length $L_{eff} = (KT\mu\tau/e)^{1/2}$. Several models have been developed to describe the influence of dislocations on L_{eff} [23,24], and some of the numerical and experimental results for silicon are shown in Figure 4.7. One can see that for dislocation densities above 10^4 cm^{-2} the diffusion length decreases and recombination velocities from $S_n \approx 10^3$ to 10^6 cm/s are obtained from the comparison with experimental measurements. The influence on L_{eff} is more pronounced if the diffusion length L_n of the undisturbed bulk is high.

4.2.5 Electron Beam-Induced Current Technique

It is evident from the preceding analysis that the density of states distribution $N(E)$ of extended defects is most important for the recombination behavior. Several

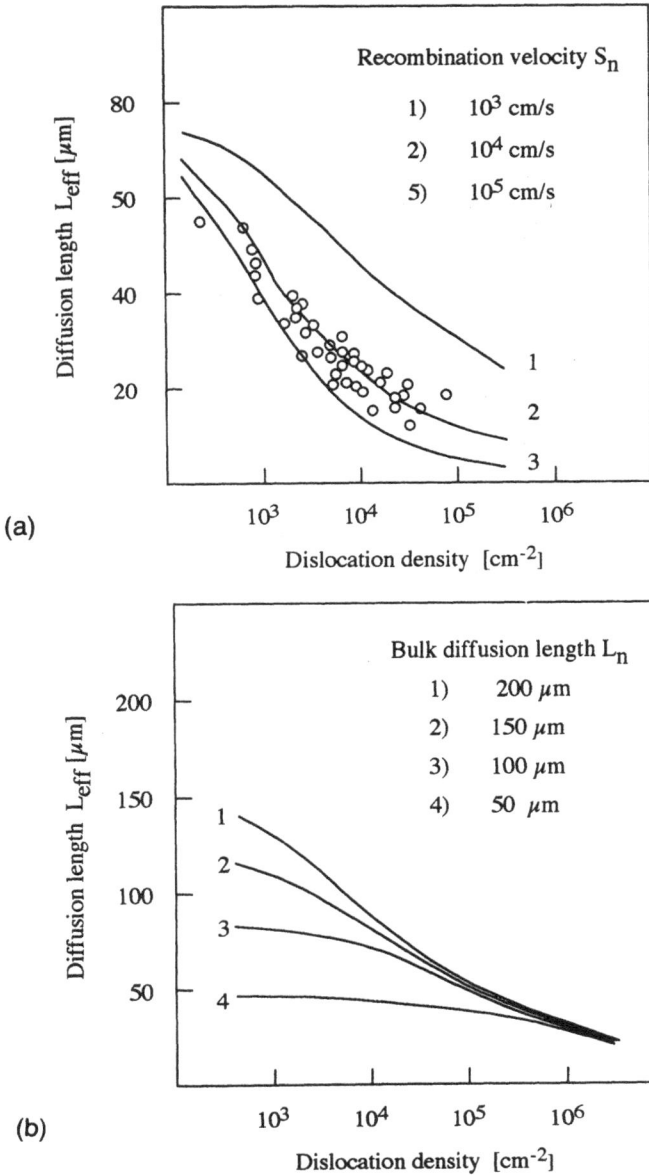

Figure 4.7 Effective diffusion length L_{eff} of polycrystalline solar silicon as a function of the dislocation density N_d: (a) for different values of the dislocation recombination velocity S_n, and (b) for different values of the bulk diffusion length L_n with $S_n = 10^3$ cm/s. Open circles are measured data [23].

experimental techniques like admittance and deep-level transient spectroscopy (DLTS) have been proved to be very useful for the characterization of single defects in semiconductors like silicon [25–27], germanium [28,29], or ZnO [30,31], and they are increasingly applied to solar cell semiconductors.

A direct determination of the recombination velocity of a grain boundary (or of dislocations) is possible with the electron beam-induced current (EBIC) technique (or light beam-induced current (LBIC) technique), which is schematically depicted in Figure 4.8. Electron-hole pairs are generated by an electron (or light) beam in a scanning electron microscope (SEM), for instance, and separated in a built-in electrical field of a Schottky contact or *pn* junction. The resulting current is amplified and used to modulate the signal of the SEM monitor. The collected current is reduced at defects because of the locally enhanced carrier recombination; therefore, the defect shows up as a dark feature in the SEM image (Figure 4.9). This techniques allows a direct and immediate observation of the recombination behavior of the grain boundaries and other lattice defects and is very useful for the characterization of electrically active defects in solar cell semiconductors. For a quantitative determination of the recombination velocity, the recombination current is measured in a single line-scan across the boundary. The analysis of the data requires the solution of the diffusion problem for the minority carries in the vicinity of the boundary. A solution for the uniform generation of electron-hole pairs under illumination was derived in the previous chapter. For the EBIC technique, the theory was developed by Donolato [32] under the assumption that the diffusion of minority carriers to the defect is the rate-limiting step for the recombination. The model takes into account the inhomogeneous generation of electron-hole pairs by the electron beam.

Typical values that have been obtained for electrically active grain boundaries in germanium and silicon vary from $S_n = 10^2$ to 10^6 cm/s [33]. Experimentally, one

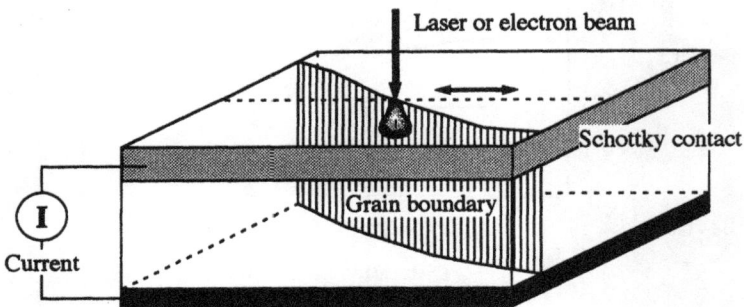

Figure 4.8 Schematic diagram of the EBIC (or LBIC) method. The recombination can be enhanced at lattice defects, which reduces locally the current through the Schottky diode.

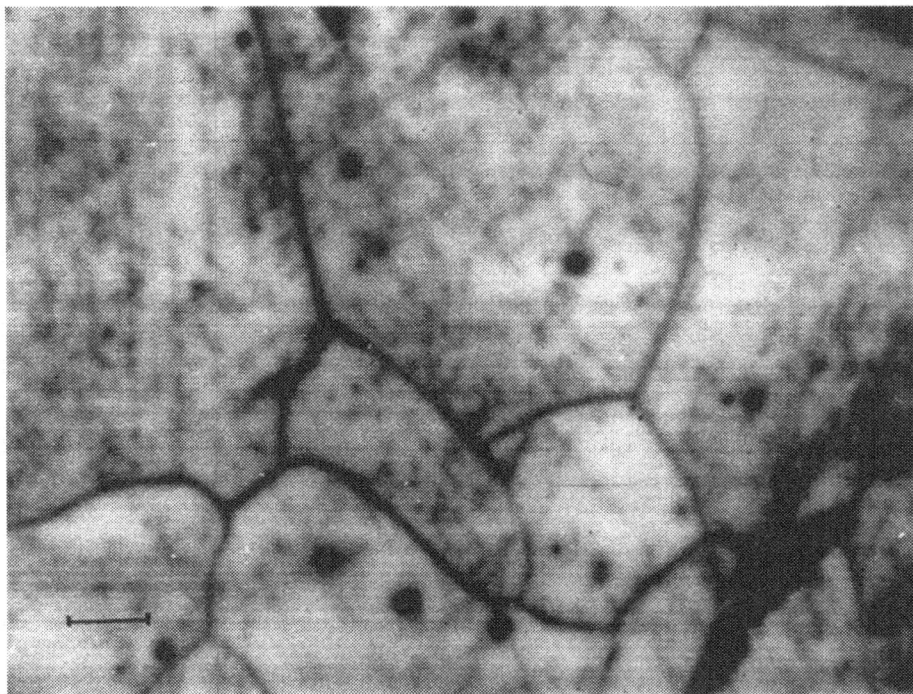

Figure 4.9 EBIC image of grain boundaries in polycrystalline silicon. Dark spots are due to the enhanced recombination at dislocations and precipitates (scale marker $\cong$ 50 μm).

finds a dependence of the recombination velocity on the injection current. This can be understood if one considers the dependence of the recombination velocity on the grain boundary charge, which is increased with higher injection levels. The potential decreases then, as can be seen from Figure 4.5, and because of the exponential dependence on V_B, the total recombination velocity decreases, too. For the comparison of different data, it is therefore necessary to measure at low excitation levels, where V_B remains essentially unchanged, although in this case the signal-to-noise ratio becomes a limiting factor.

The EBIC (or LBIC) technique is useful for the determination of the total electrical activity and recombination velocity. The quantitative determination of the distribution of the defect states is difficult, since it requires a deconvolution of the results in terms of the actual trap levels.

REFERENCES

[1] Dresner, J., *Semiconductors and Semimetals*, Vol. 21C, R. K. Willardson and A. C. Beer, eds., New York: Academic Press, 1984, p. 193.

[2] Sze, S. M., *Physics of Semiconductor Devices*, New York: Wiley & Sons, 1981, p. 790.

[3] Pankove, J. I., *Optical Processes in Semiconductors*, Englewood Cliffs, NJ: Prentice Hall, 1971.

[4] Bonc-Bruevic, V. L., and S. G. Kalasnikov, *Halbleiterphysik*, VEB Deutscher Verlag der Wissenschaften, Berlin, 1982.

[5] Brooks, H., and J. Herring, *Adv. Electr. & Electr. Phys.*, Vol. 7, 1955, p. 85.

[6] Burk, D. E., and V. De La Torre, *IEEE Electron. Device Lett.*, Vol. 5, 1984, p. 231.

[7] Buehler, M. G., and W. R. Thurber, *IEEE Trans. Electron. Device*, Vol. 23, 1974, p. 968.

[8] Verhoef, L. A., and W. C. Sinke, *IEEE Trans. Electron. Devices*, Vol. 37, 1990, p. 210.

[9] Verhoef, L. A., A. Zondervan, F. A. Lindholm, M. B. Spitzer, and C. J. Keavney, *J. Appl. Phys.*, Vol. 63, 1988, p. 4683.

[10] Labusch, R., and W. Schröter, *Dislocations in Solids*, F. R. N. Nabarro, ed., New York: North Holland, 1980, p. 129.

[11] Düster, F., and R. Labusch, *Phys. Stat. Sol.* (b), Vol. 60, 1973, p. 161.

[12] Podbielski, R., and H. J. Möller, *Poly-Micro-Crystalline and Amorphous Semiconductors*, P. Pinard and S. Kalbitzer, eds., Les Editions de Physique, Les Ulis, 1984, p. 365.

[13] Seto, J. Y. W, *J. Appl. Phys.*, Vol. 46, No. 12, 1975, p. 5247.

[14] Dziewior, J., and W. Schmid, *Appl. Physics Lett.*, Vol. 31, 1977, p. 346.

[15] Greuter, F., and G. Blatter, *Semicond. Sci. Technol.*, Vol. 8, 1990, p. 111.

[16] Panayotatos, P., E.S. Yang, and W. Hwang, *Solid State Electronics*, Vol. 25, 1982, p. 417.

[17] Romanowski, A., and A. Buczkowski, *Solid State Electronics*, Vol. 28, 1985, p. 645.

[18] Joshi, D. P., *Solid State Electronics*, Vol. 29, 1986, p. 19.

[19] Blatter, G., and F. Greuter, *Phys. Rev. B*, Vol. 33, 1986, p. 3952.

[20] Blatter, G., and F. Greuter, *Phys. Rev. B*, Vol. 34, 1986, p. 8555.

[21] Read, W. T., *Phil. Mag.*, Vol. 45, 1954, pp. 775, 1119.

[22] Schröter, W., and R. Labusch, *Phys. Stat. Sol.*, Vol. 36, 1969, p. 359.

[23] El Ghitani, H., and S. Martinuzzi, *Mater. Res. Soc. Symp. Proc.*, Vol. 106, 1988, p. 225.

[24] Pizzini, S., A. Sandrinelli, M. Beghi, D. Narducci, F. Allegretti, and F. Porchio, *J. Electrochem.*, Vol. 135, 1988, p. 155.

[25] Broniatowski, A., *Phys. Rev. B*, Vol. 36, 1987, p. 5895.

[26] Aucouturier, M., A. Broniatowski, A. Chari, and J. L. Maurice, *Polycrystalline Semiconductors*, H. J. Möller, H. Strunk, and J. Werner, eds., *Springer Proceedings in Physics*, Vol. 35, Berlin, 1989, p. 64.

[27] Werner, J., *Polycrystalline Semiconductors*, G. Harbeke, *Springer Series in Solid State Science*, Vol. 57, Berlin: Springer, 1985, p. 95.

[28] Petermann, G., and P. Haasen, *Polycrystalline Semiconductors*, H. J. Möller, H. Strunk, and J. Werner, eds., *Springer Proceedings in Physics*, Vol. 35, Berlin, 1989, p. 332.

[29] Szkielko, W., and G. Petermann, *Poly-microcrystalline and Amorphous Semiconductors*, P. Pinard and S. Kalbitzer, Les Edition de Physique, Paris, 1985, p. 379.

[30] Rossinelli, M., G. Blatter, and F. Greuter, *Br. Ceram. Proc.*, Vol. 41, 1989, p. 177.

[31] Greuter, F., G. Blatter, M. Rossinelli, and F. Schmükle, *Mater. Sci. Forum*, Vol. 10, 1986, p. 235.

[32] Donolato, C., *J. Appl. Phys.* Vol. 54, 1983, p. 1314.

[33] Tabet, N., C. Monty, and Y. Narfaing, *Polycrystalline Semiconductors*, H. J. Möller, H. Strunk, and J. Werner, eds., *Springer Proceedings in Physics*, Vol. 35, Berlin, 1989, p. 89.

Chapter 5
Structural and Electrical Properties of Lattice Defects

It was shown in the previous chapter that recombination of electrons and holes at lattice defects is a major source of recombination losses in a semiconductor solar cell. The efficiency is limited by the number of trap levels introduced in the band gap by the crystalline defects. The electrical and other physical properties (e.g., the mobility of point defects) are closely related to the atomic structure of the lattice defect and both aspects have been studied thoroughly for many defects during recent years. In a crystalline semiconductor, the lattice defects include point defects and extended defects like dislocations, grain boundaries, interfaces (including surfaces), and precipitates. In amorphous semiconductors, lattice defects are less defined, and rather unsaturated and distorted covalent bonds have to be considered. These defects are either introduced during the preparation of the starting material or later, during the various steps of the solar cell processing. The general structural and physical properties of crystalline defects as far as they are important for solar cell materials, will be discussed in this chapter. Since most of the experimental results are available for silicon, general properties of defects will be frequently discussed with reference to this material.

5.1 POINT DEFECTS

The properties of point defects that are important from the material perspective of solar cell semiconductors are their electrical properties, their diffusion and solubility behavior, and their interaction with other lattice defects. One can distinguish intrinsic point defects if they are formed from the same species as the atoms of the lattice, and extrinsic point defects if they are of a different nature. In thermal equilibrium, the concentrations of intrinsic point defects are generally low in semiconductors, but there exist many non-equilibrium processes that can generate rather high concentrations. In compound semiconductors, intrinsic point defects are also introduced to compensate for deviations from the stoichiometry. They are formed in large concentrations then and have a considerable impact on the electrical and optical properties. Typical examples are the thoroughly investigated antisite EL2-defects in GaAs and the intrinsic defects that are responsible for the doping of copper ternary compounds. Intrinsic point defects also participate in many physical processes, such as the precipitation or diffusion of impurity elements, and consequently they shall be discussed first.

5.1.1 Intrinsic Point Defects and Diffusion

5.1.1.1 Equilibrium Concentration

For a crystal consisting of one species of atoms, two basic types of intrinsic point defects can occur, namely, vacancies V (one missing atom at a regular, substitutional site) and self-interstitials I (one extra atom squeezed into the crystal). For a binary compound semiconductor with the sphalerite structure and two species of atoms occupying different sublattices, a third type of intrinsic defect can occur: the antisite defect, an atom occupying a substitutional site of the other sublattice. One has to specify the sublattice for each vacancy and interstitial (e.g., for a two-component compound V_a, V_b, I_a, I_b) here, since the atomic environment and, therefore, the physical properties are different. In ternary compounds with the chalcopyrite structure, for instance, (see Chapter 8) where two kinds of atoms occupy one sublattice in an ordered arrangement, it is also possible to form an antisite defect on the same sublattice, which leads to an even higher multiplicity of intrinsic defects (12 in this case). In addition, complexes can be formed, such as divacancies and vacancy self-interstitial pairs (Frenkel pairs), or aggregates of more than two species, which shall still be considered point defects. The definition and identification of intrinsic point defects in amorphous semiconductors require a different approach, which will be discussed in Chapter 9.

In thermodynamic equilibrium, a crystal contains a certain concentration of intrinsic point defects, since the Gibbs free energy is lowered by their formation. The thermal equilibrium concentrations (in dimensionless atomic fractions) of vacancies C_V^{eq} and of self-interstitials C_I^{eq} are given by

$$C_V^{eq} = \exp \frac{S_V^F}{KT} \exp\left\{-\frac{H_V^F}{KT}\right\} \qquad C_I^{eq} = \exp \frac{S_I^F}{KT} \exp\left\{-\frac{H_I^F}{KT}\right\} \qquad (5.1)$$

where K is Boltzmann's constant and T is the absolute temperature. S_V^F and H_V^F are the entropy and the enthalpy of formation of a vacancy, which is equivalent to the removal of an atom from its substitutional site and the deposition on the surface, and S_I^f and H_I^f are the corresponding quantities for self-interstitials. If both vacancies and self-interstitials are present, it can be expected that they react with each other according to the reaction

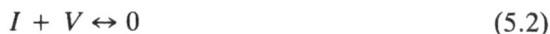

$$I + V \leftrightarrow 0 \tag{5.2}$$

where 0 represents the undisturbed substitutional site in the lattice. The equilibrium reaction between vacancies and interstitials can be described by the law of mass action:

$$C_V C_I = C_V^{eq} C_I^{eq} = \exp\left\{-\frac{G_V^F + G_I^F}{KT}\right\} \tag{5.3}$$

where C_V and C_I are the vacancy and interstitial concentrations, respectively, and $G_V^F = H_V^F - TS_V^F$ and $G_V^F = H_I^F - TS_I^F$ are the Gibbs free energies.

In compound semiconductors, reaction (5.2) can only occur between a vacancy and an interstitial of the atomic species that belongs to the same sublattice; otherwise, an antisite defect would result. Therefore, one has to formulate an equivalent equation for the vacancy and interstitial concentrations in each sublattice. In addition, because of the conservation of the stoichiometry of the compound, the concentrations of vacancies and interstitials in each sublattice are not independent of each other, which leads to supplementary constraints.

In doped semiconductors, the formation enthalpies depend on the position of the Fermi level E_F if the vacancy or interstitial introduces a trap level E_T in the band gap. In this case, the system can gain energy by the occupation of the level with an electron or hole when the defect is formed. For instance, for a vacancy, one has the reaction $V^0 \leftrightarrow V^+ + e^-$ for a donor level and $V^0 \leftrightarrow V^- + h^+$ for an acceptor level. For the formation energies $H_{V^+}^F$ and $H_{V^-}^F$ of the charged defect in the two cases, one obtains

$$H_{V^+}^F = H_{V^0}^F - (E_T - E_F) \qquad H_{V^-}^F = H_{V^0}^F + (E_T - E_F) \tag{5.4}$$

$H_{V^0}^f$ is the formation energy and E_T the trap level of the neutral vacancy. The concentrations of the charged vacancy $C_{V^+}^{eq}$ and $C_{V^-}^{eq}$ are then given for a non-degenerate trap level by

$$C_{V^+}^{eq} = C_{V^0}^{eq} \exp\left\{\frac{E_T - E_F}{KT}\right\} \qquad C_{V^-}^{eq} = C_{V^0}^{eq} \exp\left\{-\frac{E_T - E_F}{KT}\right\} \tag{5.5}$$

where $C_{V^0}^{eq}$ is the concentration of the neutral vacancy given by (5.1). These equations can be derived from the general expression (4.30) for the occupancy of a trap level, which was given in Chapter 4. Equation (5.5) is completely general and applies to any defect that introduces a trap level in the gap. The total concentration of the defect is the sum of charged and neutral defects $C_V^{eq} = C_{V^0}^{eq} + C_{V^-}^{eq} + C_{V^+}^{eq}$ (see (4.31)).

5.1.1.2 Diffusion

In general, the thermally generated concentrations of vacancies and interstitials in semiconductors are very low ($<10^{16}$ cm^{-3}), even at the melting point, and the experimental determination of the formation enthalpies and entropies are difficult and usually indirect. The most direct information about the behavior of vacancies and self-interstitials can be obtained from investigation of the self-diffusion. The self-diffusion coefficient D^{SD} depends on the concentrations of vacancies and interstitials as given by

$$D^{SD} = D_V^{SD} + D_I^{SD} = D_V C_V^{eq} + D_I C_I^{eq} \tag{5.6}$$

where D_V and D_I are the corresponding diffusion coefficients. Their temperature dependence can be written

$$D_V = D_{V^0} \exp\left\{-\frac{H_V^M}{KT}\right\} \tag{5.7a}$$

$$D_I = D_{I^0} \exp\left\{-\frac{H_I^M}{KT}\right\} \tag{5.7b}$$

which yields, together with (5.1), for D_V^{SD} and D_I^{SD},

$$D_V C_V^{eq} = D_{V^0}^{SD} \exp\left\{-\frac{H_V^{SD}}{KT}\right\} \quad \text{with} \quad H_V^{SD} = H_V^F + H_V^M \tag{5.8a}$$

$$D_I C_I^{eq} = D_{I^0}^{SD} \exp\left\{-\frac{H_I^{SD}}{KT}\right\} \quad \text{with} \quad H_I^{SD} = H_I^F + H_I^M \tag{5.8b}$$

where H^F and H^M are the formation and migration enthalpies, respectively. Intrinsic point defects can also have a strong impact on the diffusion behavior of impurities which require the participation of either vacancies or interstitials as a diffusion vehicle. Different diffusion mechanisms have been proposed for elements that occupy either substitutional or interstitial sites, or can exist in both configurations, such as the transition metals. The different mechanisms are schematically depicted in Figure 5.1.

For substitutional elements, the diffusivity D_α is composed of a contribution D_α^V involving vacancies (vacancy mechanism) and a contribution D_α^I involving self-interstitials (interstitialcy mechanism), where the relative contributions depend on the specific element and the temperature. In a non-equilibrium situation, where the equilibrium concentrations of vacancies and interstitials are disturbed, the diffusion coefficient is given by

$$D_\alpha = \frac{C_I}{C_I^{eq}} D_\alpha^I + \frac{C_V}{C_V^{eq}} D_\alpha^V \qquad (5.9)$$

where C_V and C_I are the actual vacancy and self-interstitial concentrations.

For impurities that can occupy both substitutional and interstitial sites, the diffusion involves an exchange of atoms between the two sites. Basically, two mechanisms have been proposed in the literature for the diffusion of these impurities in semiconductors: the dissociative and kick-out mechanisms [1], both of which require the participation of native point defects. The dissociative mechanism describes the transition of a substitutional atom A_s to an interstitial site A_i by the reaction

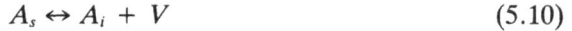

$$A_s \leftrightarrow A_i + V \qquad (5.10)$$

which involves the generation of a vacancy V. Alternatively, the kick-out mechanism requires the presence of a self-interstitial I in the reaction

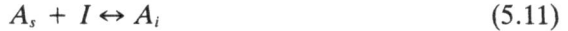

$$A_s + I \leftrightarrow A_i \qquad (5.11)$$

Both mechanisms are controlled by the concentrations and diffusivity of either vacancies or self-interstitials, and hence lead to different predictions about the diffusion behavior. For instance, the basic equations for the diffusion via the kick-out mechanism describe the transport of the interstitial impurities (diffusivity D_i and concentration C_i) and of the self-interstitials (D_I, C_I), and the interchange between the different species by a law of mass action for the reaction in (5.11).

$$\frac{\partial C_I}{\partial t} = D_I \frac{\partial^2 C_I}{\partial x^2} + \frac{\partial C_s}{\partial t} + K_I \left\{ 1 - \frac{C_I}{C_I^{eq}} \right\} \qquad (5.12)$$

$$\frac{\partial C_i}{\partial t} = D_i \frac{\partial^2 C_i}{\partial x^2} - \frac{\partial C_s}{\partial t} \qquad (5.13)$$

$$\frac{C_s}{C_I C_i} = \frac{C_s^{eq}}{C_I^{eq} C_i^{eq}} \qquad (5.14)$$

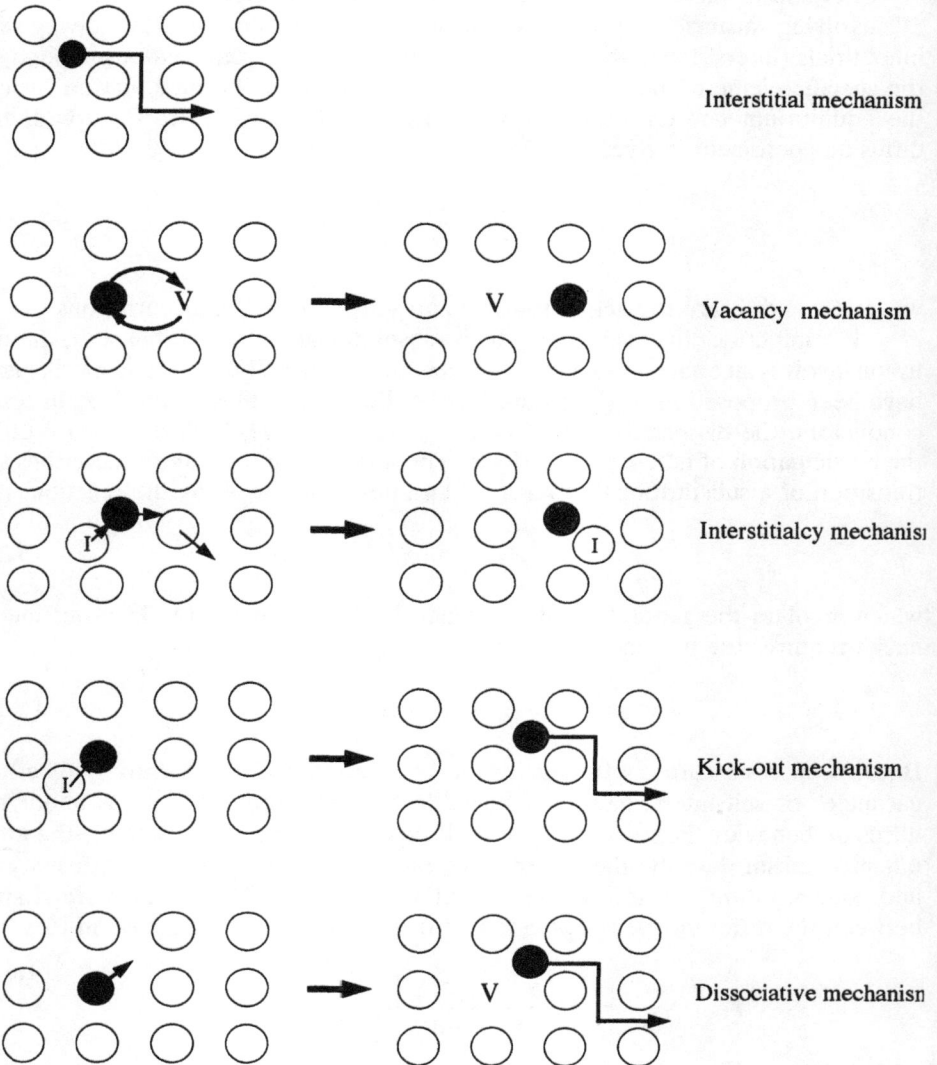

Figure 5.1 Schematic representation of different diffusion mechanisms in semiconductors: (*V*) vacancy and (*I*) self-interstitial atoms. For the interstitialcy mechanism, it is assumed that the impurity-self-interstitial pair adopts a split interstitial configuration. The diffusive jump involves the correlated motion and rotation of the self-interstitial-impurity complex.

C_s and D_s are the concentration and diffusion constants for the impurity on substitutional sites. The quantity K_I describes the strength of sinks or sources for self-interstitials, which are frequently related to dislocations by

$$K_I = \gamma_I N_D D_I C_I^{eq} \qquad (5.15)$$

where N_D is the dislocation density and γ_I is a parameter that characterizes the geometrical dislocation arrangement and is in most models on the order of one. In general, numerical methods have to be applied to solve this set of equations, but in some cases approximate solutions can be derived.

With low-cost solar cell semiconductors with a rather high density of dislocations or grain boundaries, one can assume that the sink density is high enough so that the equilibrium of the self-interstitial concentration can be locally established almost instantaneously ($C_I = C_I^{eq}$). In this case, (5.12) and (5.15) lead to the normal diffusion equation for the total impurity concentration $C_{tot} = C_s + C_i$:

$$\frac{\partial C_{tot}}{\partial t} = D_{eff} \frac{\partial^2 C_{tot}}{\partial x^2} \qquad D_{eff} = D_i \frac{C_i^{eq}}{C_i^{eq} + C_s^{eq}} \qquad (5.16)$$

where D_{eff} is an effective diffusion constant. The same equation is obtained for substitutional and interstitial components, since both concentrations are proportional according to (5.14). If the impurity is mainly dissolved interstitially ($C_i^{eq} \gg C_s^{eq}$), the diffusion constant is directly determined by the diffusion coefficient of the interstitial species $D_{eff} \approx D_i$. The same result is also obtained for the dissociative mechanism if an equilibrium for the vacancy concentration is assumed. The diffusion profiles are the well-known error-function solutions.

The diffusion behavior becomes more complex if vacancies and self-interstitials are not in equilibrium locally. This situation can arise, for instance, in dislocation-free silicon and has been observed for the gold diffusion. Gold is supposed to occupy mainly substitutional sites, where it is rather immobile, but diffuses rapidly in the interstitial lattice. The diffusion via the kick-out mechanism is described by an effective diffusion coefficient,

$$D_{eff} = \frac{C_s^{eq} D_I C_i^{eq}}{C_s^2} \qquad (5.17)$$

which is strongly dependent on the concentration of the substitutional component. This equation shows that it is the diffusion of the self-interstitial that dominates the gold transport. The diffusion profiles obtained in this case are different from the usual error function solution. The analysis of experimental results for gold are in good agreement with the predictions of the kick-out model [2].

5.1.1.3 Vacancies and Self-Interstitials in Silicon

There exists a long-standing controversy of whether self-interstitials are present at all in silicon in concentrations comparable to the vacancies. Today, most authors seem to agree that both species coexist and have to be taken into account for understanding diffusion and other processes in silicon, such as impurity precipitation or complex formation of point defects. The most direct information is obtained from the investigation of the self-diffusion coefficient D^{SD} as a function of the temperature. Results obtained from various methods indicate that a nonlinearity in the Arrhenius plot for D^{SD} occurs. At temperatures near the melting point, the activation energy is around 5 eV with a prefactor $D_0^{SD} \approx 10^3$ to 10^4 cm²/s, which decreases below 1270K to about 4.0 to 4.6 eV and $D_0^{SD} \approx 1$ to 100 cm²/s. This change in the parameters indicates a change in the self-diffusion mechanism, and it has been suggested that it is the vacancy mechanism (D_V^{SD}) that dominates at low temperatures and the self-interstitial diffusion (D_I^{SD}) that dominates in the high-temperature regime.

Further experimental evidence comes from the analysis of a large number of impurity diffusion experiments, such as doping or some transition elements where it is assumed that the impurity diffusion requires the participation of vacancies or self-interstitials. In particular, the analysis of gold and platinum diffusion (via the kick-out mechanism) has led to activation enthalpies and entropy factors which seem to agree quite well with the self-diffusion data obtained by tracer measurements [3].

Much less is known, however, about the formation and migration enthalpies that determine the concentration and diffusion coefficients of vacancies and self-interstitials according to (5.1) and (5.7). Since the self-diffusion data do not allow separation of both terms, further experimental methods are necessary. For instance, the analysis of the diffusion profiles after gold and platinum diffusion yielded the following results for the self-interstitial (atomic) concentration and diffusion coefficient [3]:

$$C_i^{eq} = 2 \times 10^4 \exp\left\{-\frac{3.8 \text{ eV}}{KT}\right\} \quad D_I = 0.2 \exp\left\{-\frac{1.2 \text{ eV}}{KT}\right\} \text{ [cm}^2\text{/s]} \quad (5.18)$$

Annealing experiments following electron irradiation at helium temperature also indicate a rather low migration energy both for the vacancy and the interstitial. Since there is no direct evidence for the self-interstitial, the indirectly obtained data remain controversial, and reliable information on C_V^{eq} and C_I^{eq} or D_V and D_I are not known. However, it seems to be that the activation enthalpies for vacancies and self-interstitials are rather similar between 4 and 5 eV (being lower for the vacancy), and that the migration enthalpies vary between about 0.1 and 1.2 eV for the interstitial and 1 and 2 eV for the vacancy, respectively. A summary of some experimental results and parameters is given in Table 5.1.

According to (5.5), the concentration of vacancies and interstitials depends on the Fermi level if the defects can be charged. The experimental investigation of the doping dependence of the diffusion of shallow- and deep-level impurities, which involve the participation of vacancies or self-interstitials, has shown that it can be described correctly if one assumes an acceptor level in the lower half of the band gap for the vacancy, and donor and acceptor levels in the lower and upper half, respectively, for the silicon interstitial [1].

Direct experimental and theoretical investigations have revealed a more complex level structure for the vacancy, which shows the behavior of a negative U-center [4]. The donor level $V^{0/+}$ $(E_V + 0.057$ eV$)$ lies below the level $V^{+/++}$ $(E_V + 0.13$ eV$)$; therefore, the charge state V^+ is unstable and V^{++} changes directly to V^0 when the Fermi level is raised. Since both donor levels lie near the valence band edge and a further acceptor level near the conduction band edge $(E_C - 0.09$ eV$)$, this result cannot easily be reconciled with the doping dependence of impurities diffusing by the vacancy mechanism. Much less is also known about the level structure of the self-interstitial, which has not been observed directly so far, and further investigations are necessary to confirm the conclusion from the diffusion experiments. A summary of data for the single defects and the simplest complexes is given in Table 5.2.

Despite the remaining controversy about the concentrations of vacancies and self-interstitials in thermal equilibrium, it is generally agreed that native defects in silicon can be generated in excess by a variety of processes. Some of the important processes are the thermal oxidation [5] of a silicon surface or the intrinsic formation of oxygen precipitates (see Section 5.1.2), which produces an excess of self-interstitials, whereas the addition of a chlorine-containing compound in the oxidizing

Table 5.1

Activation Enthalpies (H^{SD}) and Prefactors (D_0^{SD}) for the Diffusion of Vacancies and Self-Interstitials as Determined From Measurements of Self-Diffusion (SD) or Foreign Atom Diffusion, Respectively, in Undoped Silicon

Defect	D_0^{SD} [cm²/s]	H^{SD} [eV]	Remarks	H^F [eV]	H^M [eV]
Vacancy	20.0	4.4	(SD)	2–2.6	1–2
	154.0	4.65	(SD)		
	0.57	4.03	(Au)		
	40.0	4.6	(Au/Pt)		
Interstitial	1460	5.02	(SD)	3.8–4.4	0.1–1.2
	9000	5.13	(SD)		
	914	4.84	(Au)		
	4000	5.0	(Au/Pt)		

Table 5.2

Electronic Levels of Vacancy, Divacancy, and Di-Interstitial in Silicon, Which Can Occur After Electron Irradiation (Included are the annealing temperatures above which the complex dissociates or the defect levels disappear)

Defect	Trap Level [eV]	Annealing Temp. [K]	Remarks
$V^{0/+}$	$E_V + 0.05$	170	Vacancy
$V^{++/+}$	$E_V + 0.13$		Negative U-Center
$V^{0/-}$	$E_C - 0.09$	60	
$(V - V)^{+/0}$	$E_V + 0.22$	570	Divacancy
$(V - V)^{0/-}$	$E_C - 0.42$	610	
$(V - V)^{-/--}$	$E_C - 0.23$	570	
$Si_i - Si_i$	$E_C - 0.49$	440	Di-interstitial

atmosphere (e.g., HCl, Cl$_2$) leads to an injection of vacancies into the crystalline silicon [6]. An even higher supersaturation of self-interstitials compared to the surface oxidation is the diffusion of phosphorus from the surface into silicon [7]. It could be shown by *high-resolution electron microscopy* (HREM) that this is partly due to the formation of phosphorus-silicide precipitates near the surface, which occupy a larger volume than silicon.

According to reaction (5.2), the equilibrium between vacancies and self-interstitials will be restored after some time, but experimental results indicate that this process can be rather slow. Measurements at 1100°C have shown that it takes about one hour to reach thermal equilibrium, and considering the estimated diffusion coefficients of the two defects, the extrapolation to 900°C leads to times of several days [8]. This means that for experiments performed in the temperature range between 900 and 1000°C, vacancies and self-interstitials can be treated as independent species and can exist in supersaturation for some time, unless efficient sinks such as dislocations for the annihilation of these defects are not present. Several typical device processing steps such as the doping diffusion lie in this temperature regime.

The supersaturation of intrinsic point defects will have a strong impact on the diffusion behavior of impurities that require the participation of either vacancies or interstitials as a diffusion vehicle. This is the case, for instance, for the substitutional groups III and V doping elements in silicon, for which the diffusivity D_α is composed of a contribution D_α^V involving vacancies (vacancy mechanism) and a contribution D_α^I involving self-interstitials (interstitialcy mechanism), according to (5.9). Generally, atoms smaller than silicon, such as boron or phosphorus, diffuse preferentially via the interstitialcy mechanism, and their diffusion can be enhanced by a supersaturation of self-interstitials. On the other hand, for antimony, which diffuses via the vacancy mechanism, the diffusion is retarded. From a combination of different diffusion experiments at 1100°C, values have been obtained for D_α^I/D_α, which are listed in Table 5.3.

A well-known phenomena in silicon is *oxidation-enhanced diffusion* (OED) or *oxidation-retarded diffusion* (ORD) of doping elements, which occurs when self-interstitials are generated by oxidation. Another important example is the diffusion of phosphorus from a surface (for instance, in the emitter region of a diode), which can be accompanied by self-interstitial generation and lead to an acceleration of the diffusion of other doping elements (emitter-push effect; see Figure 5.2).

An influence of the intrinsic point defects on the diffusion behavior is also known for the transition metals, which can occupy both substitutional and interstitial sites. Their diffusion mechanism involves the transition between substitutional and interstitial sites with the participation of intrinsic defects, which will be discussed in Section 5.1.3.

Table 5.3
Normalized Interstitial
Diffusion Coefficient D_α^I of
Some Doping Elements in
Silicon (D_α is the total
diffusion coefficient)

Element	D_α^I/D_α
Boron	0.9–1.0
Phosphorus	0.9–1.0
Arsenic	0.2–0.4
Antimony	0.01–0.02

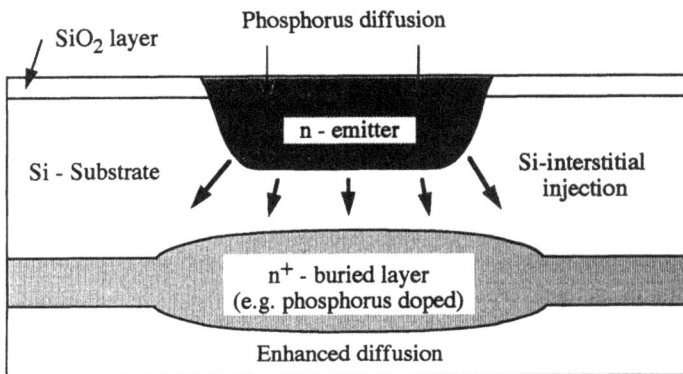

Figure 5.2 Enhanced diffusion of doping elements in a buried layer below the emitter region during phosphorus diffusion (emitter-push effect).

Non-equilibrium concentrations of intrinsic point defects can also be produced during radiation with high-energy particles or photons. This offers the possibility of studying the properties of intrinsic defects that otherwise may be too difficult to observe because of their low concentration. In this way, many results have been obtained, for instance, on vacancies and vacancy-related complexes in silicon and other semiconductors. For solar cell applications, these results are particularly useful, since exposure to high-energy radiation occurs for solar cells that are used in space. Radiation-induced defects are the dominant cause for the degradation of the solar cells in space (this will be addressed in Section 5.1.5).

5.1.1.4 Intrinsic Defects in Binary Compounds

In addition to thermally generated intrinsic point defects in binary (and multinary) compounds, intrinsic defects are also introduced to compensate for deviations from the ideal stoichiometry of the compound. Their concentrations can exceed by far the thermally generated defects, particularly in multinary compounds. The most prominent and thoroughly investigated native defect is the EL2 center, which is the dominating electrical defect in semi-insulating GaAs.

Most of the binary AB compound semiconductors crystallize in the cubic sphalerite structure where the two atomic species occupy the different face-centered cubic (fcc) sublattices (Figure 5.3). This ordered arrangement of atoms allows a larger variety of intrinsic defects. In addition to the vacancies (V_A, V_B) and the self-interstitials (A_i, B_i) that can be formed for each species, antisite defects (A_B, B_A) can occur where an A atom occupies a B-site on the second sublattice (or vice versa). In thermal equilibrium, the concentrations of these defects are determined by their formation entropies and enthalpies, but as mentioned above are more likely to be present because of an excess of A or B atoms. Because of the different numbers of valence electrons of the A and B atoms, only charge-compensating defects can occur simultaneously for a given composition. The general theory that has been worked out for non-stoichiometric ionic crystals by Kröger [9] gives for a binary AB compound the following defects, which have been summarized in Table 5.4. For a given stoichiometry that can be characterized by the ratio $\Delta s = [A]/[B]$, either one or the other defect occurs in excess. The concentration of these defects in thermodynamic equilibrium is determined by the stoichiometry and their formation entropies and enthalpies.

In non-equilibrium situations, which are the more frequent cases, the kinetics such as diffusivity are important, too. The influence on the electronic properties of the semiconductor is determined by the acceptor and donor character of the defects, which, in an ionic model, can be estimated from the excess or deficiency of the valence electrons for the particular defect. The character that can be expected for the different defects is indicated in Table 5.4; however, no information can be

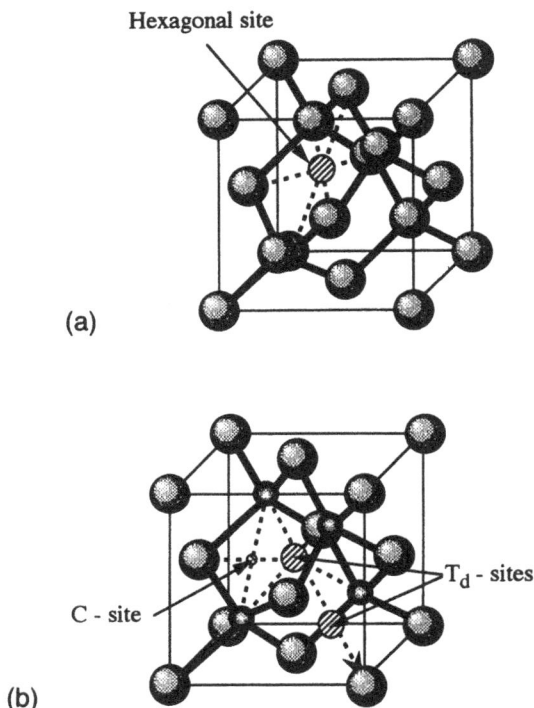

(a)

(b)

C - site

T_d - sites

Hexagonal site

Figure 5.3 Schematic representation of the diamond cubic (a) and sphalerite lattices (b). The A and B atoms in a binary AB compound occupying the two fcc sublattices are shown by black and light circles. The hexagonal (a) and two equivalent tetrahedral interstitial sites T_d (b) in the diamond cubic and sphalerite lattices, respectively, are indicated by hatched circles. A small circle shows a C-site, the center of a rhombus formed by three adjacent atoms and a T_d-site. The mid-position between two adjacent C-sites are M-sites, the equilibrium positions for hydrogen in silicon, for instance.

Table 5.4

The Majority Intrinsic Defects for a Binary AB Compound With a Given Stoichiometry $\Delta s = [A]/[B]$ (The donor (D) or acceptor character (A) is derived assuming a compound with predominantly ionic character)

Defect Type	$\Delta s > 1$		$\Delta s < 1$	
Schottky	V_B	(D)	V_A	(A)
Frenkel (A)	A_i	(D)	V_A	(A)
Frenkel (B)	V_B	(D)	B_i	(D)
Antisite	A_B	(A)	B_A	(D)

obtained about the position and the particular level structure of the defects. It was discussed in the previous section that the formation energies of a charged defect differ from the neutral one; hence, the position of the trap levels are also important for the determination of the defect concentration. Generally, p-doping lowers the formation energies of defects with donor character and n-doping favors acceptor defects. The self-interstitial defects generally seem to have the highest formation energies in many semiconductors; therefore, one is mainly left with the vacancy and antisite defects. (Whether the rather high self-interstitial concentration in silicon, which is due to the unusual large entropy factor, is an exception remains to be seen.)

In agreement with this general rule, it has been observed in undoped (LEC-grown) GaAs that the main defect in arsenic-rich material is the As_{Ga} antisite, and in gallium-rich material the Ga_{As} antisite. In the first case, the material is semi-insulating, mainly because of the donor character of the As_{Ga} antisite, which compensates residual acceptor impurities (Zn, C); in the second case the crystal is p-type. The As_{Ga} antisite defect has attracted considerable interest in the past because it seems to be related to the major electrically and optically active defect in semi-insulating GaAs. The thorough investigation of the properties of the so-called EL2 defect and the As_{Ga} antisite has yielded enormous insight into the structural and electronic properties of GaAs, and will be discussed in Chapter 7.

In doped crystals, the defect situation is different because the shift of the Fermi level can favor the occurrence of the vacancy defects in either case. In fact, it can be observed that in the arsenic-rich n-doped crystals, the V_{As} becomes the dominant defect, whereas in the gallium-rich material, the As_{Ga} acceptor remains dominant, even for p-doping. The V_{Ga} vacancy, which might be favored by p-doping, seems to be a rather unstable defect and probably disappears by the reaction $V_{Ga} \leftrightarrow As_{Ga} + V_{As}$.

5.1.1.5 Intrinsic Defects in Chalcopyrite Compounds

Some of the copper ternary compounds that crystallize in the chalcopyrite structure are currently the most promising candidates for photovoltaic applications (see Chapter 8). As with binary compounds, intrinsic defects are introduced to maintain the crystal structure for non-stoichiometric composition. A major difference is, however, that some of the intrinsic defects are shallow-level defects, which can be used to dope the crystal. In fact, the conductivity type of some of these semiconductors (for instance, for $CuInSe_2$) is usually determined by the composition of the crystals instead of a doping by shallow level impurities. Despite the importance of the intrinsic defects for the copper ternary compounds, their structural and electronic properties are still poorly understood.

There exist 36 known ternary (I-III-VI$_2$ or II-IV-V$_2$) chalcopyrites (see Table 8.2 in Chapter 8) that can be considered isoelectronic analogs to binary II-VI or III-V semiconductors with the zinc-blende (sphalerite) structure. Despite the electronic similarity, the substitution of two column II elements by column I and III elements is, in the first case for instance, accompanied by significant changes in the lattice structure. The differences can be understood by considering the chalcopyrite unit cell (Figure 5.4). Anions (X) and cations (A, B) occupy the two different sublattices, as in the zinc-blende structure. However, an ordering of A and B atoms in the cation sublattice leads to a doubling of the zinc-blende unit cell. In addition, the crystals often show a tetragonal distortion where the ratio between the lattice parameters, $\eta = c/2a$, often differs from 1 by as much as 12%. For a random distribution of the A and B atoms on the cation sublattice, the chalcopyrite again reduces to the sphalerite structure, and some ternary compounds seem to prefer the disordered structure [10] at room temperature. A third difference in the ordered structure is the displacement of the anions from their zinc-blende sites. Each anion (X) has two A and two B cations as nearest neighbors, whereas in the zinc-blende structure, each anion has four cations of the same kind as nearest neighbors (and vice versa). Because of the different bond strength, the anion X usually adopts an equilibrium position that is closer to one pair of cations than to

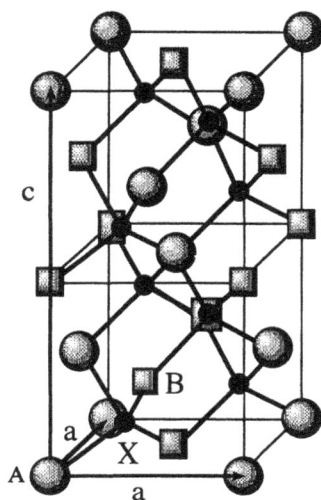

Figure 5.4 Tetragonal unit cell of the chalcopyrite lattice. For an ABX_2 compound, the positions of the cations (A, B) are represented by large circles and squares, and of the anions (X) by small black circles.

the other (Figure 5.4), which explains the frequently observed anion displacement u and the unequal bond length $R_{AX} \neq R_{BX}$ (bond alteration) related to u, η by

$$R_{AX} = a\left[u^2 + \frac{1}{16}(1 + \eta^2)\right]^{1/2} \tag{5.19a}$$

$$R_{BX} = a\left[\left(u - \frac{1}{2}\right)^2 + \frac{1}{16}(1 + \eta^2)\right]^{1/2} \tag{5.19b}$$

Theoretical calculations of the bonding between next neighbors indicates a mixture between covalent and ionic bonding. Although the lattice structure is still very similar to the sphalerite structure, which indicates a major contribution of the covalent bonding, the electronic and other properties have been frequently discussed by considering these semiconductors ionic compounds. For the following structural considerations, however, the covalent character will be emphasized.

There are twelve intrinsic point defects that can be formed in the ABX_2 chalcopyrite lattice, three vacancies (V_A, V_B, V_X) and interstitials (A_i, B_i, X_i) for each atomic species, and six antisite defects. In addition to the antisite defects, which can be formed by an exchange of anions (X) and cations (A, B) as in binary compounds (A_X, B_X, X_A, X_B), one can also form antisite defects on the same sublattice by an exchange of the cations (A_B, B_A). Structural models for some of these defects, which show the next-neighbor environment, are depicted in Figure 5.5. Since nothing is known about the structural properties of any of these defects one can only describe some of the apparent differences in comparison with similar defects in the diamond cubic or sphalerite structure.

As with the vacancy in the dc lattice, it can be assumed that the unsaturated bonds reconstruct (Figure 5.6). For the cation vacancies (V_A, V_B), only X-X bonds can form, whereas contrary to the anion vacancy (V_X), in binary compounds, two types of bond reconstructions are possible here, between A-A and B-B, or two A-B bonds. The structures of the other defects show similarly the occurrence of cation-cation and anion-anion bonds, with the exception of the antisite defects on the cation sublattice A_B and B_A. Because of the polarization of the bonding and ionic charges at the atoms in these compounds, one can assume that bonding between atoms that carry like charges (such as A-A or X-X) are rather unfavorable. One may therefore expect that the formation energies of the antisite defects A_B and B_A that avoid these bonds are lower compared to the other defects. Recent calculations of the formation energies [11,12] in $CuInSe_2$ indeed show that these two antisite defects have the lowest energy (see Table 8.3) and, hence, are likely to be the most dominant intrinsic defects in this ternary compound.

The electronic character of these defects has been determined from a simple calculation of the number of valence electrons and the assumption that the crystals

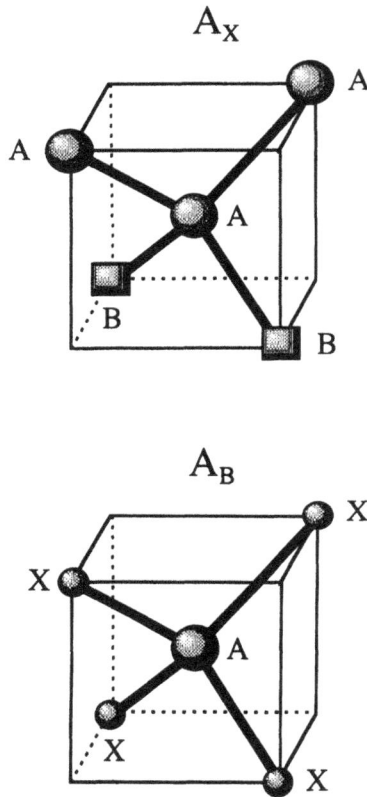

Figure 5.5 Structural models for two antisite defects in a ternary ABX_2 compound with chalcopyrite structure: A_X and A_B antisite defects.

are ionic compound. Considering the bond reconstruction for the vacancies, this approach may be too simple to actually predict the character of these defects. In fact, different conclusions have been drawn as to the acceptor or donor character of the selenium vacancy in CuInSe$_2$. The results are also summarized in Table 8.3 and show that both donor and acceptor defects occur. The positions of the energy levels remain, however, undetermined. At present, no conclusive experimental or theoretical results exist that allow the determination of the structural and electronic properties of any of these defects in CuInSe$_2$ or any other copper ternaries; therefore, this issue remains a challenge for the future.

It has been mentioned above that a difference between binary and copper ternary semiconductors is the possibility of determining the conductivity of the ternary compounds simply by changing the composition. The experimental results

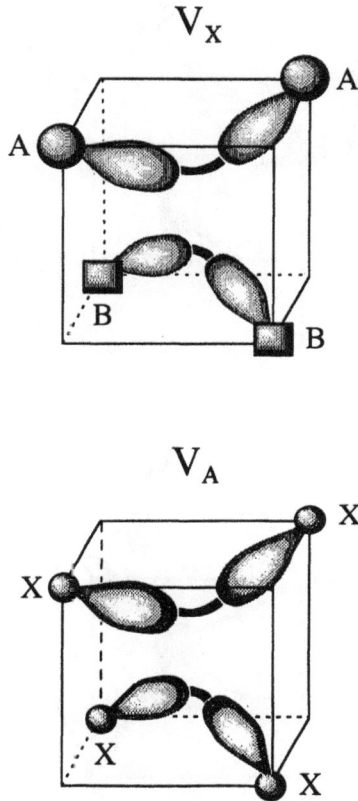

Figure 5.6 Structural models for two vacancy defects in a ternary ABX_2 compound with chalcopyrite structure. Vacancy at an anion site V_X and at a cation site V_A. Possible bond reconstruction is indicated.

indicate that at least some of the intrinsic point defects produced by the non-stoichiometry are shallow-level impurities. The same behavior has not been observed for binary compounds where, for instance, in arsenic-rich GaAs, a deep-level defect—namely, the As_{Ga} antisite (EL2 defect)—is formed. One might explain this difference between the binary and ternary compounds if one assumes that the antisite defects on the cation sublattice (A_B, B_A) are the shallow level impurities, which cannot be formed in a binary system. Photoluminescence and Hall effect measurements in $CuInSe_2$, which will be discussed in Chapter 8, indeed show the occurrence of shallow-level impurities, although no defect could be conclusively related to a particular energy level. One of the problems is which of the possible defects actually occurs for a particular composition. The parameters usually used to characterize the stoichiometry of a ternary compound are the molecularity Δm and the stoichiometry Δs, which are defined by [13]

$$\Delta m = \frac{[A]}{[B]} - 1 \qquad \Delta s = \frac{s_1[X]}{s_2[A] + s_3[B]} - 1 \qquad (5.20)$$

where s_i is the number of excess and deficiency electrons for cations (A, B) and the anion (X), respectively. Since the formation of structural defects has to maintain the neutrality of the crystal, only charge-compensating defects can be introduced simultaneously. General models have been developed for ionic crystals to relate the composition described by these parameters to the formation of the defect pairs that can be present. Unfortunately, these results are still inconclusive, partly due to the fact that the formation energies of these defects and their electrical character are not known with sufficient accuracy.

5.1.2 Carbon and Oxygen

Besides the doping elements, which are usually intentionally introduced, oxygen and carbon are the most abundant impurities in silicon. These elements also occur in compound semiconductors, but are usually not considered the most important impurities for the electronic properties, partly because other impurities or native defects occur with higher concentrations and are thus more dominant. Therefore, in the following, the properties of oxygen and carbon are mainly discussed with respect to silicon.

Apart from the introduction through the starting material, oxygen and carbon are incorporated in the melt during crystal growth through the gas ambient and the contact with the crucible. The amount of an impurity that is finally incorporated in the solid is determined by the melt concentration and the segregation coefficient $k_o = C_\alpha/C_L$ (c_α and C_L are the concentrations in the solid and the liquid). The rather high segregation coefficients for carbon ($k_o = 0.06$) and oxygen ($k_o = 1.25$) and their relatively high solubility in the solid compared to other elements are unfavorable for a refinement during solidification. Other elements such as nitrogen ($k_o = 7 \times 10^{-4}$) or the transition elements ($k_o < 5 \times 10^{-4}$), which may also be introduced at the same time, can be more easily retained in the melt. Depending on the growth technique, the pollution of the melt is unavoidable to some extent, so it becomes important to understand and reduce the influence of these impurities on the electronic properties of the material.

Dislocation-free Cz-silicon crystals contain oxygen and carbon in concentrations of about 10^{18} cm^{-3} and 10^{17} cm^{-3}, respectively, which is far above the solubility at room temperature. Due to the slow diffusion at lower temperatures and the relatively fast cooling rates during crystal growth, the solid solution remains supersaturated. However, annealing an as-grown material leads to enhanced oxygen and carbon diffusion and finally the precipitation of the equilibrium phases SiO_2 and SiC. Since annealing steps are part of the device processing, it is important to understand the precipitation behavior and the impact of the annealing on the material properties.

Cz- and Fz-silicon crystals are mainly used for electronic devices, but also for high-efficiency solar cells, and the results obtained mainly from measurements on high-quality crystals are thus immediately relevant for the solar cell fabrication, too. More important for the photovoltaic application is, however, low-cost polycrystalline silicon, which can be prepared by a variety of techniques (see Section 6.2). These are materials with high concentrations of carbon, oxygen, and other impurities, and a variety of other lattice defects, such as dislocations and grain boundaries. Therefore, they may respond differently to annealing treatments. Since much less is known about the carbon and oxygen behavior in these materials, one has to rely on extrapolations from the results that have been observed in single-crystal silicon.

5.1.2.1 Oxygen in Silicon

Oxygen in silicon forms a peritectic phase diagram with the SiO_2 phase (Figure 5.7), which explains the segregation coefficient $k_0 > 1$. The maximum solubility in the solid is about 2×10^{18} cm^{-3}, and the solubility as a function of the temperature above 750°C is given by [14]

$$C_O^{eq} = C_{Ob}^{eq} \exp\left\{-\frac{H_O^F}{KT}\right\} \qquad C_{Ob}^{eq} = 1.8 \qquad H_O^F = 1.52 \text{ [eV]} \qquad (5.21)$$

Similar values for $C_{Ob}^{eq} = 0.52$ and $H_O^F = 1.4$ eV have been given in [15]. The solubility of oxygen depends on various factors such as the doping, the presence of carbon, thermal stresses in the crystal, or the ambient atmosphere during crystal growth. Therefore, the oxygen content varies with the conditions under which a crystal is grown or has been treated later. Oxygen occupies an interstitial position between two silicon atoms in a slightly off-center position, as indicated in Figure 5.8. In equilibrium, six equivalent positions are then possible. The diffusive jump of oxygen requires the breaking of the Si-O bond without the participation of an intrinsic point defect. The interstitial migration through the silicon lattice described by the diffusivity D_O has been measured over several orders of magnitude and is given by [15]

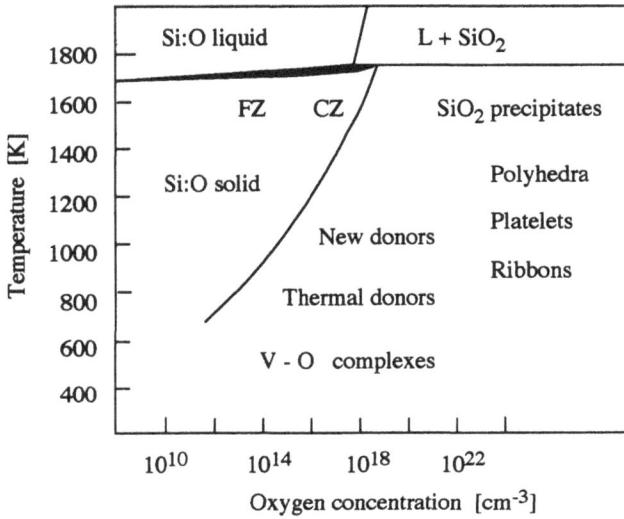

Figure 5.7 Phase diagram for oxygen in silicon and typical temperature ranges for the occurrence of oxygen related defects after annealing. Shaded area indicates the narrow liquid-solid two-phase regime.

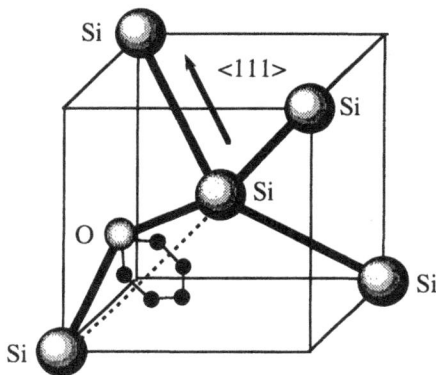

Figure 5.8 Schematic representation of the off-center bond-centered position of an interstitial oxygen in the silicon lattice. Black circles indicate equivalent oxygen sites.

$$D_O = D_{O0} \exp\left\{-\frac{H_O^D}{KT}\right\} \qquad D_{O0} = 0.11 \text{ [cm}^2\text{/s]} \qquad H_O^D = 2.51 \text{ [eV]} \qquad (5.22)$$

In agreement with these considerations, it can be observed that an excess of vacancies or self-interstitials has no influence on the diffusion behavior of oxygen.

Because of its incorporation in an interstitial position, dissolved oxygen leads to an increase in the lattice constant. Although no electrical activity has been observed for oxygen in the interstitial position itself, the presence of oxygen can give rise to a number of electrically active defects.

Vacancy-Oxygen Complexes Below 400°C

At low temperatures, oxygen can form a variety of complexes with other point defects. Of particular importance are irradiation-induced defects, which occur, for instance, after ion implantation, plasma deposition processes, or after electron or ion irradiation [16,17]. The most thoroughly investigated defects are vacancy-oxygen complexes: V-O_n. The simplest complex V-O is the so called A-center with an atomic structure depicted in Figure 5.9. A single oxygen atom occupies a slightly

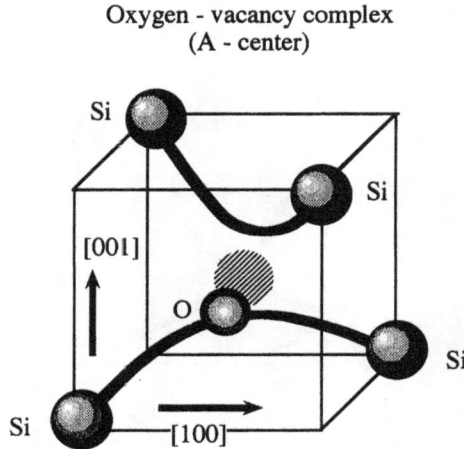

Oxygen - vacancy complex
(A - center)

Figure 5.9 Atomic configuration of the vacancy-oxygen complex (A-center). The oxygen atom is slightly displaced from the substitutional position (shaded area).

off-center bond-centered site and bridges two of the four broken bonds of the vacancy, thus forming a Si-O-Si "molecule." The negatively charged complex in *n*-type silicon is stable below 620K, whereas the neutral complex is only observed below 100K. The *A*-center has an acceptor level near the conduction band at $E_T = E_C - 0.17$ eV. The addition of further oxygen atoms yields complexes that introduce trap levels deeper in the band gap. Multivacancy-oxygen defects such as $O-V_2$, $O-V_3$, O_2-V_2, and O_3-V_3 have been identified by *electron spin resonance* (ESR), where in all cases the oxygen is trapped at the vacancy in the Si-O-Si molecule configuration. The results are summarized in Table 5.5 and show that the complexes are only stable at lower temperatures and are dissociated above 400°C.

In thermal equilibrium, the vacancy concentrations around room temperature are so low that vacancy-oxygen complexes are not important. Under non-equilibrium conditions (for instance, in spacecraft), irradiation-induced vacancies can significantly enhance the concentrations of vacancy-oxygen complexes and thus change the electronic properties of oxygen-rich silicon crystals (see also Section 5.1.5).

Primary Oxygen Agglomeration Between 400 and 650°C

Depending on the annealing temperature (and times), different stages of oxygen precipitation have been observed for as-grown oxygen-rich silicon. The first stage occurs at temperatures around 450°C. The diffusivity is so low that only the agglomeration of a few oxygen atoms can occur. The defects that form after annealing times of several hours introduce two shallow-donor levels in the band gap that are considered two different charge states $(0/+$ and $+/++)$ of the same defect. Several trap levels between $E_C - 70$ meV and $E_C - 150$ meV are observed successively,

Table 5.5

Electronic Levels of Vacancy-Oxygen Complexes in Silicon, Which Occur After Electron Irradiation (Included are the annealing temperatures above which the complex dissociates or the defect levels disappear)

Defect	Annealing Temp. [K]	Energy Level [eV]
V-O ($-$)	620	$E_C - 0.17$
V-O (0)	100	
V-O$_2$		$E_C - 0.50$
V-O$_3$		$E_V + 0.40$
O-V$_2$	620	$E_C - 0.17$
O-V$_3$	720	$E_V + 0.40$

which suggests that these so-called *thermal donors* (TD) in fact consist of a variety of different defects (named TD1 to 9). The concentrations of the different TD centers can be monitored separately by *infrared* (IR) absorption measurements. With annealing time, the concentrations increase first up to about 3×10^{16} cm^{-3}. Since the oxygen concentration in the bulk decreases at the same time with an activation energy that corresponds to the oxygen diffusion, it has been assumed that the TDs are different stages of small oxygen agglomerates that develop into each other by the addition of more oxygen atoms. The simultaneous measurement of the concentration of different TD centers as a function of temperature also shows a thermally activated behavior. A major problem is, however, that the activation energy for the formation of the TD centers is only about 1.7 eV, compared to 2.5 eV for the oxygen diffusion. Several models for the TD generation have been developed [18–20] based on the oxygen aggregation model with the common necessity of assuming an enhanced oxygen diffusion at low temperatures, which has to be about two to three orders faster than the extrapolation for the normal oxygen diffusion at high temperatures. Recently, a new approach has been tried to solve the dilemma with the suggestion that the TDs are in fact self-interstitial aggregates, and that only the first step is an oxygen complex, serving as a nucleus. The nuclei may be small oxygen complexes (presumably O_3) formed at higher temperatures and existing already in the as-grown material. Self-interstitials may be produced during the formation of O_2 complexes at lower temperatures following the reaction [20]

$$O_i + O_i \leftrightarrow O_2 + I \qquad (5.23)$$

Although this model can account for the main experimental observation, no direct proof for the nature of the TD exists, and further investigations are necessary to solve the controversy of whether oxygen or self-interstitial aggregates are the observed TD centers.

After annealing above 600°C, the TDs disappear or do not form at all. Instead, a new type of defect appears in the temperature range around 600° to 700°C with a rod-like morphology and an average size of about 20Å. *High-resolution transmission electron microscopy* (HREM) revealed crystalline precipitates that were at first identified as a high-pressure SiO_2 phase (coesite). Later measurements indicated a rather low oxygen concentration and the features were reinterpreted as hexagonal silicon [21]. The exact nature of these very small crystalline precipitates thus remains controversial.

Oxygen Precipitation Above 800°C

The annealing treatment at temperatures above about 800°C finally shows larger amorphous precipitates with an average size up to 1000Å, depending on the annealing temperature and time. The precipitates are identified as the SiO_x phase with $x \approx 2$. An important aspect is that the formation is associated with a volume change $\Delta\Omega$ given by

$$\Delta\Omega = \Omega_{SiO_2} - \Omega_{Si} - 2\Omega_{O_i} \qquad (5.24)$$

Considering the position of the oxygen interstitial, the excess volume of the interstitial is small, $\Omega_{O_i} \approx 0$ and since $\Omega_{SiO_2} \approx 2\Omega_{Si}$, the volume change corresponds approximately to the volume of one silicon atom $\Delta\Omega \approx \Omega_{Si}$. As soon as the volume increase of the growing precipitate can no longer be elastically adapted the surrounding matrix is either plastically deformed or self-interstitials are emitted (about one silicon atom for two oxygen atoms). Alternatively. the absorption of vacancies is also feasible. The self-interstitial emission or vacancy absorption can finally lead to the formation of dislocation loops or interstitial-type stacking faults, which can be observed experimentally.

The morphology of oxygen precipitates depends on the annealing temperature; and roughly three regimes can be distinguished (see Figure 5.7). Precipitates in the form of ribbons occur below 800°C. Between 800° and 1050°C, plate-like precipitates dominate, which are often accompanied by prismatic dislocation loops. Above 1050°C, the precipitates take the shape of a regular or truncated octahedron. The platelets and ribbons have the (001) habit plane, while the octahedral precipitates have (111) bounding faces. The main features of the precipitation process can be explained on the basis of conventional nucleation growth theories and are noncontroversial [22].

It has been found that precipitates formed between 600° and 900°C are electrically active and associated with donor states [23]. They are commonly termed "new donors" in order to distinguish them from the thermal donors formed at low temperatures. It is assumed that these donor states are related to the SiO_x/Si interface states and probably result from incomplete oxidation (excess silicon) in the SiO_x precipitate near the interface. These donor states may be destroyed by a high-temperature heat treatment when composition at the interface is restored.

The main aspects of the oxygen behavior described so far are valid for crystals with low carbon concentrations. Experimental results frequently show that carbon can have a significant influence on the precipitation behavior at all stages. Since carbon is frequently present in comparable concentrations, especially in solar cell silicon, these interaction processes are very important.

5.1.2.2 Carbon in Silicon

Carbon in silicon is a common impurity because it is introduced from graphite heaters or crucibles and decomposing CO_2 gas during crystal growth. Carbon dissolved in silicon is in equilibrium with SiC and forms a eutectic phase diagram

(Figure 5.10). In thermal equilibrium, the maximum solubility in the solid is about 3.5×10^{17} cm^{-3} (7 ppm) and the eutectic concentration about 5×10^{18} cm^{-3} (100 ppm). The solubility as a function of the temperature is given by [24]

$$C_C^{eq} = C_{C0}^{eq} \exp\left\{-\frac{H_C^F}{KT}\right\} \qquad C_{C0}^{eq} = 10^2 \qquad H_C^F = 2.3 \text{ [eV]} \qquad (5.25)$$

Contrary to oxygen, a carbon atom occupies a substitutional position. Because it is a smaller atom compared to silicon it leads to a decrease of the lattice constant corresponding to a lattice contraction of about one atomic volume for each carbon atom. This is contrary to the lattice expansion of oxygen; therefore, one can expect that an attractive interaction between oxygen and carbon atoms can occur, which may help to reduce the strain. If both carbon and oxygen are present in silicon in high concentration, which is frequently the case in silicon for terrestrial solar cells, the solid solubility is less defined, presumably because of this interaction between carbon and oxygen. In fact, dissolved impurity concentrations of up to 2×10^{18} cm^3 (40 ppm) have been measured exceeding the maximum solubility by an order

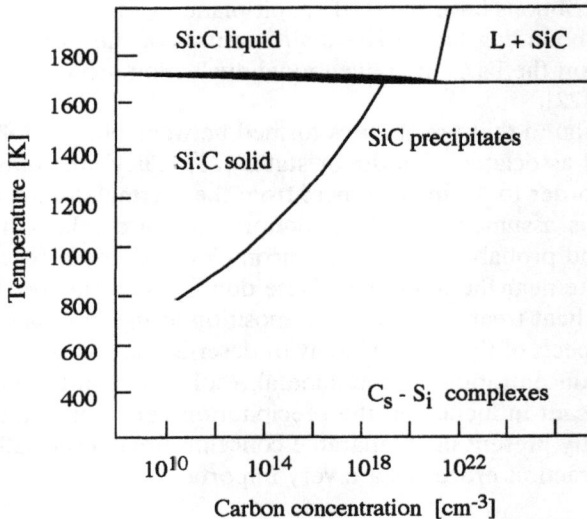

Figure 5.10 Phase diagram for carbon in silicon and temperature range for the occurrence of SiC precipitates. Shaded area indicates the narrow liquid-solid two-phase regime. The CI-complexes occur after irradiation at low temperatures (below room temperature), but may also exist at higher temperatures in crystals with a high supersaturation of self-interstitials.

of magnitude. This is therefore evidence that high concentrations of oxygen enhance the solubility of carbon considerably.

Carbon becomes mobile above about 1000°C, and the diffusion coefficient which has been measured in the temperature range 1050 to 1400°C, is given by [25]

$$D_C = D_{C0} \exp\left\{-\frac{H_C^D}{KT}\right\} \qquad D_{C0} = 1.9 \; [cm^2/s] \qquad H_C^D = 3.2 \; [eV] \qquad (5.26)$$

Since carbon occupies substitutional sites, the diffusion requires intrinsic point defects. In general, the diffusion can be described by (5.9), involving vacancies and self-interstitials. The rather high activation energy suggests a vacancy mechanism, but it has been observed that a supersaturation of self-interstitials after in-diffusion of phosphorus [26] enhances the diffusion of carbon at 900°C. Therefore, it has been proposed that the diffusion mechanism should involve a silicon self-interstitial (Si_i or I). A possible mechanism is the diffusion in the interstitial lattice where the carbon interstitial is formed through the kick-out reaction

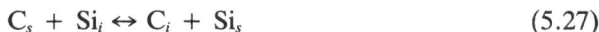

$$C_s + Si_i \leftrightarrow C_i + Si_s \qquad (5.27)$$

Carbon on a substitutional site is an isoelectronic element to silicon and electrically inactive. However, a variety of electrically active carbon-related defects can be observed after low-temperature electron irradiation, which are summarized in Table 5.6. The first defect is the carbon interstitial, which is stable below room temperature. Interstitial carbon introduces two trap levels, an acceptor level at $E_C - 0.12$ eV and a donor level at $E_V + 0.30$ eV [27]. A second defect, which occurs in the same temperature regime, also introduces two trap levels: at $E_C - 0.16$ eV and a donor level at $E_V + 0.36$ eV. It has been suggested that this defect is a carbon-carbon interstitial C_i-C_s complex. Alternatively, it is discussed that the complex

Table 5.6

Electronic Levels of Carbon and Carbon Complexes in Silicon That Can Occur After Electron Irradiation (Included are the annealing temperatures above which the complex dissociates or the defect levels disappear)

Defect	Annealing Temp. [K]	Acceptor Level [eV]	Donor Level [eV]
C_i	320	$E_c - 0.12$	$E_V + 0.30$
C_i-C_s	300	$E_c - 0.16$	$E_V + 0.36$
C_s-Si_i	340		$E_V + 0.30$
C_i-O_i	650		$E_V + 0.38$

involves a self-interstitial and two carbon atoms C_s-Si_i-C_s, so a controversy remains about the correct identification of the defect. A similar electrical behavior is observed for the carbon-silicon interstitial C_s-Si_i complex (CI), which occurs in *p*-type silicon [28] and is stable below 65°C. A split-interstitial position, as shown in Figure 5.11, has been suggested for this defect. In crystals with higher oxygen concentrations, a further defect has been reported which involves the participation of oxygen and is presumably a C_i-O_i complex [29]. This defect is rather stable up to temperatures of about 400°C and is an efficient recombination center.

Carbon-Related Microdefects

If carbon is present in concentrations exceeding the solubility, a driving force exists for precipitation in the form of SiC. However, even in carbon-rich ($>5 \times 10^{16}$ cm^{-3}) Cz and Fz-crystals, silicon-carbide precipitates are rarely observed [30]. This is partly due to the fact that carbon has a diffusion coefficient that is about an order of magnitude lower compared to oxygen, so that for typical cooling conditions diffusion becomes too slow for the precipitation of the SiC particles. Another reason may be the lack of suitable nuclei for heterogeneous nucleation. Instead, it has been shown that the presence of carbon can induce a variety of secondary defects. Although the nature of these defects has not been identified in all cases, the investigation of the conditions of their formation has led to the following model, which can account for many observations.

In a typical cooling situation, after crystal growth for instance, of a float zone (low oxygen content) and dislocation-free silicon crystal, a supersaturation of self-interstitials can build up depending on the growth condition (cooling and growth rate). Since in silicon the formation energy of interstitial-type dislocation loops is fairly high, the formation of carbon self-interstitial (CI) complexes and CI agglomerates can be a more favorable alternative, because of the volume contraction.

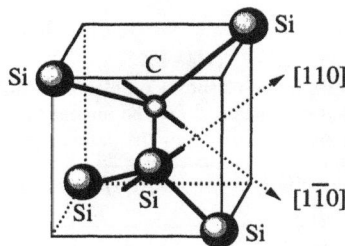

Figure 5.11 Structural model for the carbon-silicon interstitial (C_s-Si_i) complex in silicon, which occupies a split-interstitial position.

Without plastic deformation, the growth can only proceed by the further absorption of self-interstitials (or emission of vacancies). It has been proposed that these agglomerates are identical to the *B-swirls*, a defect that is observed in Fz-grown silicon crystals under certain growth conditions (see Section 6.1). More sophisticated arguments [31] lead to the conclusion that *B*-swirls may eventually collapse and transform into the so called *A-swirls*, which have been identified as interstitial-type dislocation loops with an average size of about 0.5 to 10 μm. Within this model, one can expect that a reduction of the self-interstitial concentration should also retard the carbon agglomeration or precipitation. Such a suppression of carbon precipitates is actually observed in carbon-rich ($>10^{17}$ cm^{-3}) Fz-grown silicon after annealing between 550° and 1300°C, when dislocations are present which can easily absorb self-interstitials and reduce the driving force for the precipitation.

The investigation of carbon-rich Cz- and Fz-silicon has clearly shown that the carbon is involved in the formation of microdefects, and many results can be explained by assuming the formation of CI agglomerates to be an early stage of carbon precipitation. The nature and the mechanism for the formation of these microprecipitates is, however, not yet completely understood.

Interaction of Carbon and Oxygen

Since carbon and oxygen atoms contract and expand the lattice, respectively, it seems evident that an interaction occurs. In fact, it is observed that the solubility of oxygen is enhanced when carbon is present and vice versa. In silicon with an oxygen concentration of about 10^{18} cm^{-3}, the solubility of carbon is about three times larger compared to oxygen-poor crystals. In carbon-rich crystals ($>10^{17}$ cm^{-3}), the solubility of oxygen is increased by a factor of about four. In addition, electrical measurements in carbon- and oxygen-rich crystals show a trap level at $E_V + 0.38$ eV which has been identified with a carbon interstitial—oxygen interstitial complex C_i-O_i. This defect, which can have a rather detrimental impact on the lifetime in silicon, dissociates above about 400°C.

Carbon and oxygen also interact during precipitation. Experimental results show that both the oxygen and carbon precipitation is related to the excess or deficiency of self-interstitials. The previous discussion has shown that the equilibrium concentration of self-interstitials can be disturbed in several ways. A supersaturation can occur simply by lowering the temperature because of the lower solubility, or in oxygen-rich crystals by the oxygen precipitation itself, for instance, through the reactions (5.23) or (5.24). The concentration can also be changed by external processes such as surface oxidation or phosphorus diffusion.

Considering the second external processes, only a supersaturation of self-interstitials will *promote* the carbon precipitation and *retard* the oxygen precipitation. An undersaturation has the opposite effect. These trends are in general agreement with the observation discussed before. If the two impurities are present in comparable concentrations, their interaction has to be taken into account. A supersaturation of self-interstitials will certainly again promote the carbon diffusion and the formation of CI agglomerates. Considering the different volume changes, it is likely that the CI agglomerates may act as nucleation sites for the SiO₂ precipitates, and, hence, *enhance* the oxygen precipitation, contrary to the situation where the carbon concentration is low. Experimental results in fact indicate that the level of carbon concentration may be essential for the different behavior of oxygen precipitation in cases where self-interstitials are externally generated.

At lower temperatures, where thermal donors form, the situation is different. It has been observed that carbon reduces the TD formation rate considerably. This has been used as an indication that the thermal donors may actually be self-interstitial agglomerates, and, hence, the process competes with the CI formation. In view of the controversy about the nature of the TD, the influence of the carbon on the nucleation rate remains an open problem.

A different situation is the first case, when no external generation processes occur, but a supersaturation of self-interstitials is present after cooling down the crystal. If the carbon concentration is sufficiently high, CI agglomerates can form as explained above. For high oxygen supersaturations, the driving force for precipitation may be high enough so that precipitation occurs independently of the carbon concentration. For lower oxygen concentrations, the presence of CI agglomerates can facilitate the nucleation of oxygen precipitates. This is in general agreement with the experimental observations [32]; however, more refined considerations of the detailed balance between self-interstitials, CI agglomerates, oxygen precipitates, or dislocations are required to account for all experimental situations.

Silicon-Carbide Precipitation at High Temperatures

One of the puzzling results for carbon in Cz- and Fz-grown crystals is that, in general, the SiC precipitation is suppressed. In fact, it seems to be rather difficult even in carbon-rich single crystals at high annealing temperatures (>1300°C) to obtain SiC precipitates. On the contrary, one can observe SiC precipitates quite frequently in carbon-rich silicon for solar cells [33]. Typically, SiC precipitates with sizes between 0.1 and 1 μm can be observed (see Chapter 6). TEM and HREM investigations have identified the silicon carbide cubic 3C (β − SiC) and the silicon carbide hexagonal 15R (α − SiC) phases. Since SiC can occur in a variety of different phases (polymorphism), other structures may occur as well. The approximately 50 known SiC phases differ by the stacking sequence of the close-packed (111) planes. The various SiC phases are semiconductors with band gaps between 2.2 and 3.2 eV.

Contrary to Cz- and Fz-silicon, SiC precipitation is more frequently observed in cast or ribbon-grown solar silicon. This is partly due to higher carbon concentrations compared to electronic-grade silicon. However, since the SiC precipitation is generally hampered in the solid, it has been proposed that under particular conditions precipitation can occur directly from the melt in these materials [34]. The model assumes that the SiC precipitates nucleate at the melt interface and grow by the diffusion of carbon in the melt until they are incorporated by the advancing interface (see Section 5.4.2.3).

The numerical analysis shows that slow interface velocities and high supersaturations of the carbon concentration lead to large precipitates. Typical values for directionally solidified solar silicon are $v = 0.001$ μm/s and $\Delta C_{am} \approx 150$ ppm, which yields precipitate sizes in the micron range. These results thus indicate that larger SiC precipitates (>0.1 μm) may actually have formed at the melt interface. For higher interface velocities of about 0.01 μm/s or lower supersaturations of about $\Delta C_{am} \approx 100$ ppm, precipitate sizes below 10 nm result. These conditions occur for some ribbon or sheet growth techniques (see Section 6.2.2) or in Cz- or Fz-crystals (low carbon concentrations). Whether these nanometersize SiC precipitates actually form, or if carbon stays in solution and gives rise to the CI agglomerates in the solid state, has not been decided yet, but the result can explain why the SiC precipitation is less frequently observed in these materials.

Electrical Activity of Carbon- and Oxygen-Related Defects

There is clear evidence that oxygen forms a variety of defects that are electrically active. At room temperatures, vacancy-oxygen complexes can form in irradiated crystals. If carbon is present in comparable concentrations, the C_i-O_i complex forms with a rather detrimental impact on the lifetime of silicon. These complexes anneal out between 300° and 450°C. At higher temperatures (450°C), the thermal donors are formed with maximal concentrations of about 10^{16} cm^{-3}. Since they anneal out above 600°C, they are probably not important for solar cell silicon. At higher temperatures (600° to 900°C), new donors appear which are likely to be related to SiO$_x$ precipitates, whereas SiO$_x$ precipitates formed above 1000°C are less electrically active. The origin of the electrical activity is certainly not understood yet, but may be related to the Si/SiO$_x$ interface. Another possibility is that the electrical activity of the SiO$_x$ precipitates is determined by impurities that are gettered, a process used to remove electrically detrimental impurities.

Still much less is known about the electrical activity of the carbon-related defects at higher temperatures. Since there is clear evidence that carbon has an influence on the oxygen precipitation, which can lead to electrically active defects, the main influence on the electrical properties of silicon may be indirect. It has also been observed that A- and B-swirls change the lifetime of single-crystal silicon, but there is no conclusive evidence yet that the precursors, possibly the CI agglomerates, may be responsible for the reduced lifetime in ribbon-grown silicon (see Section 6.2).

Another possibility is the occurrence of very small (nanometer size) but electrically active SiC precipitates which have not been found yet. The electrical properties of SiC precipitates, if they occur, are likely to be determined by their semiconducting bulk properties. One can expect that the electrical conductance is strongly affected by the contamination of impurities (doping) from the surrounding silicon matrix. The interaction with impurities can be seen in Figure 6.15, for instance, which shows a SiC precipitate in cast solar silicon decorated with copper precipitates. Contaminated SiC precipitates in the micron range will be rather conductive and thus detrimental to the performance of a *pn* junction if, for instance, they lie in the depletion region.

5.1.3 Transition Metals in Silicon

Deep-level impurities with high-capture cross sections for electrons and holes are efficient recombination centers in semiconductors. Transition metal impurities are known to form deep trap levels in the forbidden gap. The most extensively studied impurities are the 3d transition elements (from Ti to Cu) and some of the 4d (e.g., Mo, Pd, Ag) and 5d elements (e.g., Pt, Au) in silicon. Because of their high mobility they can be easily introduced during device processing and are thus of special concern for photovoltaic applications. A summary of the most important properties from the perspective of the photovoltaic application will be given in the following sections mainly for silicon, and are listed in Table 5.7 [35].

Transition metals are also present in other semiconductors, but are usually less investigated because they are not the most important deep-level defects that can be observed. For instance, in GaAs, the dominant deep-level EL2 center occurring in semi-insulating LEC material has a typical concentration of about 10^{16} cm^{-3}. Generally, the concentration of transition metals can be kept below this level, and hence these impurities play a greater role in rather pure semiconductors such as electronic-grade silicon.

The situation in low-cost solar cell silicon is different, since higher concentrations of impurities have to be tolerated. Therefore, the transition metals contribute significantly to the total recombination behavior of the material. After the impurities are incorporated in the semiconductor, the quality of a material can still be improved by gettering the detrimental impurities in regions that are less important for the device performance, or by passivation. Whereas gettering is a standard technique in microelectronics to remove transition metals from the active region of the device, it is not routinely applied in the silicon solar cell technology. One

Table 5.7
Experimental Data for the Diffusivity and Solubility of Transition Elements in Silicon (D_0 and Q_D are the prefactor and activation energy, respectively, of the diffusivity, and H_α^s, S_α^s the formation enthalpy and entropy of the solubility. The diffusion coefficients at 1100°C, the maximum solubilities (C_α^{max}), and the room temperature equilibrium (silicide) phases with silicon are included)

3d Elements	Ti	V	Cr	Mn	Fe	Co	Ni	Cu
Silicide Phase	$TiSi_2$	VSi_2	$CrSi_2$	$MnSi_2$	$FeSi_2$	$CoSi_2$	$NiSi_2$	Cu_3Si
H_α^s [eV]	3.05		2.79	2.81	2.94	2.83	1.68	1.49
S_α^s/K	4.2		4.7	7.3	8.2	7.6	3.2	2.4
C_α^{max} [cm^{-3}]	1×10^{15}	2×10^{16}	1×10^{16}	2×10^{16}	2×10^{16}	3×10^{16}	7×10^{17}	2×10^{18}
Q_D [eV]	1.79		1.0		0.87		0.47	0.43
D_0 [cm^2/s]	1.5×10^{-2}		10^{-2}		6.2×10^{-3}		2×10^{-3}	4.7×10^{-3}
D_{1100} [cm^2/s]	4×10^{-9}	$\approx 10^{-7}$	2×10^{-6}	3×10^{-6}	4×10^{-6}	$\approx 10^{-5}$	4×10^{-5}	1×10^{-4}

4d Elements	Zr	Nb	Mo	Tc	Ru	Rh	Pd	Ag
Silicide Phase	$ZrSi_2$	$NbSi_2$	$MoSi_2$	$TcSi_2$	$RuSi_2$	$RhSi_2$	Pd_2Si	
C_α^{max} [cm^{-3}]	5×10^{14}	5×10^{14}	2×10^{14}				3×10^{16}	5×10^{17}
Q_D [eV]								1.6
D_0 [cm^2/s]								0.002
D_{1100} [cm^2/s]					5×10^{-6}			3×10^{-9}

5d Elements	Hf	Ta	W	Re	Os	Ir	Pt	Au
Silicide Phase	$HfSi_2$	$TaSi_2$	WSi_2	$ReSi_2$	$OsSi_3$	$IrSi_3$	$PtSi$	
C_α^{max} [cm^{-3}]		1×10^{14}	9×10^{13}					1×10^{17}
Q_D [eV]					1.3	1.9		0.39
D_0 [cm^2/s]					0.042	0.1		2.4×10^{-4}
D_{1100} [cm^2/s]		1×10^{-12}			4.5×10^{-8}	7.1×10^{-7}	1.1×10^{-8}	9.0×10^{-6}

reason is that, contrary to an electronic device, the entire solar cell is an active region for the photovoltaic conversion, and, secondly, that the hydrogen passivation has been proven to be quite efficient. However, gettering techniques may become more important for thin-film devices in the future.

5.1.3.1 Solubility

The transition metals (except the noble metals) form complex phase diagrams, with a number of intermetallic phases in each case. For most of the elements, a silicon-rich $MeSi_2$ phase is formed in equilibrium with the α-phase (silicon with dissolved impurity) at room temperature. Exceptions are the 5d elements osmium and iridium, which occur in the $MeSi_3$ phase, and the elements with a complete (or nearly complete) d-shell, which form metal-rich phases (Cu_3Si, $PdSi$, and $PtSi$), or no silicide at all, as for the noble metals silver and gold. Solubility curves for the α-phase of the 3d transition elements are shown in Figure 5.12 and are characterized by an increasing solubility of the elements with increasing atomic number. In general, the solubilities of transition metals are several orders of magnitude lower compared to the doping elements. The solubility C_{a0}^{eq} below the eutectic temperature is exponentially dependent on the temperature

$$C_{a0}^{eq} = \exp \frac{S_\alpha^S}{KT} \exp\left\{-\frac{H_\alpha^S}{KT}\right\} \qquad (5.28)$$

where S_α^S and H_α^S are the formation entropy and enthalpy of solution for the neutral metal atom in solid silicon in equilibrium with the metal-silicide. Figure 5.12 shows that for the 3d elements the solubility of copper and nickel is significantly higher compared to the other elements. This is also reflected in the values for the formation enthalpies (Table 5.7), which allow separation of the elements into two subgroups, namely, 3dI (Ti, Cr, Mn, Fe, and Co) with $H_\alpha^S \approx 2.85$ eV and 3dII (Cu, Ni) with $H_\alpha^S \approx 1.6$ eV. There is little information available on vanadium, which belongs to the 3dI group, and most of the 4d and 5d transition metals.

The solubility for the transition elements above the eutectic temperature is retrograde. Empirically, it has been found that the maximum solubility of the

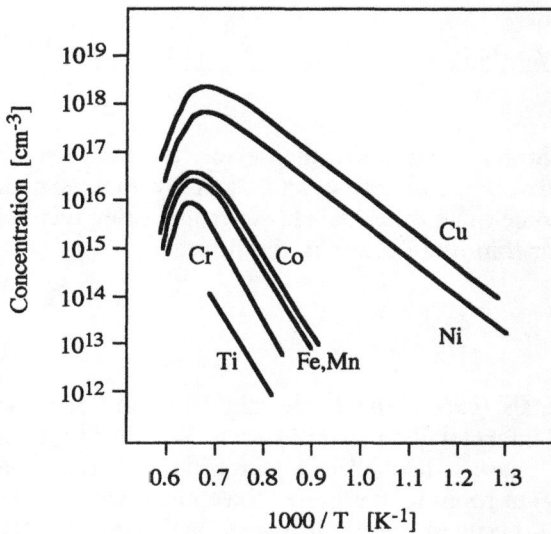

Figure 5.12 Experimental solubilities of 3d transition metals in silicon [35].

impurities (see Table 5.7) is related to the distribution (segregation) coefficient k_0 by the Fischler relation

$$k_0 = 1.92 \times 10^{-22} \, C_{max}^{eq} \tag{5.29}$$

The distribution coefficients for transition elements are generally very low; therefore, it is possible to refine silicon crystals during solidification and keep the metal impurity concentrations in the solid low. The solubility of the transition elements at room temperature is also low because of the strong temperature dependence of the solubility, less than about one atom/cm^3. Therefore, to a large extent, precipitation of the metal-silicide phases will occur during cooling.

As with the doping dependence of the concentration of intrinsic defects, the solubility of the transition metals can also be changed by doping. Since most of these impurities are multilevel defects, the occupation of the trap levels reduces the formation enthalpy H_α^S (of the neutral defect) and thus increases the solubility. If one considers a two level defect with an acceptor level E_{Ta} and donor level $E_{Td} < E_{Ta}$, which is typical for a transition metal impurity (Figure 5.13), the for-

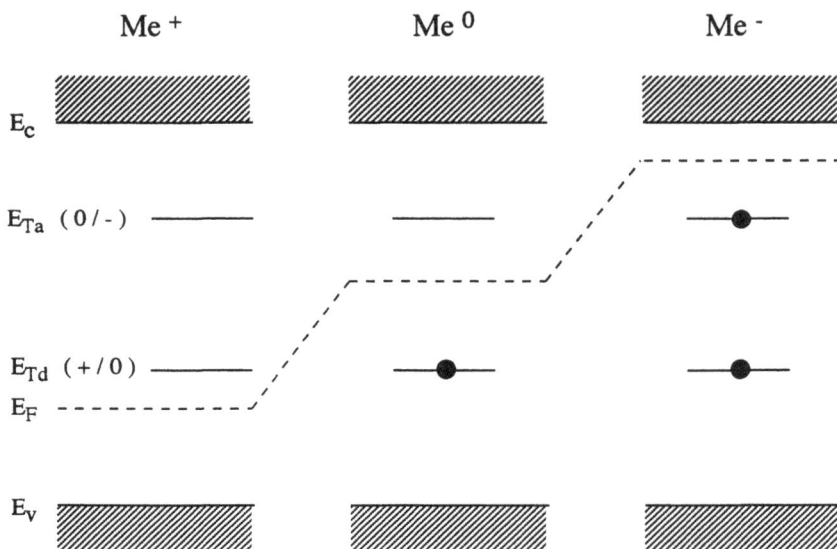

Figure 5.13 Simplified electronic level scheme for a transition metal. The occupation of the two energy levels as a function of the position of the Fermi energy is depicted.

mation enthalpies depend on the position of the Fermi energy. The solubilities of the charged defects $C_{\alpha+}^{eq}$, $C_{\alpha-}^{eq}$, according to (4.30), are

$$C_{\alpha+}^{eq} = C_{\alpha0}^{eq} \exp\left[\frac{E_{Td} - E_F}{KT}\right] \tag{5.30a}$$

$$C_{\alpha-}^{eq} = C_{\alpha0}^{eq} \exp\left[\frac{E_F - E_{Ta}}{KT}\right] \tag{5.30b}$$

where $C_{\alpha0}^{eq}$ is the solubility of the neutral defect and is given by (5.28). Due to the increase of the charged species, the total solubility

$$C^{eq} = (C_{\alpha0}^{eq} + C_{\alpha+}^{eq} + C_{\alpha-}^{eq}) \tag{5.31}$$

increases for $E_F < E_{Td}$ and $E_F > E_{Ta}$.

5.1.3.2 Diffusion

Transition metals are fast diffusers in silicon at high temperatures ($T > 0.7\,T_m$, T_m melting temperature). The diffusion constant as a function of the temperature can be expressed by

$$D = D_0 \exp\left[-\frac{H_D}{KT}\right] \tag{5.32}$$

and is shown in Figure 5.14 for the 3d elements [35]. In Table 5.7, some of the activation energies H_D and prefactors D_0 in the temperature range 900 to 1350°C are listed together with the diffusion coefficient at 1100°C. The activation energies for the 3d elements vary between 0.5 to 1.0 eV and the diffusion coefficients at the melting point are in the range of 10^{-5} cm^2/s. The comparison with the diffusion coefficients for self-diffusion ($H_{SD} \approx 4.86$ eV) and some of the doping elements ($H_D \approx 3.7$ eV for P or B) suggests that in the high temperature regime these high-diffusion coefficients can only be explained by an interstitial diffusion mechanism. Although titanium has a rather large activation energy compared to the other 3d impurities, *electron spin resonance* (EPR) results indicate that the atoms also occupy interstitial positions at high temperatures.

An important issue regarding the diffusion and solubility of transition metals is whether they occupy substitutional (C_s^{eq}) or interstitial sites (C_i^{eq}). For most of the 3d elements, it has been concluded from experimental measurements that at

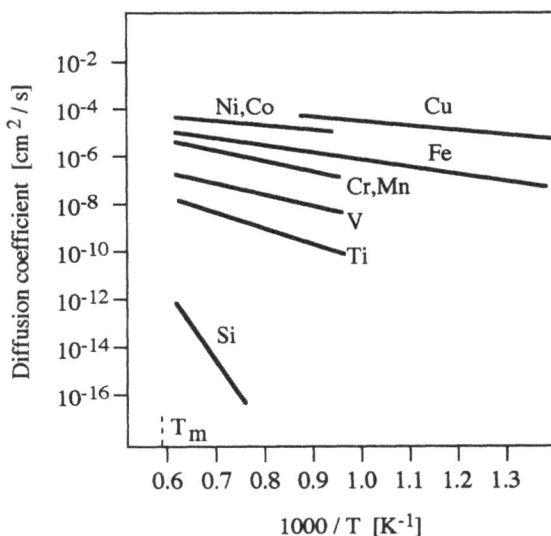

Figure 5.14 Temperature dependence of the diffusion coefficients of 3d transition metals in silicon compared with self-diffusion data for silicon [35, 52].

high temperatures they preferentially occupy tetrahedral interstitial sites (see Figure 5.3), with a smaller fraction of atoms occupying substitutional sites. In general, one has to assume that for each element the concentration ratio C_i^{eq}/C_s^{eq} is different and varies with the solvent and the temperature.

A summary of the various experimental results for silicon is that a significant fraction of the transition impurities occupy interstitial sites, and the high diffusivity is due to a high mobility of the interstitial species in the lattice. In general, the diffusion of such impurities, which occupy both substitutional and interstitial, sites occurs via either the dissociative or kick-out mechanism, which involves an exchange of atoms between the two lattice sites and the participation of native point defects. As discussed in Section 5.1.1, the diffusion can be enhanced considerably in the presence of a high concentration of vacancies or interstitials and led to complex diffusion profiles.

For most of the 3d elements, the experimental results at high temperatures indicate, however, that the interstitial diffusion is the dominant diffusion mechanism independent of native lattice defects. The high diffusion coefficient for the 4d and 5d elements indicates a similar behavior, but more experimental data are certainly necessary to establish the diffusion mechanism and the ratio of atoms occupying substitutional and interstitial sites. Nonetheless, one cannot rule out at present that for some elements the more complex diffusion processes have to be considered to account for all possible diffusion conditions. For instance, diffusion measurements for gold in silicon can only be reconciled with theoretical models by assuming that a large fraction of gold occupies substitutional sites in silicon and diffusion occurs via the kick-out mechanism [32].

Considering the low solubility and rather high diffusivity of the metals at low temperatures, it is evident that during or after cooling the major fraction of the dissolved impurities will preferentially precipitate. Alternatively, the elements can also associate with other impurities or intrinsic point defects or may form small clusters. The morphology of the defects that actually nucleate at low temperatures is determined by the kinetics of the precipitation and complex formation process and is hence dependent on the thermal history of the specimens and the presence of other defects. Of particular interest is the interaction with interfaces, dislocations, or other lattice defects (see Section 5.1.5). In some cases, the various stages of precipitate formation have been studied by HREM, and the results usually show a complex morphology (size, phase, and structure) of the precipitates (see, for instance, Figure 6.20) [36].

In solar cell semiconductors with lower perfection and purity, one can, however, expect that, after cooling, the transition metals are either precipitated in the bulk or at extended lattice defects or have formed complexes with other point defects. The ramifications of these interaction processes on the electronic properties of the semiconductors will be discussed in the following sections.

5.1.3.3 Energy Levels

Transition metals are the most important impurities forming deep energy levels in the forbidden band gap. Because of their usually high mobility, they are easily introduced during device processing and, because of their electrical activity, can be rather detrimental to the device performance. Considering the complexity of the solubility and precipitation behavior, it is not surprising that the elements can occur in a large variety of different defect states that depend on the experimental conditions. It is evident that the electronic levels associated with a complex defect such as a three-dimensional precipitate are difficult to predict and can vary with the morphology. The variability of the defect structure is reflected in a large number of different trap levels that have been associated with a particular impurity. A comparison of different results is thus only possible if the defect configuration at the atomic level can be established and related to a particular trap level.

This problem has been solved to some degree for the electronic levels of impurities when they occupy interstitial sites. Some of the transition elements with lower diffusivity can be kept in the interstitial position after quenching. The electronic levels of the interstitial 3d elements (Ti to Fe) in silicon, which are given in Table 5.8, are well established. The results show that, in general, the impurities

Table 5.8

Donor and Acceptor Levels of 3d Transition Elements on Interstitial Lattice Sites (For other transition elements, the occupation of an interstitial site has not been observed or could not be confirmed. The energy levels of the metal-boron pair are included in cases where they occur)

Impurity	Acceptor Level $M^{-/0}$	Donor Level $M^{0/+}$	Double Donor Level $M^{+/++}$	Donor Level $MB^{0/+}$
Ti	$E_C - 0.08$ eV	$E_C - 0.28$ eV	$E_V + 0.27$ eV	
V	$E_C - 0.16$ eV	$E_C - 0.45$ eV	$E_V + 0.30$ eV	
Cr		$E_C - 0.22$ eV		$E_V + 0.28$ eV
Mn	$E_C - 0.11$ eV	$E_C - 0.42$ eV	$E_V + 0.25$ eV	$E_C - 0.55$ eV
Fe		$E_V + 0.39$ eV		$E_V + 0.10$ eV

introduce donor and acceptor levels that are associated with different charge states. The lowest state is the double-positive donor state, which lies in the lower half of the band gap, and for iron and chromium is shifted into the valence band. In p-type silicon at low temperatures, the interstitial metals are therefore positively charged (M_i^+). A particularly important process that can then occur is the formation of a shallow acceptor-metal pair. The mobile ion can be captured by a negatively charged shallow acceptor (A_s^-) and form a ($M_i^+ A_s^-$) pair simply by Coulomb attraction. These acceptor-donor pairs introduce a new donor trap level which has been observed for a number of transition elements. The results are also listed in Table 5.8 for boron-doped silicon. Acceptor-metal pairs have also been observed for gallium- and aluminum-doped silicon. The binding energies are usually small and close to 0.5 eV, so the pairs dissociate at elevated temperatures above about 200°C [35].

Although, in general, fast cooling or quenching is necessary to keep the metal on interstitial sites, it is observed that a larger fraction of the impurities such as titanium, vanadium, or chromium may still remain dissolved even after slow cooling because of the decreasing diffusion coefficient with the decreasing atomic number of the 3d metals. In fact, it is observed that even small concentrations of about 10^{14} cm^{-3} of titanium or vanadium [37] can severely degrade the solar cell efficiency, likely because of the remaining interstitial component.

Cobalt, nickel, and copper in silicon cannot be kept on interstitial sites, even by rapid quenching. They precipitate or form other defects and introduce a variety of levels, as discussed above. Some results that have been reported in agreement by a number of authors are compiled in Table 5.9 [38,39]. In many cases, however, the results cannot be reconciled with measurements from other authors. The concentrations of defect levels associated with precipitates or complexes are usually rather low compared to the impurity concentration, most likely because the formation of electrically active defects requires the participation of many atoms. Therefore, higher concentrations of the metal impurity are necessary to produce

Table 5.9
Various Donor and Acceptor Levels of Some Transition Elements That Are Experimentally Observed From Hall Effect or DLTS Measurements (The atomic nature of the defects is in most cases unknown or controversial)

Impurity	Acceptor Levels			Donor Levels
Co		$E_V + 0.35$ eV	$E_C - 0.53$ eV	
Ni	$E_V + 0.22$ eV		$E_C - 0.35$ eV	
	$E_V + 0.18$ eV	$E_V + 0.21$ eV	$E_C + 0.33$ eV	
Cu	$E_V + 0.24$ eV	$E_V + 0.37$ eV	$E_V + 0.52$ eV	
	$E_V + 0.20$ eV	$E_V + 0.35$ eV	$E_V + 0.53$ eV	
Pd		$E_V + 0.34$ eV		$E_V + 0.32$ eV
			$E_C - 0.54$ eV	$E_V + 0.32$ eV
Ag			$E_C - 0.33$ eV	$E_V + 0.33$ eV
			$E_C - 0.54$ eV	$E_V + 0.29$ eV
Pt		$E_V + 0.36$ eV	$E_C - 0.25$ eV	
			$E_C - 0.28$ eV	
Au			$E_C - 0.54$ eV	$E_V + 0.35$ eV

deep-level concentrations, which can become harmful for the efficiency of silicon solar cells (see Chapter 6). Since the metal impurity concentrations can be kept below certain levels ($<10^{16}$ cm^{-3}) rather easily, the slowly diffusing transition metals are more detrimental, because even at low concentrations a substantial fraction can remain on interstitial sites and be electrically active.

The situation for the 4d and 5d transition metals in silicon is less clear. Although some of these elements are thoroughly studied, such as gold, the atomic configuration is in most cases not established. The main trap levels that have been frequently observed are given in Table 5.9 for the most common of these elements. But as with the situation for precipitating metals, such as copper and nickel, controversial measurements have also been reported for most of these impurities, which suggests that the electrical activity of these elements is also due to complexes, aggregates, or precipitates.

5.1.4 Hydrogen and Passivation

The behavior of hydrogen in semiconductors has been the object of intense experimental and theoretical investigations over the last decade [39–41]. Besides the fundamental scientific interest, the major reason is the importance of the role hydrogen plays in modern silicon technology. It has been recognized that at almost every stage of device processing, hydrogen can be introduced into the silicon lattice and affect the device properties. In the fabrication of crystalline silicon solar cells, hydrogen is intentionally incorporated because it can have beneficial effects on the performance and the efficiency. Hydrogen is even more important for amorphous silicon, where it is usually introduced during the deposition of the films and is essential for the utilization of the material. This particular aspect of the role of hydrogen will be discussed in Chapter 9. Hydrogen is also one of the four dominant neutral impurities in high-purity germanium (besides Si, O, and C), and has also been investigated in GaAs and some other compound semiconductors, such as AlGaAs and GaP. Most of the information is available for silicon; therefore, the main results will be discussed for this material.

A consistent picture of the state of hydrogen in the silicon lattice has evolved during the last years and can be summarized as follows (see also Table 5.10). At low temperatures (<500°C), hydrogen can occur in three different states:

1. It can be trapped at unsaturated covalent bonds. This Si-H site has the lowest potential energy of all states. It is also clear that trapping can occur at point defects and extended lattice defects, although it is not certain whether unreconstructed bonds or chemically induced reconstruction of bonds is the cause.
2. In the absence of defect sites to which hydrogen can be bound, molecular hydrogen (H_2) is the stable configuration. Molecular hydrogen occupies the tetrahedral interstitial site (T_d) in the diamond cubic lattice (Figure 5.3), is electrically and optically inactive, and is almost immobile.
3. Atomic hydrogen occupies an interstitial site (M-site), which is the lowest energy site for unbound hydrogen. The diffusivity for atomic hydrogen is high even at room temperatures. The calculated and experimental activation energies are 0.3 eV and 0.5 to 1.2 eV, respectively, for migration from M to M-sites across the C-sites (Figure 5.3).

The three forms of hydrogen are in equilibrium among themselves, and the bound and molecular forms dominate at low temperatures, which explains why hydrogen is then rather immobile. At higher temperatures, the bounded configurations break up and the atomic form dominates; therefore, the effective diffusivity becomes high. Theoretical calculations show that atomic hydrogen should have a ground state in the valence band, but there is no direct experimental evidence that a donor level exists in the upper half of the band gap.

Passivation of Deep Levels

The particular property of atomic hydrogen, namely, that it reacts with certain point defects, impurities, or lattice defects, such as grain boundaries and dislocations in crystalline silicon and other semiconductors (GaAs, AlGaAs, GaP, etc.), and passivates their electrical activity, is the driving force for the utilization of hydrogen in the device and solar cell technology. Typically, the passivation is achieved by exposing the specimen to a low-pressure (0.1 to 0.5 torr) rf hydrogen

Table 5.10

Donor (D) and Acceptor (A) Levels of Impurities and Lattice Defects in Silicon (a) and GaAs (b), Which Can Be Passivated by Hydrogen (For GaAs, defect levels are compiled which occur during the epitaxial growth. The origin and nature of the defects is in most cases unknown or controversial. E_D is the dissociation energy for the defect-hydrogen complex as determined from annealing measurements. A distribution of defect states $N(E)$ occurs for extended defects)

Lattice Defects of Impurities	Electron Trap [eV] $E_C - E_T$	Hole Trap [eV] $E_V + E_T$	Dissociation Energy E_D [eV]
Metals			
Au	0.54 (A)	0.35 (D)	2.3
Ag	0.54 (A)	0.29 (D)	2.2
Pd	0.22 (A)	0.32 (D)	2.4
Pt	0.28 (A)	—	2.3
Cu		0.20 0.35 0.53 (A)	2.5
Ni		0.18 0.21 0.33 (A)	2.5
Fe		0.32 (D)	
Doping elements			
B		0.045 (A)	1.1
Ga		0.065 (A)	1.6
Al		0.057 (A)	1.9
Complexes			
TD (thermal donor)	0.07 0.15 (D)		1.1
V-O (A-center)	0.17 (D)		1.9
V-V (divacancy)	0.23 (D)		1.9
Extended defects			
Dislocations	$N(E)$		3.1
Grain boundaries	$N(E)$		2.5

Lattice Defects or Impurities	Electron Trap [eV] $E_C - E_T$	Dissociation Energy E_D [eV]
Point defects		
LEC: EL2/EL6	0.82 0.22	2.5
LPE: EL5/EL3	0.38 0.54	2.2
MBE: EL6/EL5	0.28 0.45	2.9
Doping elements		
Si, S, Se, Te, Sn, Ge	Shallow level	2.1
Zn, Cd, Mg, Be		1.6
Extended defects		
Grain boundaries	$N(E)$	2.1

plasma at temperatures between 100 and 400°C. The depth to which the hydrogen penetrates and neutralizes the defects depends on the density of all sites to which the hydrogen can be bound (especially the shallow level impurities), the temperature, and the duration of the plasma treatment. It can vary between about 1 and 100 μm in single-crystal silicon. In polycrystalline material, the diffusivity is enhanced, probably along grain boundaries, and hydrogen can thus diffuse much deeper into the crystal.

An alternative method is the hydrogen incorporation by ion implantation at low energy and high currents. This technique is usually used for passivating amorphous and crystalline silicon for solar cell applications. The commonly called Kaufmann ion sources allow very good control over the penetration depth and other parameters, but they have the major disadvantage of extensive damage of the near-surface region, which can be detrimental to the device performance.

Besides lattice defects such as dislocations and grain boundaries, many of the metal-related deep trap centers can be neutralized by atomic hydrogen. As discussed in the previous chapter, the most troublesome impurities in silicon are the transition metals, which can be easily introduced during device processing. They have high capture cross sections for electrons and holes and hence degrade the performance even at low concentrations. The mechanism by which hydrogen passivates deep levels is rather clear in cases where the origins are dangling bonds. The bonding of the hydrogen atom to the dangling bond forms bonding and antibonding states that are pushed into the valence and conduction bands, respectively, and hence become electrically inactive. The mechanism for the passivation of transition metals presents a different problem. Since the microstructure of all deep-level centers in semiconductors is largely unknown, one can only speculate about some aspects of the deactivation process. Some impurities such as platinum and palladium in silicon are known to distort the lattice, so they may be bound more strongly to some of the next-neighbor atoms, whereas the remaining bonds are to some extent dangling bonds which may bond with the hydrogen. In other cases, the presence of hydrogen may well cause a distortion or even a reconstruction of bonds, so a stable bond with the hydrogen atom can be formed.

In the following sections, the origin of deep trap levels associated with grain boundaries (and dislocations) will be discussed. Despite the prevailing discussion in the literature on the origin of the electrical activity of extended lattice defects, the experimental and theoretical results seem to indicate that impurity-related trap levels play the dominant role. Alternatively, it is suggested that structurally determined unsaturated bonds are the centers of the electrical activity. Corresponding to the uncertainty about the origin of the electrical activity, the mechanism of the passivation also remains unsolved. It is evident, however, that in both cases (impurities and/or dangling bonds), a passivating effect can be expected. Certainly, a better understanding of the structure and electrical properties of grain boundaries and dislocation is necessary to understand the observed passivation effects in silicon and other semiconductors. Table 5.10 compiles some of the impurity and lattice defect states in silicon and GaAs that can be passivated [39].

It is a general observation that a passivated defect can be reactivated by annealing. The thermal stability of the deactivated defects is indicated by the dissociation energy E_D for the hydrogen-defect complex. In a simple kinetic model, the fraction of defect-hydrogen complexes N/N_0 remaining after a time t at an annealing temperature T is given by

$$\ln \frac{N_0}{N} = t\nu \exp \frac{E_D}{KT} \tag{5.33}$$

Usually, an annealing treatment above 400°C (for an activation energy of about 2.5 eV) is sufficient to reactivate the trap level. Generally, the most stable levels are introduced by dislocations and grain boundaries, followed by deep-level and shallow-level impurities, as will be discussed next. After a passivation treatment, deep and shallow levels are passivated likewise. From the perspective of practical applications, where it is required to retain the doping concentration, the lower thermal stability of doping elements offers the possibility of reactivating them at low temperatures without the loss of deep-trap passivation.

Shallow Level Passivation

It is well established now that shallow-level impurities in a variety of semiconductors can be passivated by hydrogen. In silicon, GaAs, and AlGaAs, the deactivation of both acceptor and donor impurities occurs, and in CdTe, ZnTe, and other semiconductors, the passivation of acceptors has been observed. In silicon, the acceptors boron, gallium, aluminum, and so on are almost completely passivated (over 99%) within the penetration depth of the hydrogen. By sharp contrast, the shallow donors phosphorus, arsenic, and antimony are only weakly deactivated, and in most cases the change in carrier concentration is negligible.

The following is a model that describes the essential features of the deactivation reaction of shallow impurities, at least in silicon and GaAs. The basic assumption is that hydrogen has a donor level in the upper half of the band gap. Such a level has indeed been found by DLTS at $E_T = E_C - 0.16$ eV. In p-type silicon, this level would be depleted and the hydrogen atom positively charged. The acceptor passivation can thus be described as a compensation reaction

$$A^- + H^+ \leftrightarrow (AH)^0 \tag{5.34}$$

The ion pairing occurs as a result of the Coulomb attraction [42]. The mechanism

of the donor passivation is less clear, since in an n-doped crystal the hydrogen atom will be neutral. The passivation may then occur by the reaction

$$D^+ + H^0 + e^- \leftrightarrow (DH)^0 \tag{5.35}$$

which also requires a donor level of the DH complex in the upper half of the band gap. The electronic character of the bonding and the position of the hydrogen atom in the lattice is, however, unclear. Neutral atomic hydrogen H^0 in n-doped silicon can also form the stable and rather immobile molecule H_2 via the reaction

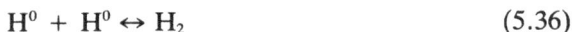

$$H^0 + H^0 \leftrightarrow H_2 \tag{5.36}$$

This reaction competes with the previous reaction and may in fact dominate at low temperatures and prevent a substantial passivation of donors in agreement with the observation. This would explain, for instance, the observed barrier effect of a highly n-doped layer in p-doped silicon [39].

Diffusion of Hydrogen

The understanding and interpretation of the diffusion behavior of hydrogen in semiconductors is complicated by the fact that trapping at impurities and lattice defects and the molecule formation can occur. Earliest high-temperature measurements in silicon determined an Arrhenius law for the diffusion coefficient (equation (5.32)), with [43]

$$D_{H0} = 9.4 \times 10^{-3} \text{ [cm}^2\text{/s]} \qquad H_H^D = 0.48 \text{ [eV]} \qquad (970 \text{ to } 1200°C)$$

whereas later studies at lower temperatures gave [44]

$$D_{H0} = 4.2 \times 10^{-5} \text{ [cm}^2\text{/s]} \qquad H_H^D = 0.56 \text{ [eV]} \qquad (400 \text{ to } 500°C)$$

The activation energy at high temperatures is close to the theoretical value of about 0.32 eV for the atomic hydrogen migration from an equilibrium M-site to another. Therefore, one can assume that at high temperatures diffusion occurs through interstitial migration of atomic hydrogen. For a hydrogen trap level in the upper half of the gap, most of the diffusion is by the neutral species, since the Fermi energy at high temperature lies in the middle of the gap and the semiconductor becomes intrinsic.

Extrapolation to low temperatures (<400°C) gives diffusivity values in both cases, which are orders of magnitude higher than what is usually observed. This impeded diffusion at low temperatures can be understood if one considers the trapping at defect sites and the formation of molecular hydrogen, which is much less mobile (activation energy 2.7 eV for migration from an interstitial T_d-site via a hexagonal site). The effect of the defect and self-trapping processes leads to an effective diffusion constant:

$$D_{\text{eff}} = \frac{C_{\text{H}}}{C_T} D_{\text{H}} \qquad C_T = C_{\text{H}} + 2C_{\text{H}_2} + C_{\text{H}-T} \qquad (5.37)$$

where C_{H}, C_{H_2}, $C_{\text{H}-T}$ are the concentrations of mobile atomic hydrogen, molecular hydrogen, and trapped hydrogen, respectively. C_T is the total hydrogen concentration and D_{H} the diffusion constant of atomic hydrogen at high temperatures. In p-type silicon, hydrogen will be present as a mixture of H^0 and H^+, whose relative concentrations depend on the temperature and the doping concentration. The charged atom can interact with acceptors at low temperatures ($<300°C$), whereas at higher temperatures the neutral species dominates, which favors the molecule formation. In both cases, the diffusion will be impeded. In n-type silicon, only the neutral species is present, and a significant pairing of molecules will occur, which leads to a much slower permeation than in p-type silicon. At temperatures below 150°C, donor-hydrogen bonds can form, which passivates the electrical activity and impedes the diffusion further.

In general, all processes take place simultaneously, and the resulting diffusion coefficient and profile are difficult to predict. One can obtain some idea of the influence of the trapping effects by considering limiting cases for steady state diffusion with a constant surface concentration of hydrogen C_{H0}. If molecule formation is the dominant process and dissociation is negligible, one obtains a steady-state diffusion profile as a function of the penetration depth x given by [45]

$$C_{\text{H}}(x) = C_{\text{H0}} \frac{1}{\left\{1 + x\left(\frac{6D_{\text{H}}}{K_m C_{\text{H0}}}\right)^{1/2}\right\}^2} \qquad D_{\text{eff}} = \frac{C_{\text{H}}}{C_{\text{H}} + 4C_{\text{H}_2}} D_{\text{H}} \qquad (5.38)$$

where K_m is the rate constant for molecule formation.

In the case that trapping at impurities dominates the diffusion process and dissociation can be ignored, the steady state solution is given by

$$C_{\text{H}}(x) = C_{\text{H0}} \exp\left\{-x\left(\frac{K_T N_T}{D_{\text{H}}}\right)^{1/2}\right\} \qquad (5.39)$$

where N_T and K_T are the trap concentration and rate constant for trapping, respectively. Without trapping, the diffusion would be described by the normal error function complement profile

$$C_H(x) = C_{H0} \text{ erfc} \frac{x}{(4D_H t)^{1/2}} \tag{5.40}$$

It is evident that experimental diffusion data differ considerably depending on the experimental variables, such as surface condition, trap concentrations, and plasma parameters. A summary of the more reliable diffusion data is given in Figure 5.15, along with the calculated diffusion constant using the experimental values mentioned above.

Finally, it should be mentioned that the solubility of hydrogen is also greatly affected because of the trapping of hydrogen, which can lead to much higher hydrogen concentrations, as one can expect from the extrapolation of high-temperature data given by

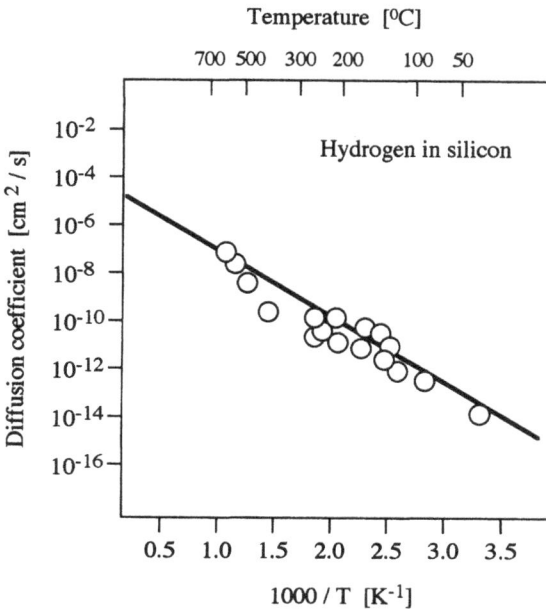

Figure 5.15 Diffusion coefficient of hydrogen in crystalline silicon, collected from [39]. Solid line calculated from the Ichimiya and Furuichi relations [44].

$$C_H = C_{H0} \exp\left\{-\frac{H_H^s}{KT}\right\}$$ (5.41)

with $C_{H0} = 2.4 \times 10^{21}$ cm^{-3} and $H_H^s = 1.88$ eV [43], or $C_{H0} = 1.6 \times 10^{20}$ cm^{-3} and $H_H^s = 1.46$ eV [44]. Extrapolation to 300°C yields in the first case $C_H \approx 2 \times 10^7$ cm^{-3}, which has to be compared with measured data of about 10^{18} to 10^{20} cm^{-3}.

5.1.5 Radiation-Induced Defects

Radiation-induced damage in semiconductors occurs through ionization and the displacement of atoms. The incoming beam of high-energy particles or photons creates ionization along the path in the solid. Electrons are excited into higher energy states and descend after sufficient time into the ground state and release the energy, mainly in the form of heat, to the lattice. Some electrons may be trapped at defects and alter their charge state, which can change the recombination behavior of this defect or lead to the dissociation of complexes. The ionization may also cause redistribution of defects by a process known as *ion-enhanced diffusion* [46]. Despite these secondary effects, which can have an influence on the lifetime of the semiconductor, ionization is primarily the cause of the energy loss of the radiation in the solid and is transient in nature. Only the remaining 0.1% of the particle energy is responsible for the permanent damage in the crystal structure. The radiation damage depends on the energy distribution and intensity of the particle beam. For solar cells used in spacecraft, details of the space radiation environment and its effect on the solar cell performance are important and are compiled and discussed in [47,48].

Radiation-induced atomic displacement by collision results, initially, in the formation of a vacancy-interstitial (Frenkel) pair. The energy transferred to the struck atom depends on the velocity of the beam particles. The maximum energy T_m for relativistic and non-relativistic particles with velocity v that can be transferred is

$$T_m = E \frac{4M_1 M_2}{(M_1 + M_2)^2} \qquad (v/c \ll 1)$$ (5.42a)

$$T_m = E \frac{2M_1}{M_2}\left(2 + \frac{E}{M_1 c^2}\right) \qquad (v/c \gg 1)$$ (5.42b)

where E and M_1 are the kinetic energy and mass of the incident particle, respectively, and M_2 is the mass of the struck atom. Kinetic energies of protons in space range from 1 to 100 MeV, with high energies occurring during sun flares. Electrons may have lower energies of about 0.1 MeV, but because of the lower mass they can have high velocities, so relativistic effects have to be considered. The threshold displacement energy T_d is defined as the minimum energy to cause displacement in the solid. Higher energies in excess of T_d are transmitted to the displaced atom, which in turn can then displace further atoms and lead to a more complicated defect structure. A relationship exists between the threshold energy and the lattice constant of the semiconductors, which is shown in Figure 5.16. Although the threshold energy is useful for determining whether radiation-induced Frenkel pairs can occur, it is usually not possible to relate them to the final defect structure. The primary reason is that these defects are mobile even at low temperatures and can thus migrate and combine with other lattice defects or impurities. The secondary defects are the primary cause for the degradation of the solar cell performance in space. The defects that are recombination-active contribute to the reduction of the lifetime of the material and thus to the degradation of the solar cell performance. According to (4.10), the different lifetimes τ_i, which are related to a particular radiation-induced defect with the concentration N_{Ti}, can be summarized and yield the total lifetime τ of the material:

$$\frac{1}{\tau} = \frac{1}{\tau_0} + \sum \frac{1}{\tau_i} = \frac{1}{\tau_0} + \sum \sigma_i v_T N_{Ti} \tag{5.43}$$

where τ_0 is the lifetime before irradiation. The individual lifetimes τ_i are expressed here by (4.29), which was derived in the SRH theory in Section 4.2.2. σ_i is the capture cross section of the defects and v_T is the thermal velocity of the minority carriers. This equation holds for low injection and defect concentrations, which are significantly less than the majority carrier concentrations. In many cases, the concentrations N_{Ti} are proportional to the fluence ϕ of the irradiation $N_{Ti} = C_i \phi$, so that (5.43) and the corresponding equation for the diffusion length $L = (D\tau)^{1/2}$ reduce to

$$\frac{1}{\tau} = \frac{1}{\tau_0} + \phi \sum \sigma_i v_{th} C_i = \frac{1}{\tau_0} + K_\tau \phi \tag{5.44a}$$

$$\frac{1}{L^2} = \frac{1}{L_0^2} + K_L \phi \tag{5.44b}$$

where K_τ is the lifetime damage coefficient, $K_L = K_\tau/D$ is the related diffusion length damage coefficient, and D is the diffusion coefficient of the minority carriers. The use of K_L rather than K_τ implies that the diffusivity does not vary with the fluence. Since diffusion lengths of a semiconductor are relatively easy to measure, these constants are extensively used to characterize radiation damage in solar cells.

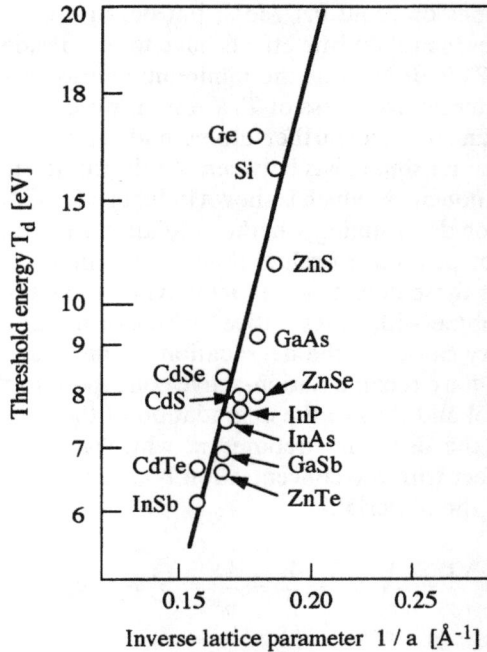

Figure 5.16 Correlation of threshold energies T_d with lattice parameters (a^{-1}) of several semiconductors [46].

In general, they depend on the energy of the incoming radiation and also on the majority carrier concentration (or conductivity) of the semiconductor. Most of the solar cells used for spacecraft are based on silicon and recently also on GaAs. A third cell type based on InP and currently in the research and development phase shows very good radiation resistance compared to silicon and GaAs and may replace these semiconductors for space applications. For 1-MeV electron radiation with fluences between 10^{13} and 10^{16} cm^{-2}, a typical value of K_L for a p-based 0.1-Ωcm silicon solar cell is $K_L \approx 8 \times 10^{-10}$, and a corresponding value for a p-based GaAs cell is $K_L \approx 4.5 \times 10^{-8}$. InP solar cells with similar cell characteristics are significantly more radiation resistant than both silicon and GaAs, a fact that is demonstrated in Figure 5.17, where normalized efficiencies are depicted as a function of the fluence [49]. In silicon and GaAs, the radiation resistance decreases (or K_L increases) with increasing base doping concentration or conductivity, whereas the reverse is observed in InP cells.

Since the damage coefficients give only an average description of the radiation resistance of a semiconductor, no insight into the damage mechanism itself is obtained that would allow understanding of the obviously complex behavior of the

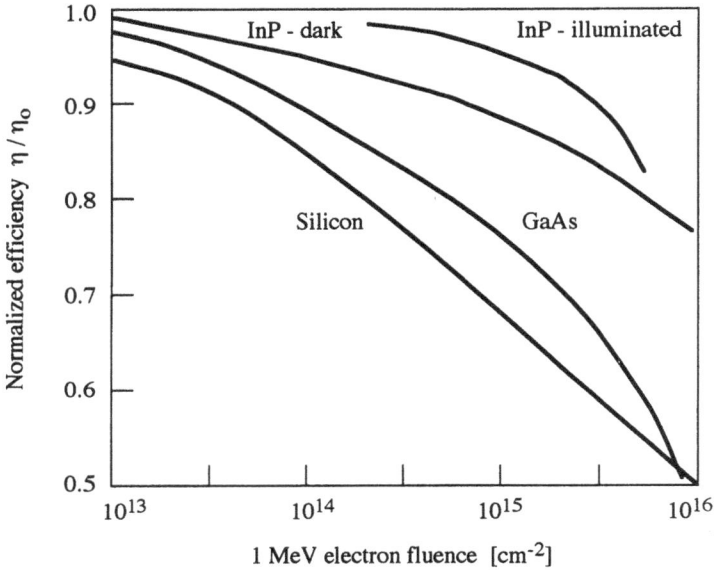

Figure 5.17 Normalized efficiencies of InP, GaAs, and silicon solar cells after 1 MeV of electron radiation [49].

semiconductors. The most valuable tools for an identification of the radiation-induced defects are the ESR and DLTS methods, which in principle allow the determination of structural and electronic properties of the defects.

Relatively little is known so far about the nature of radiation-induced defects in GaAs and InP, which can be observed, for instance, by DLTS. The major radiation-induced trap levels seem to be associated with intrinsic defects on the arsenic sublattice. The evidence favoring arsenic atom displacement is obtained from energy-dependent orientation studies. The method is based on the comparison of the introduction rates after irradiation along two opposite [111] directions. The situation is depicted in Figure 5.18 and shows that there is an "easy" and a "hard" direction for the arsenic displacement. Considering, for instance, the $[111]_{As}$ direction, it is evident that the probability of the gallium atom displacement is much less than for arsenic displacement because the gallium recoil path is obstructed by the neighboring arsenic atom, while the lattice is open in front of the arsenic atom. In the opposite direction, the result is reversed. The argument is, however, only valid for low beam particle energies and small scattering angles; therefore, energy-dependent measurements are necessary to determine the correct energy range. The anisotropy studies in connection with electrical (e.g., DLTS) measurements are a

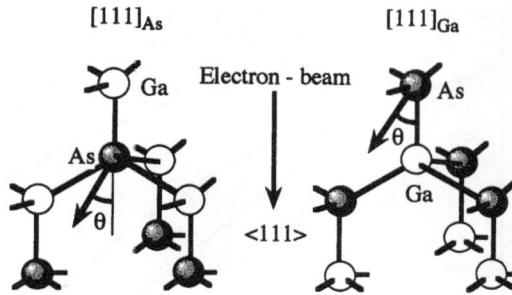

Figure 5.18 Displacement due to irradiation along the $[111]_{As}$ and $[111]_{Ga}$ direction. θ is the recoil angle of the arsenic atoms [53].

general method for studying the nature of displaced atoms after irradiation in III-V and II-VI semiconductors [49].

A further important observation in n-type GaAs is an enhanced annealing of the radiation-induced defects around 100°C under forward bias conditions, and in p-type GaAs around room temperature under reverse bias. Although the exact nature of the mechanisms responsible for the enhanced annealing are not quite clear yet, the experimental result is of particular importance for GaAs solar cells, since it offers the possibility of extending the useful lifetime by annealing treatments at rather low temperatures.

The experimental situation for silicon is more advanced because many of the intrinsic defects observed by DLTS have also been detected by ESR and other methods that can give information about the structural properties. The primary defects after irradiation are vacancies and self-interstitials. Since the self-interstitial appears to be mobile even at helium temperature and recombines with other defects quickly, it has never been observed directly. Therefore, very little is known about the isolated defect and most of the properties that have been attributed to the self-interstitial are obtained indirectly, as discussed in Section 5.1.1. A silicon di-interstitial is, however, identified after neutron irradiation and consists of two self-interstitials separated by a substitutional atom and oriented along the ⟨001⟩ axis. This defect anneals out at 170°C.

The majority of the radiation-induced defects are related to the vacancy. A partial list of the vacancy-related defects relevant to solar cell silicon, their trap levels, and the annealing temperatures above which the particular defect disappears has been given in Table 5.2. In addition, the single-defect divacancies and higher-order complexes (three and four vacancy atoms) have been observed in irradiated silicon. The divacancy can be formed either directly when the energy of the irradiating particle is high enough to displace two neighboring atoms or by migration and association of single vacancies. The divacancy in its different charge states introduces three trap levels in the band gap and is stable below about 300°C.

The vacancy can also form a considerable number of complexes with other impurity atoms, particularly the doping impurities, some of which are listed in Table 5.11 [50]. In fact, in phosphorus-doped Fz-grown silicon, the P-V complex (*E*-center) is the dominant defect observed by ESR that is stable below 130°C. In boron-doped silicon, the corresponding B-V complex occurs, although it could not be confirmed directly by ESR measurements.

Besides the doping elements, carbon and oxygen are the major impurities in silicon, and a number of complexes can occur. In oxygen-rich Cz-grown crystals, vacancy-oxygen complexes are observed, which were discussed in Section 5.1.2 and are summarized in Table 5.5. The most dominant defect is the complex V-O, which in its negative charge state is referred to as the *A*-center. Carbon is usually present in various degrees even in high-quality silicon, and one can expect that radiation-induced defects containing this atom are also of primary importance. The various defects that have been observed are also discussed in Section 5.1.2 and listed in Table 5.6.

It has been suggested that the disappearance of the silicon interstitial after irradiation in *p*-type silicon is due to trapping at shallow acceptor impurities, which exchange their site then, so that the interstitial impurities and vacancies remain in equal proportions. An example in *p*-type silicon is the formation of the boron interstitial, which can occur in different charge states. The boron interstitial is unstable at room temperature but appears to be stabilized by the complex formation with interstitial oxygen in Cz-grown crystals: B_i-O_i. This defect introduces a trap

Table 5.11
Electronic Levels of Point Defect Complexes in Silicon That Can Occur After Electron Irradiation and Are Related to Some Doping Elements (Included are the annealing temperatures above which the complex dissociates or the defect level disappears)

Defect	Annealing Temp. [K]	Trap Level [eV]
P-V	420	$E_C - 0.4$
As-V	450	$E_C - 0.47$
Sb-V	460	$E_C - 0.44$
B-V	260	
B_i^+	50	$E_C - 0.13$
B_i^-		$E_C - 0.45$
B_i-O_i	470	$E_C - 0.27$
Li_i		$E_C - 0.034$
Li_i-O_i		$E_C - 0.039$

level at $E_C - 0.27$ eV and is stable below 200°C. Despite the many defects that can be observed after irradiation, it has been assumed that the boron-oxygen complex B_i-O_i is the primary cause of the radiation-induced degradation in boron-based np silicon solar cells.

It has also been observed that doping with lithium can increase the radiation resistance [51]. Therefore, lithium doped p^+n and counterdoped p^+n solar cells have been mainly used for space applications. Lithium is a shallow donor and occupies interstitial sites in the silicon lattice. It has been inferred from infrared spectroscopy that interstitial lithium forms a complex with interstitial oxygen. It was therefore argued that this process competes with the boron-oxygen formation process and hence reduces the formation rate for this complex. This tentative conclusion on the beneficial influence of lithium on the radiation resistance is supported by DLTS measurements on irradiated lithium-counterdoped cells.

5.2 DISLOCATIONS

Today's single-crystal growth techniques for silicon and GaAs either eliminate dislocations or reduce their density to levels at which they are usually no longer detrimental to device properties. Although the general interest in the investigation of their properties also declined, a few groups continued to study the behavior of dislocations in covalently bonded semiconductors, particularly now in germanium and silicon, for several decades. These investigations have provided a very thorough understanding of the structural, electronic, and mobility properties of dislocations [54–56], which turns out to be very useful from the perspective of the solar cell application.

The requirement of low-cost techniques for terrestrial solar cells has led to the development of unconventional crystal growth techniques. Of particular interest are ingot and ribbon growth techniques for silicon (see Chapter 6). These methods produce polycrystalline crystals, which can contain so many dislocations that they are the most detrimental defects for the solar cell. Dislocations are also incorporated during heteroepitaxial growth of thin films, particularly for materials with a large lattice mismatch. This problem is both relevant for the electronic device technology and for thin-film solar cells which are grown using epitaxial techniques.

Like other extended lattice defects, dislocations are efficient recombination centers, and their recombination behavior is similar to grain boundaries, as discussed in Section 4.2.4. Since dislocations in low-cost processes are to some extent unavoidable, it is important to understand how they nucleate and multiply. In the following, the discussion will be mainly limited to dislocations in silicon because it is the most promising low-cost solar cell material so far and because the results provide the basis for the understanding of the dislocation dynamics of other semiconductors as well.

5.2.1 Mobility

The mobility and electronic structure of dislocations in covalently bonded semi-conductors is closely related to the atomic structure of the core of the dislocations. The essential results are the following. The dislocations glide on {111} planes and are dissociated into partial dislocations that bound a stacking fault (Figure 5.19). The dissociation width varies between about 10Å and 100Å, depending on the material and the dislocation type. Since the {111} glide planes consist of a double layer of atoms in the diamond cubic and sphalerite lattice, the extra half plane of the dislocation can terminate between the widely (shuffle set) or narrowly spaced (glide set) planes. There is evidence now from HREM in silicon and germanium that dislocations usually belong to the glide set, although one cannot completely rule out that shuffle set configurations occur, too. In binary AB compound semi-conductors with the sphalerite structure, the core atoms of the partial dislocations can be either all A or B atoms, which leads to two different types of dislocations (α and β), usually with different mobilities. The core region of the partials is a few atomic distances wide and geometrical models show that it contains a number of unsaturated bonds. There is general agreement now that the atomic structure is mostly reconstructed. A dissociated dislocation contains, however, even in equilibrium a number of defects such as constrictions, kinks, jogs of various heights,

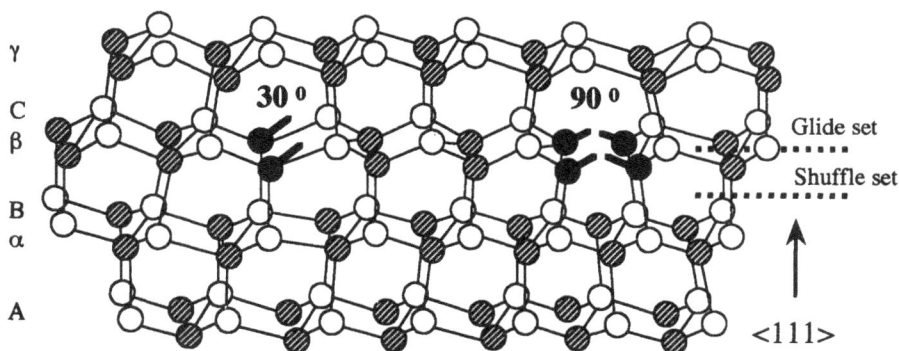

Figure 5.19 Projection of the sphalerite structure on a (011) plane showing a dissociated glide set dislocation with a stacking fault between the 30- and 90° partials. A- and B-sites are indicated by hatched and white circles, respectively. Core atoms with dangling bonds are emphasized by black circles. Exchange of A and B atoms leads to a second type of dislocation because the core atoms either terminate with a row of A (α dislocation) or B atoms (β dislocation). The stacking sequence of (111) planes for the sphalerite (and diamond cubic) lattice is $A\alpha B\beta C\gamma$. For the glide set configuration, the extra half plane of the complete dislocation ends between the narrowly spaced layers (e.g., $\alpha - B$) for the shuffle set configuration between widely spaced layers (e.g., $A - \alpha$).

or reconstruction defects (Figure 5.20). Dangling bonds that introduce deep levels in the forbidden gap may thus exist locally at these point-like defects or also at segregated impurities. Besides these details, which are important for the electrical properties of the dislocations, it is evident from the microscopic picture that the motion of a dislocation is connected with the breaking and reconstruction of bonds, a process that requires high energy. This is described in terms of a high periodic (Peierls) potential parallel to the $\langle 110 \rangle$ direction, which has to be overcome during motion. Therefore, in contrast to dislocations in metals, the motion of dislocations in covalently bonded crystals is a strongly thermally activated process. Experimental results have confirmed that the velocity v is a function of the temperature T and the stress τ acting upon the dislocation in the glide plane, and is given by the general relationship

$$v = v_0 \tau^m \exp\left\{-\frac{Q}{KT}\right\}$$ (5.45)

where Q is the activation energy for the dislocation motion. In most cases, the stress exponent m is slightly temperature-dependent and the activation energy $Q(\tau)$ is a function of the applied stress. The prefactor is a material constant. Equation (5.45) has been found for all covalent semiconductors studied so far. The velocity depends on the character of the dislocation and usually experimental results are presented for the screw and 60° dislocation, which are the fundamental dislocations types in covalent semiconductors. A general curved dislocation can be considered a mixture of screw and 60° components, and their velocity may be estimated from the two velocity values. As mentioned above, in compound semiconductors, one has to distinguish further between α- and β-type dislocations. In general, during plastic deformation both types are formed; however, their velocities are usually very different.

The experimental results obtained for the dislocation velocities in silicon are summarized in Figure 5.21 and Table 5.12. The velocities of screw and 60° dislocations are usually measured in the temperature range between 450 and 800°C and turn out to be rather similar. For undoped silicon and high stresses, an activation energy of about $Q_0 \approx 2.2$ eV and a stress exponent $m \approx 1.3$ to 1.5 have been determined. For stresses below about 20 MPa, the activation energy increases as shown in Figure 5.22 for the 60° dislocation and reaches values of about 2.5 eV at 2 MPa. The dislocations are therefore slowed down in the low-stress regime.

The dislocation velocity is considerably changed in doped crystals. Whereas isoelectronic impurities such as tin and germanium in silicon do not influence the velocity, the typical doping elements such as boron or phosphorus increase the velocity if the doping concentrations exceed the intrinsic free carrier concentration for the particular temperature where the velocity is measured. Since most of the measurements are carried out in the temperature range below 800°C, the doping

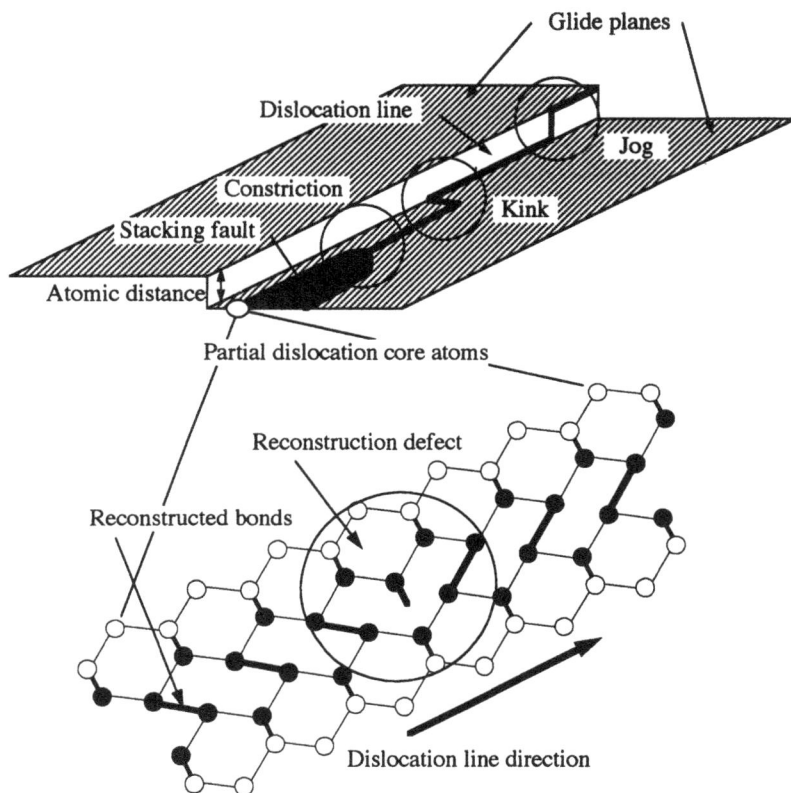

Figure 5.20 Schematic diagram depicting possible structural defects of a dislocation. Reconstruction defects occur at core atoms of dislocations.

concentrations have to exceed about 10^{17} to 10^{18} cm^{-3} in order to observe the doping effect. The experimental results depicted in Figure 5.23 and summarized in Table 5.12 show that for both *n*- and *p*-doping the activation energy Q decreases in the extrinsic doping range with increasing doping level, whereas the stress exponent and the prefactor remain essentially unchanged. Since the effect is independent of the doping element, it is evident that it is the change in the Fermi level that causes the enhanced mobility of the dislocations. There are only a few results for the stress dependence of Q in doped crystals, but they indicate that it resembles the stress dependence of the undoped crystals, so that one can summarize the doping dependence of the dislocation velocity with the relationship

$$Q = Q_0(\tau) - \Delta Q(E_F) \qquad (5.46)$$

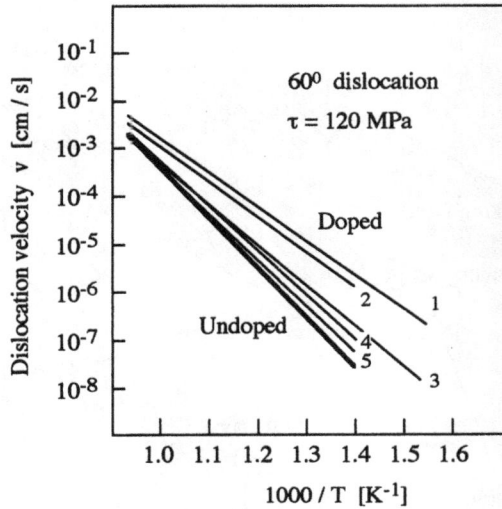

Figure 5.21 Experimental dislocation velocities of 60° dislocations as a function of the temperature for a shear stress $\tau = 120$ MPa in pure and doped silicon: (1) 8×10^{18} cm^3 (Sb); (2) 6×10^{18} cm^{-3} (As); (3) 6×10^{17} cm^{-3} (P); (4) 8×10^{18} cm^{-3} (B); (5) 2×10^{17} cm^{-3} (Al). Data collected from [56].

Table 5.12
Experimental Results for the Activation Energies Q, Stress Exponents m (stress τ in MPa), and Prefactors v_0 Defined in (5.45) for the Dislocation Velocities of 60° Dislocations in p- and n-Doped Silicon

Doping [cm^{-3}]	v_0 [cm/s MPa]	m	Q [eV]
2.0×10^{13} (P)		1.5	2.20
1.0×10^{14} (P)	9.0×10^5	1.35	2.15
1.0×10^{16} (P)	1.1×10^6		2.16
1.0×10^{17} (As)		1.35	2.05
1.2×10^{18} (P)	2.9×10^3		1.67
8.0×10^{18} (Sb)			1.40
1.4×10^{19} (P)	5.3×10^2		1.41
1.0×10^{15} (B)	9.0×10^5	1.4	2.15
2.7×10^{17} (P)	9.3×10^6		2.34
6.8×10^{17} (B)	1.5×10^6		2.20
8.0×10^{18} (B)		1.38	1.85
3.0×10^{19} (B)		1.38	1.60

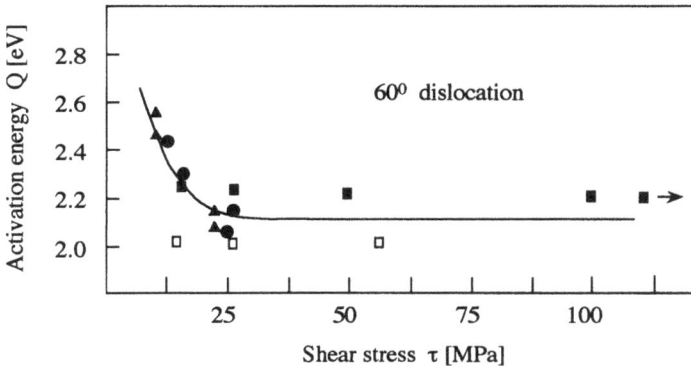

Figure 5.22 Activation energy $Q(\tau)$ of the dislocation velocity as a function of the shear stress for a 60° dislocation in pure silicon. Data summarized in [56].

where Q_0 is the activation energy of the undoped crystal.

It should be mentioned that impurities can also interact with dislocations via the elastic strain field, which has the opposite effect and slows down the dislocation motion. This usually occurs for rather high impurity concentrations or low velocities when impurities can accumulate at or near the dislocation and form a Cotrell cloud or even precipitates. For instance, a Cotrell cloud can develop during crystal growth when impurities are rather mobile, and, in fact, dislocations tend to be "pinned" in the as-grown state. A starting stress is necessary then to break the dislocation away from the pinning impurities, but afterwards the dislocation moves with the normal velocity. The pinning effect thus depends on the thermal history of the crystal. It has also been observed for slow moving dislocations at rather low stresses.

Several theoretical models based on atomic models of the dislocation structure have been developed to explain the thermally activated motion and the doping dependence. Dislocations in covalent structures have to overcome the rather high Peierls potential during motion, which would require a very high activation energy if the entire dislocation would proceed in a single activated step into the next Peierls valley. The basic assumption is therefore that the dislocation moves by the local formation of double kinks on the dislocation line and the sideways motion of single kinks under the applied stress. The thermally activated steps are thus the double kink formation and the kink migration (Figure 5.24). The activation energy Q for the dislocation velocity is thus a combination of the activation energies for these two processes H_{dk} and H_{km}, respectively. The doping dependence is explained by assuming that either H_{dk} or H_{km} are changed when the Fermi energy shifts towards the band edges. Most of the models assume that, in thermal equilibrium, point-like defects at the dislocation, such as the kinks, jogs, or reconstruction defects in

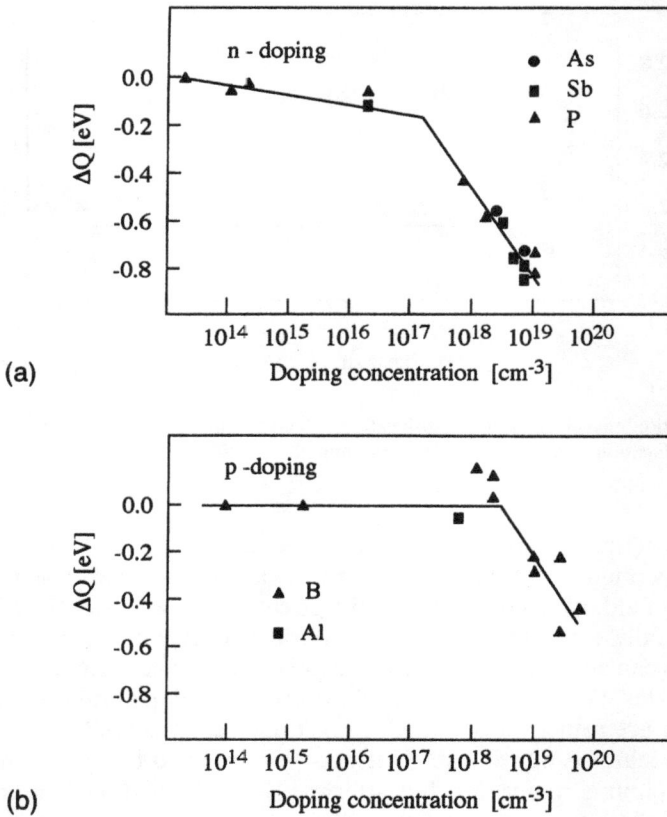

Figure 5.23 Change of the activation energy ΔQ of dislocation motion with doping for n-type silicon (a), and for p-type silicon (b). Data collected from [56].

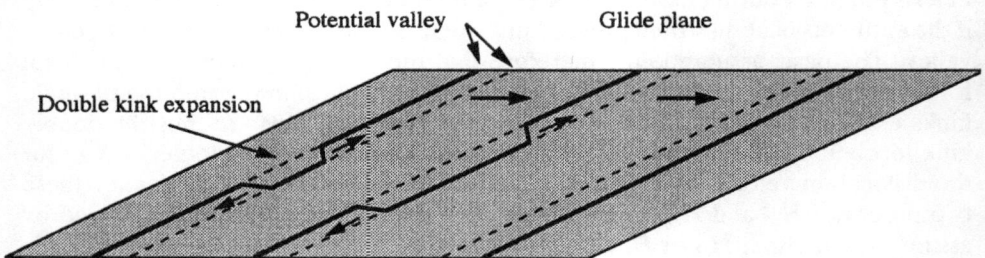

Figure 5.24 Schematic representation of the motion of a dislocation by the double kink mechanism and lateral migration of kinks along the dislocation line.

the core of the dislocation, introduce energy levels in the band gap. If, for instance, the concentration of these defects determines the double kink formation rate, the energy for this process can be reduced by the gain in electronic energy when these levels become charged. The energy change for the formation energies can be described by the same formalism as was used for the calculation of the formation energies of charged vacancies (equations (5.4) and (5.5)). For a single donor (charge state E_T^+) or acceptor level (E_T^-) connected with the dislocation defects, one obtains

$$H_{dk}^+ = H_{dk}^0 - (E_T^+ - E_F) \qquad H_{dk}^- = H_{dk}^0 + (E_T^- - E_F) \tag{5.47}$$

where H_{dk}^+ and H_{dk}^- are the new double kink formation energies when the donor or acceptor level becomes charged. Similar arguments have also been used to explain a reduction of the migration energy. In any case, the change of the double kink formation or kink migration energies translates directly into a change of the activation energy ΔQ of the dislocation velocity. Although the details of the model for the dislocation velocity are important here (see [56]), in most models, for either an acceptor or donor level, one obtains

$$\Delta Q(E_F) = (E_T^+ - E_F) \qquad \Delta Q(E_F) = (E_F - E_T^-) \tag{5.48}$$

The comparison with experimental results shows that, for silicon, a donor E_T^+ and an acceptor level E_T^- have to be assumed to account for the observed decrease of the activation energy in n- and p-type silicon. Although there is still some controversy about the identification of the defects that introduce the trap levels and whether H_{dk} or H_{km} are reduced, the essential features of the thermally activated motion and the doping dependence can be described quantitatively within these models.

5.2.2 Multiplication

The doping effect is of particular importance for the nucleation of dislocations during growth of low-cost solar cell silicon, since the crystals are usually doped. The enhanced mobility of the dislocations when the material becomes extrinsic can lead to dislocation multiplication, even at lower temperatures. Although so far no attempts have been made to describe the dislocation nucleation during growth of polycrystalline crystals quantitatively, the theoretical background for these processes has been developed.

The plastic deformation is described quantitatively from a microscopic point of view using the knowledge of the mobility of dislocations (equation (5.45)) and taking into account the mutual interaction of the dislocations. At the early stages of deformation, when slip takes place predominantly on a single glide plane, the average stress τ_{eff} acting upon a dislocation is the macroscopic shear τ, reduced by a backstress that is due to the presence of other dislocations and depends on their mean distance according to the relationship

$$\tau_{\text{eff}} = \tau - (AN)^{1/2} \tag{5.49}$$

where A is a proportionality factor that can be calculated approximately for a given arrangement of dislocations. With increasing dislocation density, the effective stress for dislocation motion thus decreases continuously. For higher stresses, when a denser network of dislocations develops, other interaction mechanisms begin to operate (for simplicity, these shall not be taken into account here). During deformation, the density of dislocations increases, and several mechanisms have been proposed by which dislocations can multiply. They can be summarized by the following dislocation multiplication law:

$$\frac{\partial N_m}{\partial t} = BN_m v \tau_{\text{eff}}^n \tag{5.50}$$

where B is a constant and n an exponent that depends on the model for the multiplication process. Values for n can be $n = 0$ or 1. The experimental observation shows that, in general, only a fraction of all dislocations contribute to the shear and the plastic deformation ϵ_{pl} of the crystal. Therefore, N_m is the density of mobile dislocations here, but at the early stages of deformation, one can assume that it is equal to the total density $N_m = N$. The second basic equation for the plastic deformation expresses the total strain rate using the sum of the elastic and plastic strain rates and is given by

$$\frac{\partial \epsilon}{\partial t} = \frac{1}{G} \frac{\partial \tau}{\partial t} + N_m b v \tag{5.51}$$

where b is the Burgers vector of the dislocations and G is the elastic shear modulus on the glide plane. The second term is the plastic strain rate $\partial \epsilon_{pl}/\partial t$, which is proportional to the mobile dislocation density and velocity (Orowan relationship). The expressions (5.49) to (5.51) together with the expression for the dislocation velocity (5.45) are the fundamental equations describing the plastic deformation at the early stages of deformation. For higher deformations, when slip occurs on several glide planes, additional assumptions have to made about the deformation process. These equations have been solved for two standard deformation conditions, namely, the dynamical deformation, where the strain rate is kept constant ($\partial \epsilon/\partial t = \text{const}$), and the creep test experiment, where the stress is kept constant

(τ = const). Good agreement with experimental results has been obtained in both cases.

The stress situation during crystal growth is similar to the creep test condition, since it is the stress that is controlled by the experimental conditions. The major problem is to calculate the thermal and local stresses in a real growth situation, which requires, in general, sophisticated numerical simulations. Once the thermal stresses have been determined, one can apply these equations iteratively and determine the dislocation density as a function of time. Computer simulations of the crystal growth process have been used already to calculate the temperature profiles during single-crystal growth; however, realistic simulations of the developing dislocation structure during growth, particularly in ingot and ribbon-grown material, are not available yet. There is no doubt, however, that numerical simulations will be used more extensively in the future, making use of the basic knowledge that is available for the dislocation motion and multiplication in covalent crystals.

5.3 GRAIN BOUNDARIES

Grain boundaries are the main crystal defects in polycrystalline thin films with grain sizes in the micron range. They also play an important role in polycrystalline bulk and sheet semiconductors, where the grain sizes are usually larger (0.1 to 10 mm). Many of the electrical and optical properties of these materials are determined by the corresponding properties of the grain boundaries. The interactions with free charge carriers, intrinsic point defects such as vacancies and interstitials, and impurities are the most important physical processes that have to be considered. They are closely related to the atomic structure and defect chemistry of the grain boundaries. The structural aspects have been studied experimentally and theoretically in recent years, primarily in silicon and germanium, whereas the most advanced concepts for electrical transport phenomena at grain boundaries have been developed for compound semiconductors like ZnO or $SrTiO_3$, which are used for varistor devices. Despite the structural and compositional differences between these materials, a rather general description of the grain boundary properties has emerged [57–61]. The general picture allows a qualitative and quantitative understanding of the electrical behavior of grain boundaries in a variety of polycrystalline semiconductors, although in special cases the particular properties of the material have to be taken into account.

The basis for an understanding of the behavior of grain boundaries is the description of the atomic structure and chemistry. The structural and chemical disorder determines the electronic states in the forbidden gap primarily responsible for the electrical activity of the grain boundaries. Therefore, the main experimental and theoretical results on the atomic structure and the distribution of the electronic states will be discussed first for elemental and compound semiconductors. The grain boundary states interact with the free charge carriers in the bulk, and two different processes can be distinguished. The permanent trapping of charge carriers leads to a grain boundary charge and the formation of a potential barrier that hinders the flow of free carriers through the grain boundary. The electrical transport properties at single grain boundaries will be discussed in the second part of the chapter. In polycrystalline semiconductors, with grain sizes below a few microns, which is typical for most polycrystalline thin films, the influence of the grain boundaries becomes so dominant that the conductance properties may differ considerably from a single crystal. The electrical transport properties in polycrystalline semiconductors will be reviewed in Section 5.5. The electronic states can also enhance the recombination of electrons and holes. Since this process is of particular importance for the photovoltaic properties of semiconductors, the recombination behavior was discussed separately, in Section 4.2.3.

The atomic structure of grain boundaries is also important for the interaction with intrinsic point defects, impurities, and dislocations. Since these defects are responsible for the electronic properties of the bulk, which can be described by the resistance and diffusion length, for instance, their interaction with grain boundaries can also have a profound impact on the bulk properties of the material. For instance, the gettering of undesirable impurities may have a beneficial effect on the electronic properties of the semiconductor. The accumulation of impurities at grain boundaries can also change the electronic behavior of the boundaries. Of practical importance is the interaction with hydrogen or other elements, which can passivate the grain boundary states to some extent. Though these interaction processes are less studied so far, they are rather important in polycrystalline semiconductors and will be discussed in Section 5.4.

5.3.1 Atomic Structure and Electronic States

Grain boundaries separate regions of different crystallographic orientation. In general, a grain boundary is defined by the orientation relationship between the two adjacent crystals and the orientation of the grain boundary plane [62,63]. The geometrical relationship between the two crystal lattices can be described by a rotation about a common axis and the rotation angle. For cubic lattices, it is convenient to describe the rotation with a coordinate system parallel to the rotation axis $[hkl]$, and the plane perpendicular to it given by the tetragonal unit cell:

$$\mathbf{a_1} = a[\overline{k^2 + l^2}, hk, hl] \qquad \mathbf{a_2} = a[0\overline{l}k] \qquad \mathbf{a_3} = a[hkl] \qquad (5.52)$$

The rotation of a vector $\mathbf{x}$ of lattice 2 in the plane perpendicular to $\mathbf{a_3}$ with respect to lattice 1 can be described by

$$y = Ax \qquad A = \left\{ \begin{array}{ccc} \cos\theta & \dfrac{1}{R}\sin\theta & 0 \\ -R\sin\theta & \cos\theta & 0 \\ 0 & 0 & 1 \end{array} \right\} \qquad (5.53)$$

with $R = (h^2 + k^2 + l^2)^{1/2}$. In the case of non-cubic lattices, such as the chalcopyrite structure, the rotation matrix becomes a more complicated expression. Experimental evidence shows that it is generally necessary to include a relative translation **d** of the two lattices with respect to each other: $y = Ax + d$. This is equivalent to a rotation about an axis parallel to the a_3-axis, but displaced with respect to the origin by

$$\mathbf{u} = (E - A)^{-1}\mathbf{d} \qquad (5.54)$$

This equation can also be used to determine the translation invariance of the atomic structure of a particular grain boundary and is important if dislocations in grain boundaries are considered. Additional degrees of freedom are required to determine the orientation of the grain boundary plane. Particular orientations occur for twist boundaries where the grain boundary plane is perpendicular to the rotation axis, and tilt boundaries with a plane parallel to the rotation axis (Figure 5.25). In the latter case, one degree of freedom remains for the orientation of the plane: a rotation about the a_3-axis. Symmetric tilt boundaries occur for orientations where each lattice is rotated by $\pm\theta/2$ with respect to the grain boundary plane. Geometrically, a general grain boundary can be considered to consist of a mixture of tilt and twist components. In fact, the experimental observation shows that in many cases this separation is a useful concept for the description of a random grain boundary.

The microscopic and atomic structure of grain boundaries depends on the misorientation angle [64–66]. For small angular deviations ($\theta < 5$ deg), the grain boundary structure consists simply of an array of dislocations (Figure 5.26) whose character and arrangement is determined by the misorientation. For a simple tilt grain boundary, a set of edge dislocations is sufficient to account for the angular deviation. The Burgers vector (length b) and the mean distance d of the dislocations determine the misorientation angle $\theta = b/d$.

For larger misorientation angles, the mean distance of the dislocations becomes so small that a description of the grain boundary structure in terms of a dislocation model is not useful anymore. A typical example of a random large angle grain boundary in silicon is given in Figure 5.27, which shows an irregular atomic structure. On a macroscopic scale, a characteristic feature of random grain boundaries is the occurrence of facets or straight segments of the boundary plane.

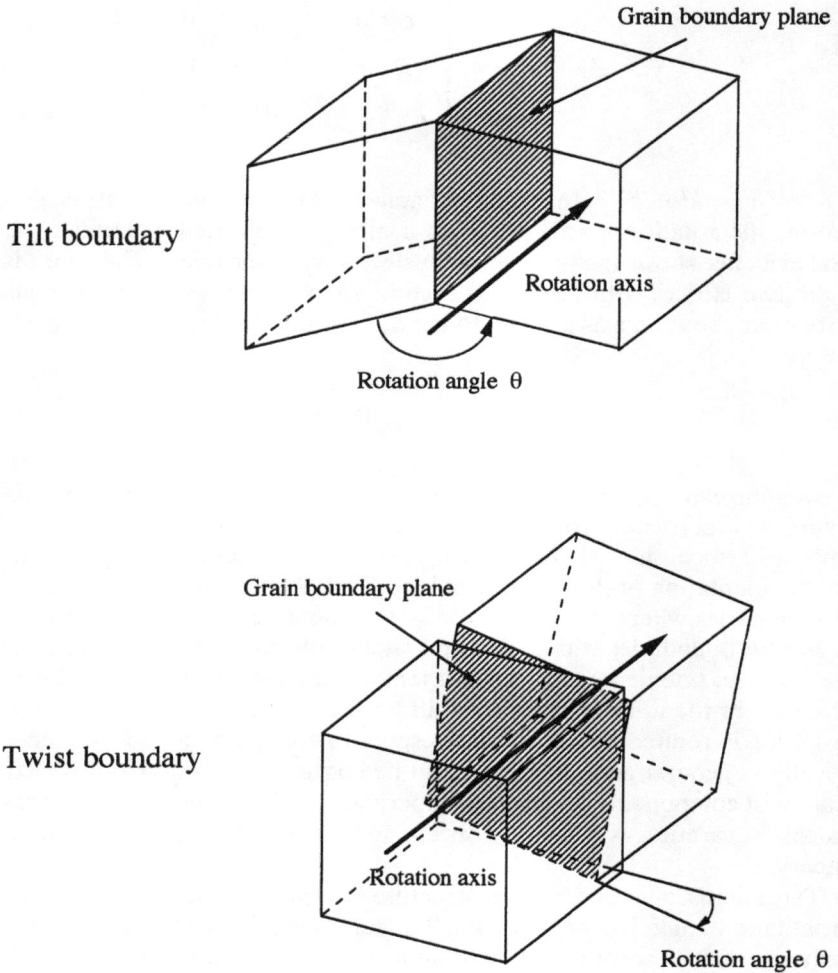

Tilt boundary

Twist boundary

Figure 5.25 Schematic representation of tilt and twist grain boundaries, defining rotation axis, rotation angle θ, and grain boundary plane.

In general, these facets, which may be only a few atomic steps wide, lie along particular crystallographic (low-index) directions, which indicates a lower grain boundary energy for this particular orientation. It has been shown experimentally that in many cases these low-energy boundaries occur for orientations in which the

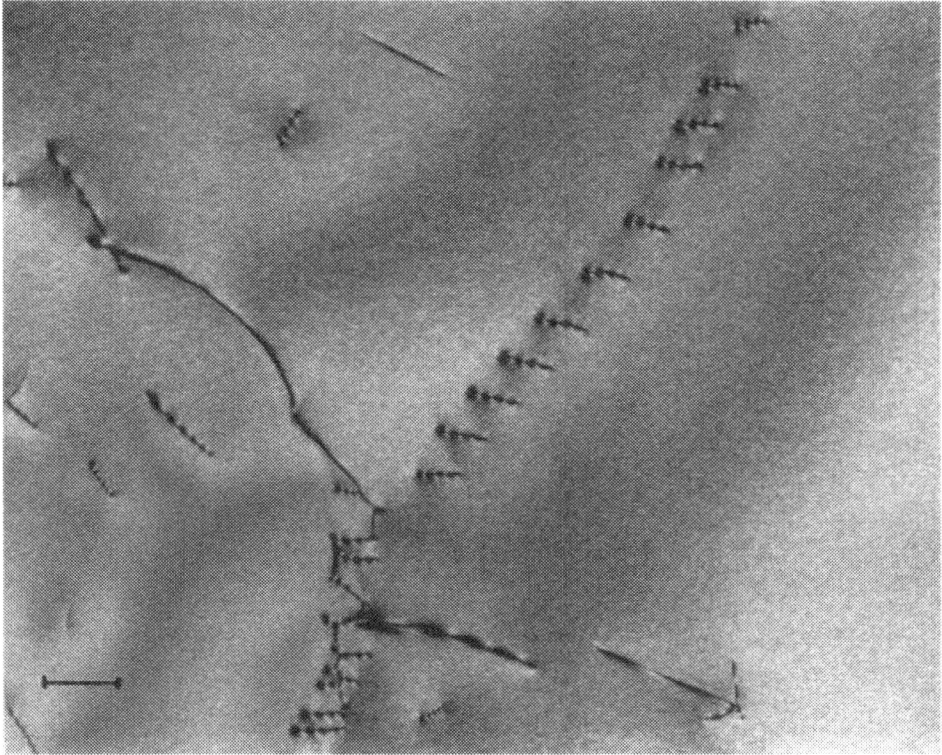

Figure 5.26 TEM micrograph of a small angle grain boundary in solar cell silicon showing a set of parallel dislocations (scale marker $\cong 0.1\ \mu$m).

two crystal lattices form a superlattice: the *coincidence site lattice* (CSL), or, more generally, the O-lattice [62]. Geometrically, the CSL is characterized by the relationship that some lattice sites are coincident in both lattices, which can be described formally by the relationship $\mathbf{y} - \mathbf{x} = \mathbf{t}$, where $\mathbf{t}$ is a translation vector of the reference lattice. Inserting this into (5.53) yields a set of vectors that form the CS-lattice:

$$\mathbf{y} = B\mathbf{t} \qquad B = (E - A^{-1})^{-1} = \frac{1}{2}\left\{\begin{array}{ccc} 1 & -\dfrac{1}{R}\cot\theta/2 & 0 \\ R\cot\theta/2 & 1 & 0 \\ 0 & 0 & 0 \end{array}\right\} \qquad (5.55)$$

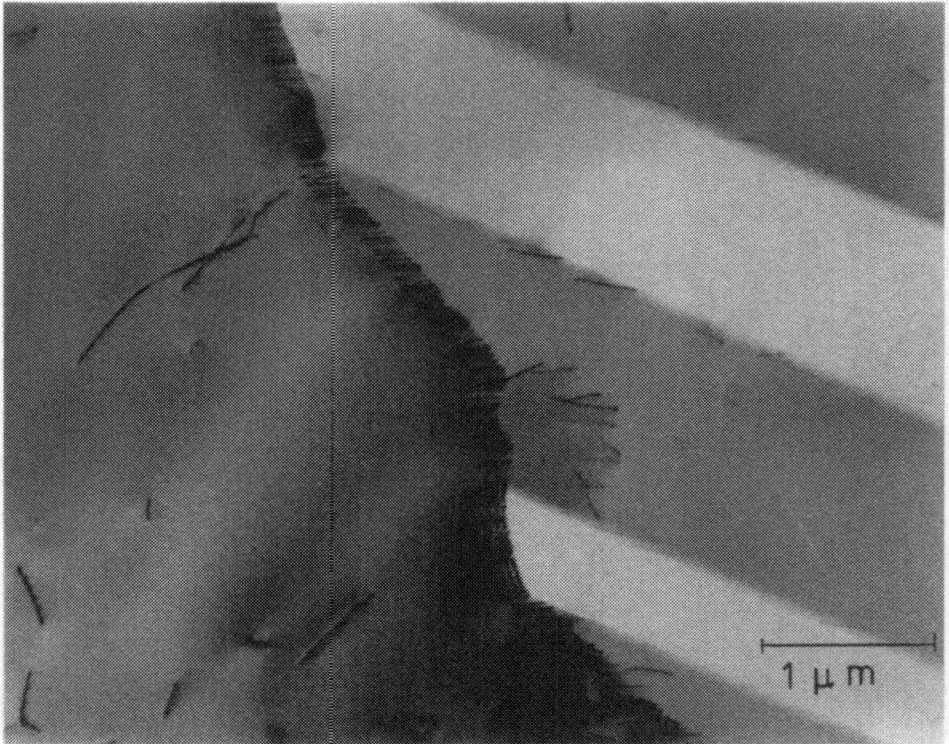

(a)

Figure 5.27 TEM micrograph of a random large-angle grain boundary in polycrystalline solar cell silicon, showing steps, dislocations, and intersecting twin boundaries (a). High-resolution TEM micrograph of the same boundary (b) (scale marker ≅ 50Å).

    The particular misorientations for which a CSL occurs can then be easily derived from the equation

$$\theta = \arctan \frac{Rk_1}{k_2} \qquad \sum = R^2k_1^2 + k_2^2 \qquad (5.56)$$

where k_1 and k_2 are integer numbers. The CSL boundaries are usually characterized by the rotation axis and the Σ-value, which is a measure of the number of coincidence points within the unit cell of the CSL. Low values indicate a higher density of coincidence points. The orientation of the straight segments of grain boundary planes frequently follows a direction with a high density of coincidence points,

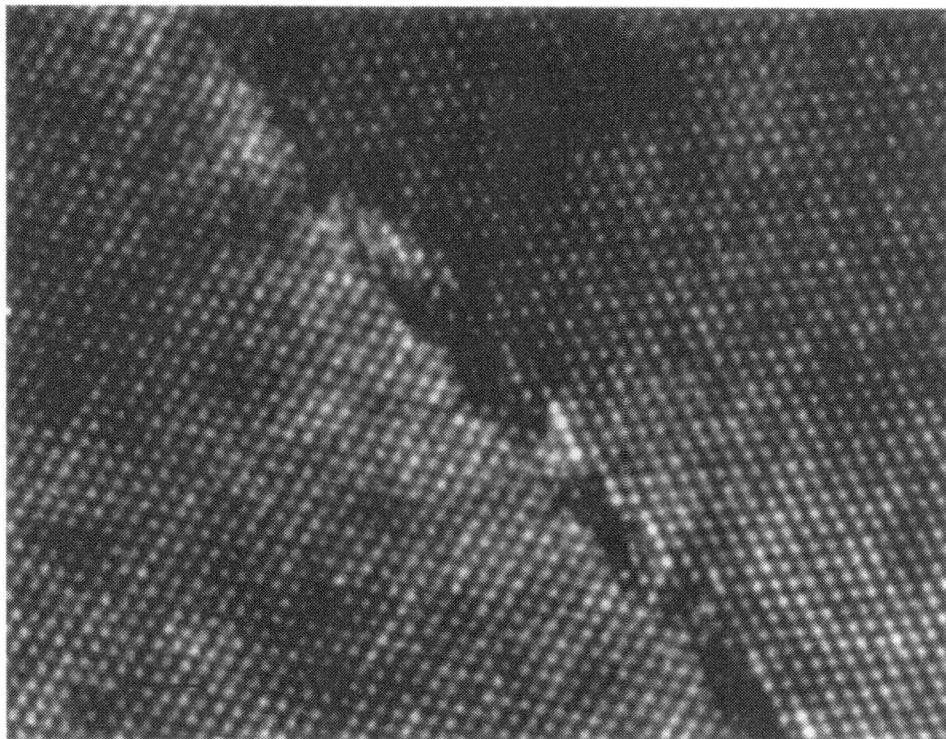

(b)

Figure 5.27 Continued.

which suggests a good atomic lattice match in these cases. The O-lattice concept is very useful in terms of a geometrical description of the orientation relationship between the two lattices but does not allow the determination of the actual atomic structure of a grain boundary. Although the geometrical modeling demonstrates that CSL boundaries are frequently characterized by a good reconstruction of the covalent bonds across the boundary plane, experimental HREM results have shown that, except for a few boundaries such as the Σ3 and Σ9 (Figure 5.28), the real atomic structures differed from the simple geometrical models. This shows that in most real boundaries the local atom and bond relaxations are not predictable from the geometrical model.

Nonetheless, geometrical models give some insight into structural aspects of grain boundaries in general. The picture that has evolved so far from the investigation of CSL grain boundaries is that the atomic structure can be described in terms of small structural units [67–69]. They are characterized, for instance, for

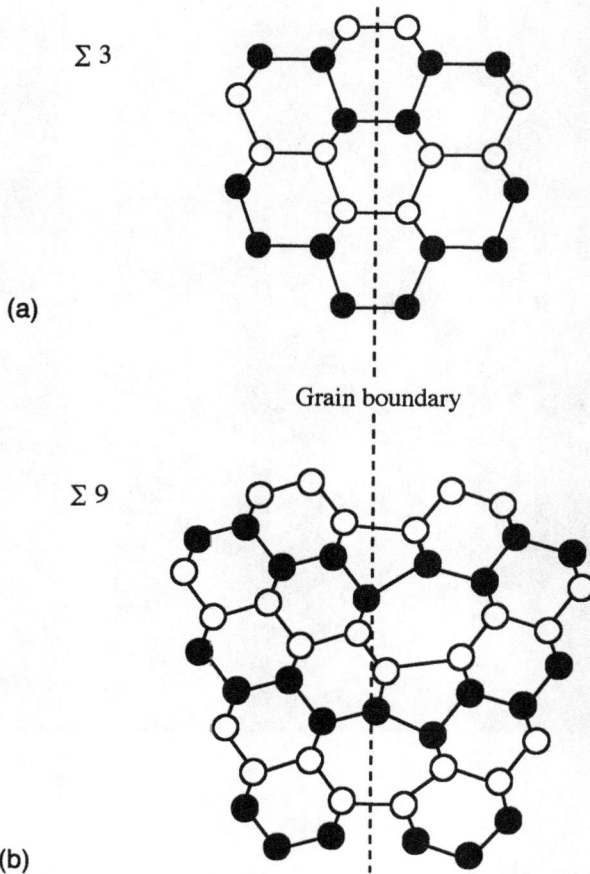

Figure 5.28 Geometrical models of the atomic structure of Σ3 and Σ9 grain boundaries in the diamond cubic lattice for rotation angles (a) $\theta = 70.5$ deg and (b) $\theta = 38.9$ deg, respectively. In the projection on a (011) plane, white and black circles indicate atoms on the paper plane and above at a distance $a\sqrt{2}$, respectively (a: lattice constant).

the ⟨011⟩ boundaries by five, six, and seven-membered rings of fourfold coordinated atoms (Figure 5.29), which occur in various arrangements in the different structures. As mentioned previously, it can be observed experimentally that a random grain boundary tends to facet into segments that have a low energy. It is generally assumed that the faceted segments can be described by the same structural units that have been derived from the atomic structures of low-energy grain boundaries. A more complex atomic configuration probably only occurs at the edges of steps and facets.

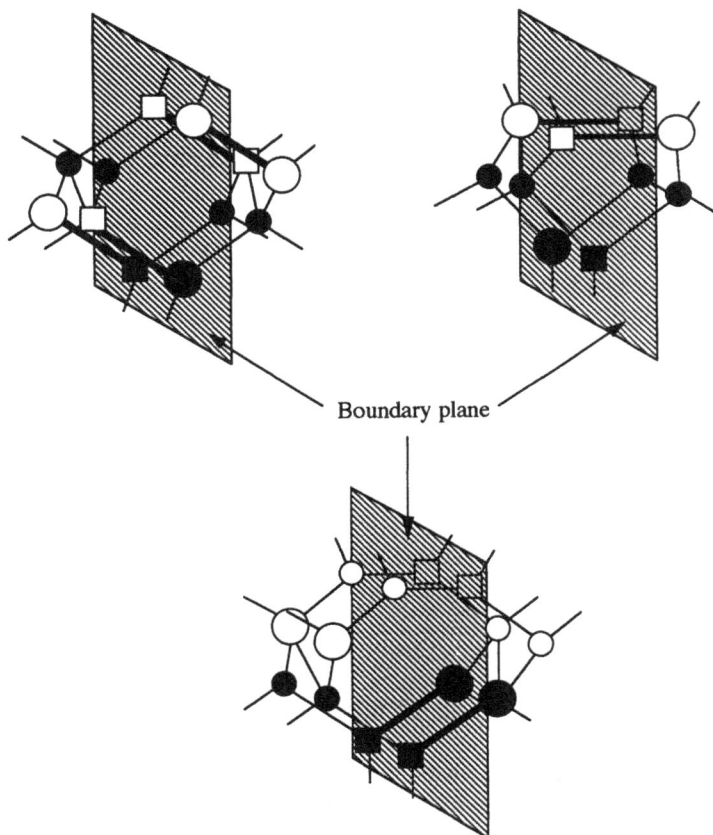

Figure 5.29 Geometrical models of five-, six, and seven-membered structural units, which can occur at grain boundaries in the chalcopyrite lattice [68]. The units are projected on a (011) plane. Considering an ABX_2 compound, the positions of the cations (A, B) are represented by large circles and squares, and the anions (X) by small circles. Black and white atoms are on neighboring (011) planes. Incorrect bonds are emphasized by bold lines. In a binary compound AX with the sphalerite structure, the A- and B-sites are occupied by the same kind of atoms. In the diamond cubic structure, all atoms are identical.

A fundamental issue from the perspective of the electrical activity of grain boundaries is the question of whether unreconstructed covalent ("dangling") bonds, which are supposed to introduce electronic levels in the band gap, occur in the grain boundary structure. Geometrical models suggested that atomic structures with unreconstructed bonds can occur and are more likely for random boundaries.

Computer calculations of particular grain boundary structures indicate, however, that at least low-energy structures are totally reconstructed and no deep electronic levels are introduced in the forbidden gap [70–72]. Experimental investigations of the electrical activity of low-energy CSL boundaries indeed show that they are electrically inactive, which is in agreement with the structural investigation. Therefore, considering the concept that random boundaries consist of the same structural units as the special lowenergy boundaries, one would rather expect reconstructed atomic structures and no electrical activity.

In a real crystal, most of the grain boundaries have random orientation and show electrical activity, which raises the questions about the origin. For instance, ESR measurements on small-grained polycrystalline silicon have shown a spin density of 10^{12} cm^{-2}, which has been attributed to dangling bonds in the grain boundaries [73]. At the present level of structural characterization, the microscopic origin of the electronic properties of grain boundaries is still far from being understood. Basically, two possibilities are feasible. First, the atomic structure of random grain boundaries cannot be described completely by reconstructed structural units. Possible sites for dangling bonds could be in the core of grain boundary dislocations and at edges of steps and facets.

Secondly, impurities (and intrinsic point defects in non-stoichiometric compound semiconductors) segregate or precipitate at grain boundaries and are responsible for the electrical activity. The experimental difficulties in the determination of the usually very low point defect concentrations in semiconductors and particularly at grain boundaries have prevented a systematic investigation, but there is increasing experimental evidence of the influence of point defects on the electrical activity.

In polycrystalline semiconductors, the majority of grain boundaries grow randomly and are usually low-symmetry, large-angle boundaries. Depending on the growth conditions, preferential orientations of the grains may occur with the possibility of the formation of coincidence (e.g., twins) and near-coincidence grain boundaries. Nonetheless, the problem remains that the numerous results on special grain boundaries so far are of limited use for the understanding of the electrical activity of random grain boundaries in polycrystalline semiconductors.

Compound Semiconductors

Systematic studies on the structure and electrical properties of grain boundaries in compound semiconductors are not available. Scattered results have been reported for GaAs, GaP, InSb, and SiC [74–78], but almost no experimental results are available for grain boundaries in CuInSe$_2$ or other ternary compounds that are important for photovoltaic applications.

The atomic structures of grain boundaries in covalently bonded compound and elemental semiconductors are similar, but the larger number of atomic species adds more complexity. In addition to the deformation of the covalent bonds (bending and stretching) at the interface (*structural disorder*), one has to consider the different occupation of lattice sites at the interface (*occupational disorder*) in a compound crystal.

The differences become evident if one compares geometrical models of atomic structures of the same CSL boundary in the sphalerite [79] and chalcopyrite lattices [80]; for instance, a $\Sigma 9$ $\langle 011 \rangle$ tilt boundary (Figure 5.30). In the diamond cubic structure, the same CS-lattice occurs at four misorientation angles, θ_1, $\theta_2 = 180 - \theta_1$, $180 + \theta_1$, $180 + \theta_2$, and the corresponding grain boundary structures are all identical. In AB compounds with sphalerite structure, where the A and B atoms occupy different fcc sublattices of the diamond cubic lattice, the atomic boundary structures for θ_1, θ_2 are different now. A characteristic feature of the grain boundary

Sphalerite lattice $\Sigma 9$ Chalcopyrite lattice

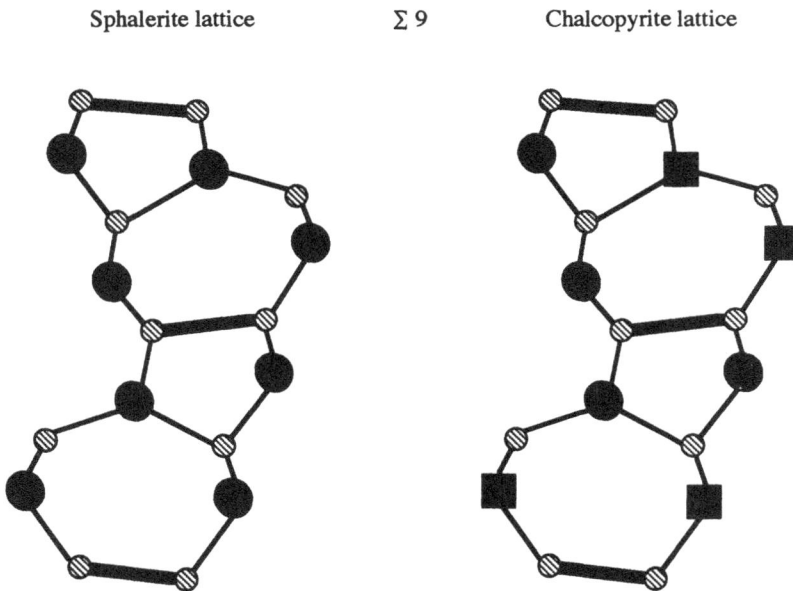

Figure 5.30 Structural unit models for a $\Sigma 9$ grain boundary in the chalcopyrite and sphalerite lattices. Four non-equivalent variants occur at the rotation angles $\theta_1 = 39.7$ deg, $\theta_2 = 141.7$ deg, $180 + \theta_1$, and $180 + \theta_2$ for the chalcopyrite lattice, and two variants at $\theta_1 = 38.5$ deg, $\theta_2 = 180 - \theta_1$, $180 + \theta_1$, and $180 + \theta_2$ for the sphalerite lattice. Half a period is shown in all cases. Bold lines indicate incorrect cation-cation or anion-anion bonds.

structure is the occurrence of a number of incorrect bonds, which can be between either A-A or B-B atoms. The same type of wrong bonds occurs in other grain boundary structures as well. although their number and distribution varies with the misorientation.

In a ternary system ABX_2 with the chalcopyrite structure, the grain boundary structures for the rotation angles θ_1, θ_2, $180 + \theta_1$, and $180 + \theta_2$ are all different now [80]. (In general, the angles θ_1 and θ_2 also differ slightly from the corresponding angles in the diamond structure, depending on the c/a ratio of the chalcopyrite unit cell. See Figure 5.4.) The complete model in Figure 5.30 consists of a period of four equidistant layers (of which only one is shown). The structure also shows the occurrence of wrong bonds, but because of the two different cations A and B, three different types of cation-cation and three types of anion-anion bonds can occur in a general boundary.

Incorrect bond sites are probably favorable nucleation sites for the formation of intrinsic point defects. This is important for non-stoichiometric compound semiconductors, where the lattice structure is maintained over a wider range of compositions by the formation of intrinsic point defects such as vacancies, interstitials, and antisite defects. Hence, in a polycrystalline compound with a high density of grain boundaries and wrong bonds, intrinsic point defects may preferentially nucleate at these grain boundary sites instead of in the bulk. Since intrinsic point defects can introduce deep and shallow levels in the band gap, their modified occurrence can have an important influence on the electrical behavior of the polycrystalline semiconductor. This is especially important for ternary compounds, where some intrinsic point defects introduce shallow levels and the doping level is usually determined by the composition of the crystals (see Chapter 8).

Experimental results, mainly for polycrystalline compounds, clearly indicate the importance of the grain boundaries for the electrical properties of the materials. However, at the present level of investigations, the knowledge about the structural and electrical properties of individual grain boundaries in compounds is still insufficient to understand all aspects of the behavior of these materials.

5.3.2 Electrical Transport Over Barriers

The previous discussion has shown that still little is known about the origin of the electronic grain boundary states in the band gap, but there is no doubt that they determine the electrical activity of the grain boundaries. Although the electrical activity of a grain boundary is not clearly defined, it can roughly be separated into two categories: the recombination of electrons and holes at the grain boundary states and the permanent trapping of charge carriers, which leads to the build-up of a grain boundary potential. There is also the possibility of an enhanced conductivity parallel to the boundary plane which may occur for high contamination levels, but is probably of minor importance in photovoltaic semiconductors.

The recombination of electron-hole pairs at extended lattice defects was discussed in Section 4.2.3. In the following, only charged grain boundaries will be considered. From the results of electrical measurements, a model for the electronic properties of grain boundaries has emerged that can account for most of the experimental data and allows the determination of the intrinsic electronic properties of the grain boundary, such as the density of the interface states $N_B(E)$. This conceptually simple model has found wide acceptance for the description of transport properties at grain boundaries and in polycrystalline semiconductors [57].

The basic idea is that the grain boundary is charged when the interface states are occupied, and thus represents a repulsive potential for the majority carriers in the bulk. In the bulk, a compensating space charge lies symmetric to the boundary plane for a homogeneously doped crystal (Figure 4.3). The resulting potential barrier (a *double Schottky barrier*) and the corresponding depleted region in the vicinity of the boundary plane are regions of enhanced resistance. The calculation of the potential barrier depends on the details of the electronic structure at the grain boundary and in the adjacent bulk of the semiconductor. Since the origin of the electronic interface states is not known, some assumptions have to be made for the modeling, which are, however, plausible considering the experimental observations.

1. The grain boundary is a planar two-dimensional interface. The electronic states in the band gap thus form a two-dimensional band, which, in general, may be distributed over the whole gap and be described by an a priori unknown density of states $N_B(E)$. The states of a neutral boundary are partly filled up to the level E_{B0}. The further occupation of the states with electrons or holes yields an interface charge Q_B.

2. The grains are homogeneously doped with one shallow impurity (for instance, a donor with an energy level E_d and a concentration N_d) and contain, in addition, several deep bulk traps (E_{Ti}, N_{Ti}).

3. The occupation of the traps is described within the depletion approximation.

In general, one has to consider that under an applied voltage U and illumination, the occupation of the grain boundary states changes, and therefore the recombination and electrical transport across the grain boundary changes as well. The general problem is thus to calculate the potential barrier under bias U and illumination (Figure 5.31). The calculation of the potential barrier $\varphi_B(x)$ from the charge density in the vicinity of the grain boundary requires the solution of Poisson's equation. In the depletion approximation, a trap in the band gap is charged between the "screening lengths" x_{li} and x_{ri}, and neutral outside. The screening lengths are determined by the position of the trap level E_{Ti} and the quasi-Fermi energy E_{FB}.

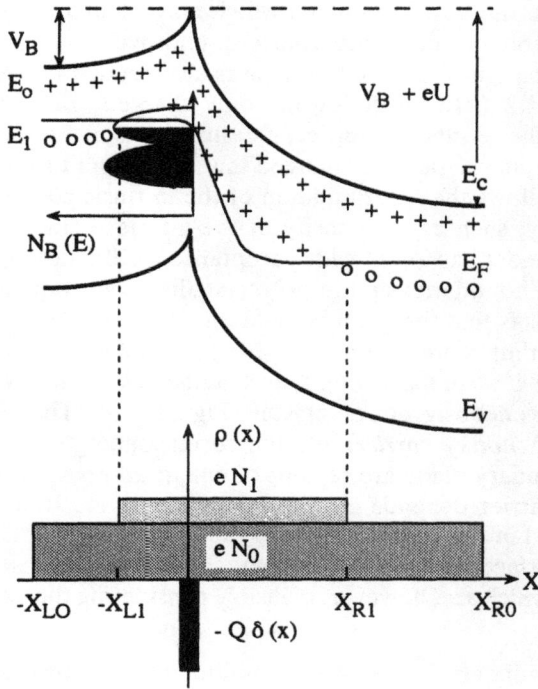

Figure 5.31 Energy band diagram and spatial charge distribution for a grain boundary barrier under bias. E_0, N_0 and E_1, N_1 are the energy levels and concentrations of a shallow and deep donor level impurity, respectively. The density $N_B(E)$ and occupation of the interface states is shown in the insert.

Assuming, for instance, an n-type semiconductor with $i = 0, \ldots, n$ bulk traps, where the first level ($i = 0$) refers to the donor level, the charge distribution is given by

$$\frac{\partial^2 \varphi_B}{\partial x^2} = \frac{e}{\epsilon\epsilon_0}\{N_B\delta(x) - \sum_i N_{Ti}[\Theta(x + x_{li}) - \Theta(x - x_{ri})]\} \tag{5.57}$$

with the boundary conditions $\varphi_B(-x_{l0}) = 0$ and $\varphi_B(x_{r0}) = U$. The expression for the barrier height $V_B = -e\varphi_B(0)$ as a function of the applied voltage U yields

$$V_B = \frac{V_C}{4}\left(1 - \frac{eU}{V_c}\right)^2 + \frac{\sum N_{Ti}\{E_{Ti} - E_{FB}\}}{e(N_d + \sum N_{Ti})} \tag{5.58}$$

where the potential V_c is given by

$$V_c = \frac{Q_B^2}{8\epsilon\epsilon_0(N_d + \Sigma\, N_{Ti})} \tag{5.59}$$

Only trap levels with $E_{Ti} > E_{FB}$ contribute to the sum in (5.58), since a trap level that is completely neutralized does not contribute to the screening of the barrier charge. The potential barrier V_B is positive in an n-doped semiconductor and negative in a p-doped semiconductor. The expression for the potential is the generalized form of (4.33) in Section 4.2.3 and is valid when deep traps are present in the vicinity of the barrier. The potential barrier at the interface depends strongly on the grain boundary charge Q_B, which can be calculated from the density of states distribution of the interface $N_B(E)$:

$$Q_B = e \int N_B(E) f_B(E + V_B) dE - e N_{Bn} \tag{5.60}$$

N_{Bn} is the electron concentration (per unit area) of the neutral grain boundary, and $f_B(E)$ is the occupation statistics for the interface states. In thermodynamic equilibrium, $f_B(E)$ is equal to the Fermi-Dirac distribution, but has to be determined in a non-equilibrium situation (for instance, under illumination).

It was shown, however, in Section 4.2.3 that using the thermionic emission model in a doped crystal, even under applied bias, f_B, maintains the form of the Fermi-Dirac statistics with a quasi-Fermi level E_{FB} for the grain boundary states given by

$$E_{FB} = E_F - KT \ln \frac{2}{1 + \exp(-eU/KT)} \tag{5.61}$$

The grain boundary potential has to be determined self-consistently from (5.58) to (5.61) for a given density of states distribution. Figure 5.32 shows calculated barriers as a function of the applied voltage for a single level and a Gaussian density of states distribution [57]. The dominating term in (5.58) is the first one, which depends strongly on the grain boundary charge Q_B and the bias U. As the bias increases, the barrier height decreases. The increase of the boundary charge with increased voltage, however, counteracts and slows down the decrease of the barrier; thus, for small voltages, the potential barrier is pinned by the grain boundary states. A highly localized density of states can thus stabilize the barrier for moderate voltages. The barrier breaks down rapidly when all grain boundary states are filled, and Q_B essentially remains constant.

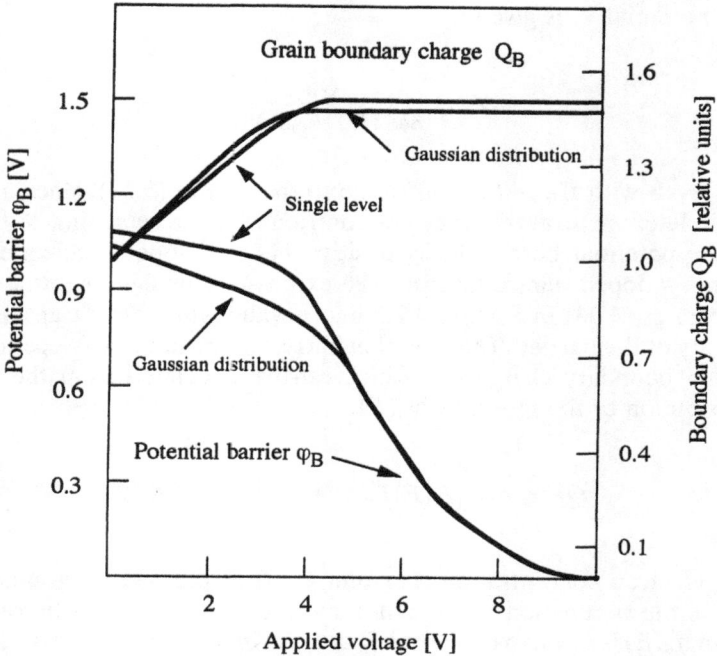

Figure 5.32 Calculation of the barrier height and grain boundary charge as a function of the applied voltage for a single level and a Gaussian density of states distribution [87].

In n-type crystals, which are considered here, the presence of donor-like deep bulk defects (charge states $+/0$) further decreases the barrier height because of the additional screening charge (see Figure 5.31). In p-type crystals, acceptor-like states (charge states $0/-$) have to be considered. For typical values, for a deep level with a concentration of $N_T = 10^{17}$ cm^{-3} and $N_T/N_d = 0.1$, the barrier can be reduced by as much as 0.1 eV, which is large compared to KT at room temperature. A second effect is that with increased voltage more ionized bulk traps are shifted below the Fermi level and do not contribute anymore to the screening. In n-type crystals, a donor-like deep level at E_{Ti} becomes completely neutral if the potential barrier reaches the value $V_B(U) = E_{FB} - E_{Ti}$, which leads to a transient stabilization of the barrier with increasing voltage.

Time-Independent Electrical Transport (Current-Voltage Characteristics)

The barrier height V_B determines the current flow of majority carriers through the grain boundary. Under dc conditions, the total current density J_{DC} for majority carriers (e.g., electrons) is determined by the currents from the left- and right-hand sides into the barriers J_L and J_R, the electron current into the grain boundary states J_{TL} and J_{TR}, and the current emitted from these states into the conduction band $2J_E$ (Figure 5.33). The total current can be calculated on either the left- (J_1) or right-hand side (J_2), respectively, [81] and is given by

$$J_1 = J_L - J_R + J_{TR} - J_E \qquad J_2 = J_L - J_R - J_{TL} + J_E \qquad (5.62)$$

Symmetric emission into the left and right grains is assumed here. The trap and emission currents determine an interface current J_B into the grain boundary states:

$$J_B = J_{TL} + J_{TR} - 2J_E \qquad (5.63)$$

J_B is related to the previously calculated recombination rate R_0 (see Section 4.2.3) for the grain boundary by the relationship $J_B = eR_0$. It has been shown there that without illumination or other hole generation processes no electron-hole recom-

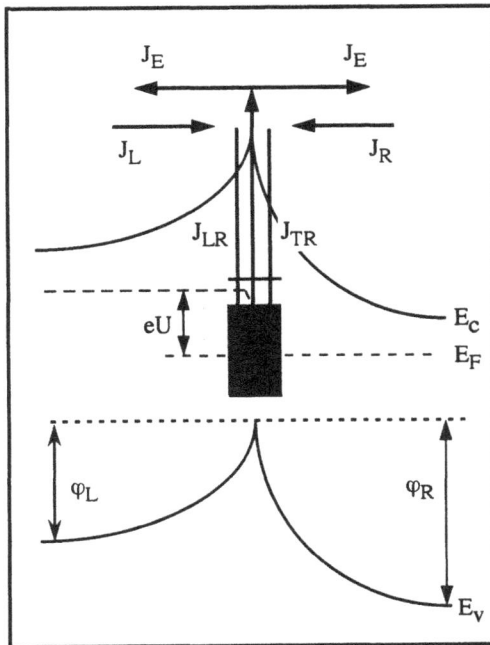

Figure 5.33 Schematic representation of the different contributions to the total current at a grain boundary.

bination occurs at the interface; therefore, the interface current is zero and the contributions of minority carriers to the total current through the barrier can be ignored here. It should be noted that under non-steady-state conditions the interface current J_B becomes non-zero and is accompanied by a change in the grain boundary charge and barrier. Combining (5.62) and (5.63) yields for $J_{DC} = J_1 = J_2$

$$J_{DC} = J_L - J_R + \frac{1}{2}(J_{TR} - J_{TL})$$
(5.64)

Generally, it is assumed that the currents can be calculated within the thermionic emission model [81], which yields

$$J_L - J_R = env_T\left\{\exp\left(-\frac{e\varphi_L}{KT}\right) - \exp\left(-\frac{e\varphi_R}{KT}\right)\right\}$$
(5.65)

$$J_{TR} - J_{TL} = e\int s_n N_B(1 - f_B)\left\{\exp\left(-\frac{e\varphi_R}{KT}\right) - \exp\left(-\frac{e\varphi_L}{KT}\right)\right\}dE$$
(5.66)

Substituting J_{DC} in (5.60), one obtains

$$J_{DC} = env_T(1 - c)\exp\left(-\frac{V_B}{KT}\right)\left\{1 - \exp\frac{eU}{KT}\right\}$$
(5.67)

where n is the electron concentration determined by the doping in the vicinity of the grain boundary $N_d = n = N_C\exp\{-(E_C - E_F)/KT\}$. The prefactor is also frequently expressed by the effective Richardson constant A^* defined by $A^*T^2 = eN_Cv_T$. The constant c is the capture probability for electrons at the interface and is given by

$$c = \frac{1}{2v_T}\int s_n(E)N_B(E)\{1 - f_B(E)\}dE$$
(5.68)

$f_B(E)$ is the Fermi-Dirac distribution with the quasi-Fermi level E_{FB} for the grain boundary states given in (5.61). Usually it is assumed that the capture cross section $s_n(E)$ is constant so that c becomes proportional to the number of empty grain boundary states N_B, with $c = S_nN_B/2v_T$. Typical values for the capture cross section in silicon and other semiconductors are $S_n \approx 10^{-16}$ to 10^{-14} cm^2, and for the density of states $N_B \approx 10^{12}$ to 10^{13} cm^{-2}. In most cases, c is thus small ($c \ll 1$) and can be neglected.

For small voltages $eU \ll KT$, the potential barrier is almost constant, and the current density J_{DC} is linearly dependent on the voltage. Therefore, one can define the zero-bias conductance σ_B for the grain boundary with $\sigma_B = J_{DC}/U$ and obtain

$$\sigma_B = \frac{e^2}{KT} n v_T (1 - c) \exp\left(-\frac{V_B}{KT}\right) \qquad (5.69)$$

σ_B as defined here has the dimension $\Omega^{-1}\text{cm}^{-2}$ and is exponentially dependent on the grain boundary potential. For larger voltages $eU \gg KT$, the current density J_{DC} depends critically on the voltage dependence of the grain boundary barrier V_B. Since V_B always decays rapidly at higher voltages when all traps are filled, the current shows a sharp rise above a certain bias, as can be seen from the calculated curves in Figure 5.34. The steepest increases occur for single-level boundary states. Deep majority traps in the space charge region tend to smooth the decay of the barrier and, hence, the increase of the current. It should be mentioned that with increasing decay of the barrier at high voltages the finite conductivity of the grains has to be taken into account, which can be seen for $U > 4V$. The current-voltage characteristic is critically dependent on the electronic structure of the grain boundary and hence can vary from sample to sample.

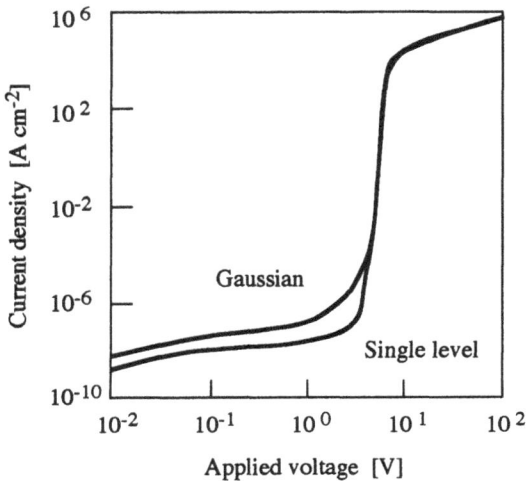

Figure 5.34 Calculated I-V characteristic for a grain boundary barrier, assuming a single level and a Gaussian density of states distribution [57].

Solar cells are usually used under dc conditions, so the grain boundary response can be described by the previous model. For the investigation of the emission and capture processes of electrons or holes and the distribution of the interface states at the grain boundary, time-dependent experimental methods such as admittance or DLTS are more advantageous. These techniques have been increasingly applied during recent years to study grain boundaries in polycrystalline solar cell semiconductors.

The time-dependent electrical behavior of grain boundaries cannot be described by the previous equations. A realistic description of the dynamical electrical activity of a grain boundary has to take into account the rearrangement of electrons and holes at the interface and bulk states. An essential difference is that the recombination current at the interface $J_B = eR_o$ (equation (4.47)) is not zero anymore. Correspondingly, the grain boundary charge, the barrier height, and the screening charge of shallow- and deep-level traps on either side of the interface become time-dependent. This leads to a more complicated description of the grain boundary response, but the resulting equations are in agreement with experimental results for a variety of semiconductors [82,83]. The theoretical description of the electrical behavior of grain boundaries is thus solved to a large extent.

Grain Boundary States

Electrical measurement data on individual grain boundaries are available for germanium [83–85], silicon [86,87], and GaAs [74]. The most comprehensive results have been reported for polycrystalline and bicrystal silicon. Localized grain boundary states have been observed in n-type Fz- and Cz-bicrystals of particular misorientations at $E_{B1} \approx E_V + 0.35$ eV, $E_{B2} \approx E_C - 0.29$ eV, or $E_{B2} \approx E_C - 0.42$ eV. In addition, most measurements show a rather broad distribution of states in the middle of the band gap, as can be seen in Figure 5.35 for a single grain boundary in silicon. The values for the density of states near midgap vary between $N_B \approx 10^{-11}$ to 10^{-13} cm^{-2} eV^{-1}. The occupation limit of the neutral grain boundary is usually found to be near the midgap level at $E_{B0} \approx 0.55$ eV. Therefore, the charging of the grain boundary states is almost the same in n- and p-doped silicon crystals.

The results for grain boundaries in germanium and GaAs indicate that the interface states are rather localized. Electrically active ⟨011⟩ tilt grain boundaries in germanium show trap levels in the lower half of the band gap at about $E_B \approx E_V + 0.1$ eV, whereas the well-reconstructed twins ($\Sigma 3$ and $\Sigma 9$) are electrically inactive. In n-doped GaAs, ⟨011⟩ tilt grain boundaries with a variety of misorientation angles have been studied by DLTS and I-V analyses. Independent of the orientation, two grain boundary trap levels at about $E_{B1} \approx E_C - 0.65$ eV and $E_{B2} \approx E_C - 0.8$ eV have been observed, whereas the density of states was reported to depend on the

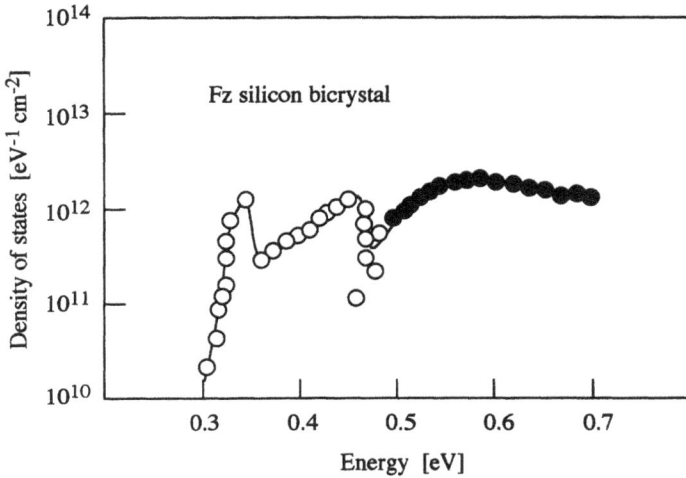

Figure 5.35 Experimental density of states distribution for a single grain boundary in Fz-silicon bicrystal with n-doping concentrations of 5×10^{12} cm^{-3} (open circles) and 1.4×10^{15} cm^{-3} (dark circles). In the latter case, the specimens were annealed at 730°C for 1 hour [84]. The energy is measured from the valence band edge.

misorientation angle in particular. Summarizing the results in compound semiconductors, the grain boundaries are characterized by:

1. Narrow localized states at energy levels typical for other lattice defects;
2. A density of states that depends on the processing and is affected by lattice point defects;
3. Deep bulk defects near the grain boundary, which can reach rather high concentrations and thus contribute to the electrical activity.

The localized states observed for the grain boundary states are also frequently found for other extended lattice defects, such as dislocations, heterostructure interfaces, and surfaces. Whereas in elemental semiconductors this coincidence is not understood, in compound semiconductors this result supports the *unified defect model* [88], which assumes that the localized states are point defects of the host lattice rather than structural imperfections at the grain boundary. Candidates for the localized states are anion or cation vacancies, antisite defects, or impurities. Since the electronic states observed for grain boundaries, heterostructures, and surfaces are similar, one may generalize the results and use the extensive knowledge from surface and heterostructure investigations to predict the positions of grain boundary states in polycrystalline compound semiconductors. Since an important class of photovoltaic materials are polycrystalline compound semiconductors, such as the copper ternaries for which almost no electrical properties of the grain boundaries are known, this generalization may be of particular importance.

5.4 DEFECT INTERACTION

Unlike semiconductors used for microelectronic devices where defect-free, high purity material is used, solar cell semiconductors are usually less defect-free and contain higher concentrations of impurities because of the constraints of low production costs. Impurities are incorporated during crystal growth and device fabrication. Segregation and precipitation processes lead to an inhomogeneous distribution of impurities in the bulk and can occur at all stages of device fabrication. Of particular importance are surfaces, grain boundaries, and other extended lattice defects that are preferential sites for the segregation and precipitation of impurities.

The interaction of the impurities with lattice defects affects the electrical properties of the semiconductor in several ways. First, it changes the distribution of the impurities in the crystal. This can be beneficial if detrimental impurities are removed from the electrically active regions of the device ("gettering"). Typical examples are the intrinsic gettering at oxygen precipitates in silicon or back surface gettering in microelectronic devices. On the other hand the electrical activity of the lattice defect itself can be changed, which may be detrimental for the performance of the solar cell if, for instance, the recombination velocity of interfaces and the surface is increased. Topographic evaluations of the grain boundary activity in ingot silicon, for instance, by four-point probe resistivity mapping (Figure 5.36) demonstrate the electrical activity of grain boundaries and show their non-uniform behavior in a crystal.

The utilization of gettering techniques in solar cell devices is limited because all regions of the semiconductor (bulk and surface) contribute to the performance of the cell. Nonetheless, it has been recognized that the investigation of the interaction of lattice defects with impurities is important from the perspective of the optimization of the solar cell performance.

The impurities that are of main interest in silicon are the doping elements (mainly B, Al, P), carbon and oxygen, which can occur in rather high concentrations, and fast diffusing transition elements (for instance, Cu, Fe, Ti), which introduce deep trap levels. In compound semiconductors, one has to consider in addition intrinsic point defects, which can act as shallow- and deep-level traps. Of particular importance also are impurities like hydrogen, which can passivate the electrical activity of lattice defects to some extent.

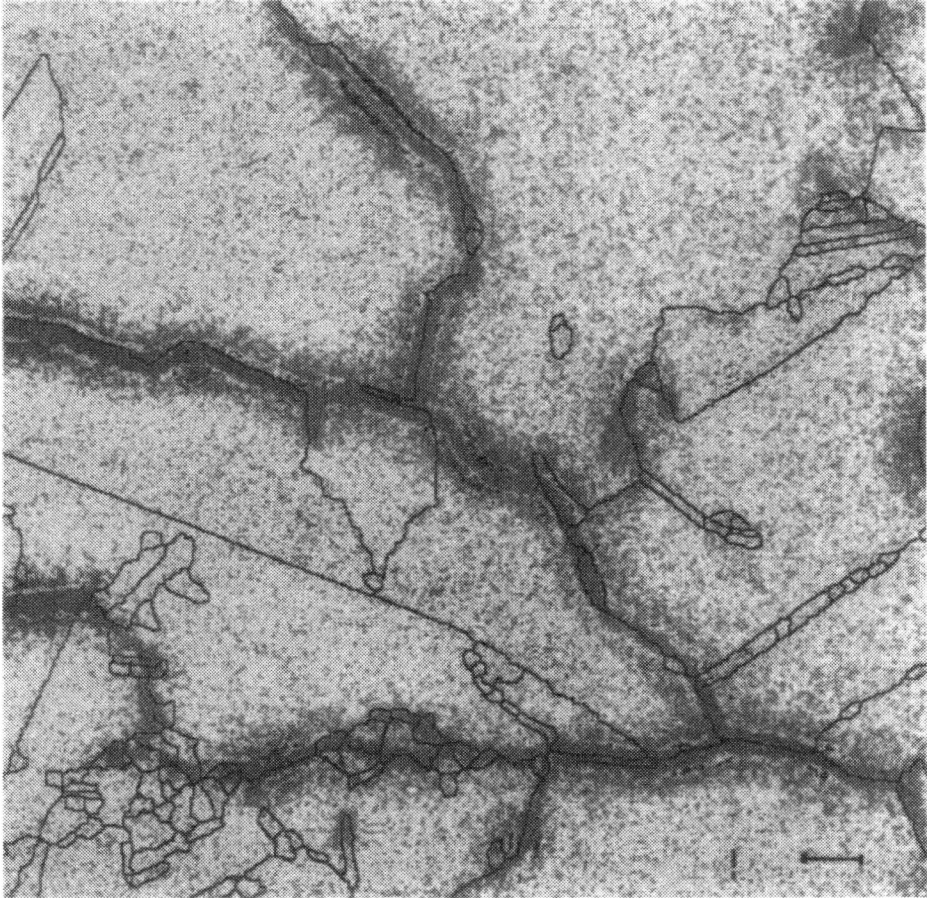

Figure 5.36 Four-point probe resistivity mapping of a polycrystalline silicon wafer showing the enhanced resistivity of some grain boundaries. The resistivity values are color-coded and increase from blue to red. The grain boundary pattern is inserted and indicated by solid lines (scale marker $\cong$ 5 mm). (Courtesy of W. Koch, Bayer Co.)

5.4.1 Segregation

An inhomogeneous distribution of impurities in the solid due to segregation is the result of a difference in the impurity solubility. In thermal equilibrium, different solubilities usually occur between the solid (C_a^{eq}) and the liquid phase (C_L^{eq}) of a material, but can also occur between different regions of the solid, such as the bulk and an interface region (C_G^{eq}) in the solid.

In the first case, segregation occurs during solidification and is determined by the segregation coefficient $k_0 = C_\alpha^{eq}/C_L^{eq}$, which is correlated with the maximum solubility of the impurity in the solid (see Fischler relation (5.29)). For $k_0 < 1$, the solid is purified and the melt enriched with impurities. Typically, in silicon and other semiconductors, the transition metals have very low segregation coefficients ($k_0 < 10^{-4}$) and can be removed efficiently from the solid, whereas doping elements and carbon are less affected during solidification because of a segregation coefficient closer to 1. There are also impurities with $k_0 > 1$, such as oxygen in silicon, which are enriched in the solid.

5.4.1.1 Fermi-Level Effect

At temperatures below the melting point, segregation of an impurity in the solid can occur when the solubility differs in various parts of the crystal. For instance, different solubilities can occur in semiconductors with space charge regions for impurities with trap levels in the band gap. Due to the band bending, the positions of the trap levels with respect to the constant Fermi level vary in the crystal, and the impurity can occur in different charge states. For instance, an impurity with an acceptor level E_{Ta} in the band gap has a solubility C_α^{eq} given by (see (5.30) and (5.31))

$$C_\alpha^{eq} = C_{\alpha0}^{eq}\left\{1 + \exp\left(\frac{E_F - E_{Ta}}{KT}\right)\right\} \qquad (5.70)$$

where $C_{\alpha0}^{eq}$ is the equilibrium solubility of the neutral species (equation (5.28)). If the Fermi-level lies above the trap level, the solubility is increased. Therefore, if the solubility is higher in a certain region of the crystal, the crystal can lower its total free energy G when impurities diffuse into the region of enhanced solubility. The equilibrium is reached when a uniform chemical potential is established in the entire crystal. This phenomenon shall be discussed in more detail in the following.

For simplicity, we separate the crystal into two regions in which the impurity occurs, in either the neutral or the charged state, respectively. For an impurity with an acceptor-like state E_{Ta}, the formation enthalpy of solution for the charged defect is given by (see (5.4))

$$H_{\alpha-}^s = H_{\alpha0}^s + (E_{Ta} - E_F) \qquad (5.71)$$

where $H_{\alpha0}^s$ is the formation enthalpy of the neutral impurity. For low impurity concentrations, which are typical for semiconductors, the entropies in the region of neutral and charged defects, respectively, are given by

$$S^s_{\alpha 0} = n_0 S^s_0 + K n_0 \ln \frac{N_0}{n_0} \qquad S^s_{\alpha -} = n_1 S^s_0 + K n_1 \ln \frac{N_1}{n_1} \qquad (5.72)$$

S^s_0 is the entropy contribution associated with the disorder of the lattice vibrations in the vicinity of the defect, and for simplicity it is assumed to be equal for the neutral and charged impurities. The second term in each equation is the contribution from the configurational disorder (the number of distributions of impurity atoms over the possible sites). N_0 and N_1 are the available lattice sites, and n_0 and n_1 are the number of neutral and charged impurities in region 0 and 1, respectively. n_0 and n_1 are related by the constraint that the total number of impurities in the crystal n is constant: $n = n_0 + n_1$. For a given temperature T, the change in free energy is given by

$$\Delta G = n_0(H^s_{\alpha 0} - TS^s_0) + n_1(H^s_{\alpha -} - TS^s_0) - KT\left(n_0 \ln \frac{N_0}{n_0} + n_1 \ln \frac{N_1}{n_1}\right) \qquad (5.73)$$

$\Delta G(n_0, n_1)$ is a function of the impurity numbers n_0 and n_1 and is at minimum in thermodynamic equilibrium. Under the constraint $dn_0 + dn_1 = 0$, differentiation with respect to n_0 and n_1 yields

$$d(\Delta G) = 0 = dn_1\left\{KT \ln \frac{N_0 n_1}{n_0 N_1} + (E_{Ta} - E_F)\right\} \qquad (5.74)$$

Introducing the concentrations $C_0 = n_0/N_0$ of the neutral species in region 0, and $C_1 = n_1/N_1$ of the charged species in region 1, one obtains

$$C_1 = C_0 \exp \frac{E_F - E_{Ta}}{KT} \qquad (5.75)$$

Since $(E_F - E_{Ta}) > 0$ for a charged acceptor-like state, the concentration of the charged impurities in region 1 is enhanced in thermal equilibrium. A corresponding equation can be derived for an impurity with a donor-like trap level in the band gap. The increase in concentration depends on the position of the Fermi energy E_F and the trap level E_{Ta} or E_{Td}, respectively.

Segregation due to the Fermi-level effect is used, for instance, in the phosphorus gettering technique, which is a standard gettering process in the microelectronics device technology to remove transition metals from the active regions of a device. Phosphorus is diffused in a *p*-type crystal and a highly doped *n*-type region is produced near the back surface. Due to the strong band bending in the space charge region, trap levels (particularly deep) are shifted below the Fermi level. In the case of an acceptor-like state, the corresponding impurity becomes charged and the solubility increases in this region. Since most of the transition metals have both deep donor and acceptor states (see Section 5.1.3), they can be accumulated in the n-doped region and effectively removed from the active front region of the device. (It should be noted that in addition to the Fermi-level effect other processes not yet completely understood also contribute to the phosphorus gettering effect [89].)

In pn-junction solar cells, the emitter region is usually also a highly n-doped region and thus attracts transition metals. Since most of the transition metals are fast diffusers, they are easily accumulated in the emitter region during device processing. In a solar cell, however, this is an undesirable effect, since it will decrease the diffusion length in the emitter region. Although a shorter diffusion length in this region compared to the base region can be more easily tolerated, this will, in general, degrade the performance of the cell.

5.4.1.2 Segregation at Interfaces

A thermodynamic description of the impurity segregation at interfaces is possible if one separates the bulk from the interface regions [90,91]. The basic idea is that the interface can be considered a two-dimensional system for which thermodynamic variables analogous to the bulk can be defined. For several atomic species ($i = 0$ to n), these variables are, in particular, the atomic concentrations in the interface C_{Gi}, the activities $a_{Gi} = f_{Gi}C_{Gi}$, and the chemical potentials μ_{Gi}. The role of the conjugate bulk quantities, pressure and volume, is held in the interface by the surface tension σ and the specific area per atom ω_i. The chemical potentials for each species in the interface are then given by

$$\mu_{Gi} = \mu_{Gi}^0 + KT \ln a_{Gi} - \sigma\omega_i \qquad (5.76)$$

The corresponding variables in the bulk are $a_{\alpha i} = f_{\alpha i}C_{\alpha i}$ and $\mu_{\alpha i}$, which are related by

$$\mu_{\alpha i} = \mu_{\alpha i}^0 + KT \ln a_{\alpha i} \qquad (5.77)$$

In semiconductors, the impurity concentrations $C_{\alpha i}$ are generally very low, so the bulk phase can be considered ideal, and the activity coefficients are $f_\alpha \approx 1$. In thermal equilibrium, the chemical potentials for each species in the interface and the bulk are equal:

$$\mu_{Gi} = \mu_{\alpha i} \qquad (i = 0, \dots, n) \qquad (5.78)$$

Experimental results show that the concentrations of impurities at the interface can become quite high, so the interaction between atoms cannot be neglected anymore. In a simple model, one can assume that the driving force for segregation is the relaxation of elastic misfit energy when a bulk atom occupies an interface site. In this case, one can express the excess free enthalpy ΔG^E and the enthalpy of mixing ΔH^M by the atomic concentrations and interaction coefficients α_{ij}:

$$\Delta G^E = \Delta H^M = -\sum_{i<j} \alpha_{ij} C_{Gi} C_{Gj} \qquad (5.79)$$

The coefficients α_{ij} are positive for an attractive interaction and negative for a repulsive interaction. In this approximation, only the interaction energies ϵ_{ij} between nearest neighbor atoms of species i and j are taken into account. The activity coefficients are related to the interaction coefficients. For a three-component system with two solutes ($i = 1, 2$) in a solvent $i = 0$, one obtains [90]

$$-KT\ln f_{G1} = \alpha_{10}(1 - C_{G1})^2 + \alpha_{20}C_{G2}^2 + (\alpha_{12} - \alpha_{20} - \alpha_{10})C_{G2}(1 - C_{G1}) \qquad (5.80a)$$

$$-KT\ln f_{G2} = \alpha_{20}(1 - C_{G2})^2 + \alpha_{10}C_{G1}^2 + (\alpha_{12} - \alpha_{20} - \alpha_{10})C_{G1}(1 - C_{G2}) \qquad (5.80b)$$

$$-KT \ln f_{G0} = \alpha_{10}C_{G1}^2 + \alpha_{20}C_{G2}^2 - (\alpha_{12} - \alpha_{20} - \alpha_{10})C_{G1}C_{G2} \qquad (5.80c)$$

The solvent concentration C_{G0} has been eliminated by using the boundary condition that the number of available sites at the interface and in the bulk are constant and can only be occupied by one atom at a time. This is equivalent to

$$\sum_i C_{Gi} = 1 \qquad \sum_i C_{\alpha i} = 1 \qquad (5.81)$$

Combining (5.76) with (5.81), one obtains a general expression for the concentration of the solute atoms i at the interface:

$$C_{Gi} = \frac{C_{\alpha i} \exp \dfrac{\Delta G_i}{KT}}{1 + \sum_j \left(C_{\alpha j} \exp \dfrac{\Delta G_j}{KT} - 1 \right)} \qquad (5.82)$$

where the free enthalpies are ($i = 0$ for the solvent atom)

$$\Delta G_i = \Delta G_{i0} + \Delta G_{im} \qquad \Delta G_{i0} = \mu_{Gi} - \mu_{\alpha i} + \mu_{G0} - \mu_{\alpha 0} \qquad (5.83)$$

ΔG_{im} accounts for the interaction with other atoms in the interface and can be expressed by the interaction coefficients. In the case of a single impurity ($i = 1$) and a low bulk solubility ($C_{\alpha 1} \ll 1$), one obtains (index omitted)

$$C_G = \frac{C_\alpha \exp \dfrac{\Delta G}{KT}}{1 + C_\alpha \left(\exp \dfrac{\Delta G}{KT} - 1 \right)} \tag{5.84}$$

where the free enthalpy takes the simple form [90]

$$\Delta G = \Delta G_0 + \alpha_0 (1 - 2C_G) \tag{5.85}$$

α_0 is the interaction coefficient between the impurity and solvent atoms. As with the solid-liquid segregation, a segregation ratio $k_G = C_G/C_\alpha$ can be defined. At a given temperature the interface concentration C_G increases with the bulk concentrations C_α and reaches a saturation value C_G^0, the maximum solubility in the interface. The saturation value may not be reached if the bulk solubility is low. In this case, the maximum solubility limit at the interface becomes temperature-dependent. For attractive segregation ($\Delta G > 0$), the enrichment at the interface k_G increases with decreasing temperature, and increasing the temperature leads to impurity desegregation.

In the case of two impurities ($i = 1, 2$), where both impurities can occupy the same interface sites, the expressions for ΔG_i assume the following form:

$$\Delta G_1 = \Delta G_{10} + \alpha_{10}(1 - 2C_{G1}) + (\alpha_{12} - \alpha_{20} - \alpha_{10})C_{G2} \tag{5.86a}$$

$$\Delta G_2 = \Delta G_{20} + \alpha_{20}(1 - 2C_{G2}) + (\alpha_{12} - \alpha_{20} - \alpha_{10})C_{G1} \tag{5.86b}$$

These equations for a ternary system describe the influence of a second impurity on the segregation behavior of impurity 1 (cosegregation). An attractive interaction between the impurities increases the segregation of impurity 1, but it can also lower the maximum interface solubility. Cosegregation phenomena have been observed in metal alloy systems.

In semiconductors, the solubilities of most impurities are rather limited and strongly temperature-dependent, as has been discussed, for instance, for the transition metals (Section 5.1.3). The thermodynamic description of the interface as a two-dimensional system also shows that the interface has a limited solubility C_G^{eq} for a particular impurity. Precipitation at the interface will occur in thermodynamic equilibrium if the solubility limit C_G^{eq} for a given temperature is exceeded. Since, according to (5.82), the interface concentration is a function of the bulk concentration, the solubility limit is reached for a particular bulk concentration C_α^{max}. The interface solubility is also exponentially dependent on the temperature. If $C_\alpha^{\text{max}} < C_\alpha^{\text{eq}}$ for a given temperature, these two values define a concentration range for which precipitation occurs at the interface, but the impurity is still dissolved in the bulk. An example may be the formation of oxide micro-precipitates in the cores of dislocations of small angle grain boundaries, which has been observed in silicon and germanium [93]. In the second case $(C_\alpha^{\text{max}} > C_\alpha^{\text{eq}})$, if the maximum solubility in the interface has not been reached, precipitation may already occur in the bulk if the concentrations exceed the solubility limit, while the impurities are still dissolved in the interface.

5.4.2 Growth Kinetics

For most impurities in semiconductors, the solubility limits both in the bulk and the interfaces are exceeded during cooling, and in thermodynamic equilibrium, precipitation will occur in the bulk and at the interface. Experimental situations are usually characterized by a thermodynamic non-equilibrium; therefore, it is more likely that the precipitation behavior is determined by the kinetics of the segregation and precipitation processes rather than by considering the equilibrium situation.

5.4.2.1 Bulk Precipitation

In general, the redistribution of impurities by segregation is controlled by diffusion processes. The segregation kinetics changes when the solubility limits in the bulk (C_α^{eq}) or at the interface (C_G^{eq}) are exceeded and precipitation occurs. Once the precipitates are nucleated, the growth of the second-phase particles (composition C_β) can be controlled by several rate-limiting steps. In many cases it is assumed that the growth is limited by the atomic diffusion of the impurity to the precipitate. In general, one also has to consider the volume change between the particle and the matrix, which may require the transport of vacancies or self-interstitials towards or away from the precipitate. The particle growth may also be controlled by a surface reaction rate when the impurity atoms at the particle-matrix interface must surmount a barrier in order to enter the precipitate. A number of theories have been developed to account for the different situations.

    The main results of the diffusion-limited growth for simple geometrical situations shall be discussed here for a supersaturated solution with concentration $C_\alpha^0 > C_\alpha^{\text{eq}}$. A supersaturation can occur, for instance, by the quenching of a specimen or, if the nucleation of the second phase during cooling is hampered, by a lack of nucleation sites. Solutions of the diffusion problem have been derived by Ham [94] under the assumption that the nucleation is not rate-limiting and that a periodic array of spherical particles has formed in the bulk. If the radius $r(t)$ is small compared to the mean distance r_m between the particles, the time dependence of the particle radius and the maximum radius are approximately given by

$$r(t) = \left\{\frac{2\Delta C_\alpha D_\alpha t}{C_\beta}\right\}^{1/2} \qquad r_{max} = r_m\left\{\frac{\Delta C_\alpha}{C_\beta}\right\}^{1/3} \qquad (5.87)$$

where $\Delta C_\alpha = C_\alpha^0 - C_\alpha^{eq}$. The first equation is valid for short times, where the bulk concentration C_α remains essentially constant and the critical radius of the nucleated particle $r_c = r(0)$ is negligible. The initial growth follows a $t^{1/2}$ law and is determined by the diffusion coefficient of the impurity and the supersaturation ΔC_α. The expressions for the time dependence become more complex for longer times and particles with nonspherical shapes. The equations are also applicable for precipitation at interfaces if the bulk diffusion is rate-limiting and interface diffusion can be ignored.

More refined theories have been developed for situations where the bulk concentration between precipitates decreases or smaller particles disappear at the expense of larger ones (Oswald ripening) [95–97]. In semiconductors with low impurity concentrations compared to metals and large distances between precipitates, these models may not be required to describe the precipitation kinetics.

It was discussed in Section 5.1.3 that the diffusion of transition elements in general may require the participation of native point defects. Since the equilibrium concentrations of vacancies and self-interstitials can be disturbed significantly during the precipitation, one cannot rule out that the growth rate of precipitates is controlled by the transport of these intrinsic defects. In order to estimate the influence of the intrinsic defect transport on the growth rate, the following simple model is assumed: a supersaturated solid solution with a homogeneous distribution of the impurity concentration $C_\alpha^0 = C_s^0 + C_i^0 > C_\alpha^{eq}$, where C_α^{eq} is the solubility limit for the impurity (the subscripts s and i denote substitutional and interstitial sites for the impurity). At high temperatures, the interstitial impurity concentration shall exceed the substitutional component $C_i \gg C_s$, which is typical for most transition elements. For the initial growth phase (short times), the total concentration remains almost constant; therefore, one can assume that $C_i^0 \approx C_\alpha^0 = $ constant. During precipitate formation, self-interstitials are generated and injected into the lattice with a generation rate K_G. In addition, it is assumed that, in the bulk, self-interstitials disappear at sinks such as dislocation (or grain boundaries) with a rate K_I. K_G and K_I (see also (5.15)) may be given by

$$K_G = \gamma_P N_P D_I C_\alpha^{eq} \qquad K_I = \gamma_I N_D D_I C_I^{eq} \qquad (5.88)$$

γ_P is a geometry factor and N_P is the density of precipitates, with $N_P = 1/r_m^3$, where r_m is the mean distance between the precipitates. The generation of self-interstitials at the precipitate shall be proportional to the number of impurity atoms entering the precipitate, and is thus proportional to the supersaturation $\Delta C_\alpha = C_\alpha^0 - C_\alpha^{eq}$. It is assumed now that the impurity diffuses via the kick-out mechanism, so one can use (5.12) to (5.15). The results are the same if one assumes the dissociative mechanism and applies the corresponding equations for the vacancy concentrations. Ignoring the spatial variation of the concentrations, the set of equations can be simplified, leading to a single differential equation for the substitutional concentration $C_s(t)$:

$$\frac{\partial C_s}{\partial t} = K_I\left\{\frac{C_s^{eq}}{C_s} - 1\right\} - K_G\left\{\frac{C_\alpha^0}{C_\alpha^{eq}} - 1\right\} \tag{5.89}$$

The short time solution for this equation is

$$C_s^0 - C_s(t) = \gamma_P N_P D_I \Delta C_\alpha t \tag{5.90}$$

The conservation of matter requires that the radius of a growing spherical precipitate be given by

$$C_\beta\{r(t) - r(0)\} = r_m\{C_\alpha(t) - C_\alpha^0\}^{1/3} = r_m\{C_s(t) - C_s^0\}^{1/3} \tag{5.91}$$

Combining (5.90) and (5.91) and assuming $r(0) = 0$, one obtains for the initial growth

$$r(t) = \left\{\frac{\gamma_P \Delta C_\alpha D_I t}{C_\beta}\right\}^{1/2} \tag{5.92}$$

The comparison with (5.87) shows that the time dependence for the self-interstitial-controlled growth is weaker, $t^{1/3}$, and determined by the diffusion coefficient of the self-interstitial D_I. Although this quantity is not well known, as was discussed in Section 5.1.1, estimations show that it may be much smaller than the diffusion coefficient for many transition metal impurities. Therefore, the precipitation process should proceed slower, as one might expect from the high diffusivity of these elements if the growth rate is diffusion-limited. Experimental results of the copper precipitation at grain boundaries in silicon indeed suggest that the growth of the precipitate is not determined by the diffusion coefficient of the copper according to (5.87), but by a slower process [98]. Since the formation of the Cu_3Si phase is accompanied by a large volume change, the injection of self-interstitials (or, alternatively, the absorption of vacancies) is necessary. Although the simplified calculation given here explains a slower growth rate, for a quantitative comparison, more experimental data and a refined calculation that, for instance, includes the particular geometrical situation at a grain boundary are necessary.

5.4.2.2 Segregation and Precipitation at Interfaces

A number of diffusion models have been developed to describe the interaction kinetics of impurities with grain boundaries and interfaces [99]. In thermodynamic equilibrium, impurities can accumulate at the interface up to the solubility limit. Considering the atomic structure and the two-dimensional nature of an interface, it is evident that the interface offers only a finite number of interface sites that can be occupied. One can assume that the dissolved impurities are distributed statistically among the available sites, at least for lower concentrations. Impurities arriving at the interface can only be trapped in empty sites and thus may not immediately enter the interface. The occupation and distribution of impurities also differs if the interaction between impurity atoms has to be taken into account, as discussed in the previous section.

In the simple case of a single impurity in a binary system with no interaction, the time dependence of the interface concentration $C_G(t)$ is given by

$$\frac{C_G(t) - C_G(0)}{C_G(\infty) - C_G(0)} = 1 - \exp(s^2)\,\text{erfc}(s) \qquad s = \frac{2(D_\alpha t)^{1/2} C_\alpha}{d C_G(\infty)} \qquad (5.93)$$

where d is the width of the interface region and D_α is the diffusion coefficient of the impurity in the volume. Since in semiconductors the amount of segregated impurities at grain boundaries is probably less than an atomic monolayer, in most cases one can assume that the interface width is of the order of the lattice constant. For short times and $C_G(0) = 0$, the equation reduces to

$$C_G(t) = 4\frac{C_\alpha}{d}\left\{\frac{D_\alpha t}{\pi}\right\}^{1/2} \qquad (5.94)$$

More general models [100] take into account the interactions between different atomic impurities in a binary or ternary system and the different values for the diffusion coefficients of the various species. The segregation kinetics becomes rather complex then and may lead to an accelerated or retarded segregation.

Once precipitates have formed at the interfaces, the kinetics can be described analogously to the precipitation in the bulk. Differences occur if one takes into account the special geometrical arrangements of precipitates that have formed in the interface plane, or the accelerated diffusion of impurities or point defects along the interface plane.

5.4.2.3 Precipitation From the Melt

The precipitation processes that have been described so far are kinetic reactions in the solid state, which occur during cooling when the solubility limits of the impurities are exceeded. For some impurities, such as carbon and oxygen in metallurgical silicon for terrestrial solar cells, the solubility limits may already be exceeded in the melt, and precipitation may occur in the melt. The model that has been developed for this particular situation [101] assumes that the precipitates nucleate at the melt-interface and grow until they are incorporated by the advancing interface (Figure 5.37). The growth of the precipitate is driven by the diffusion of the impurity in the melt. Since the interface moves during solidification and impurity atoms are continuously rejected into the melt due to the lower solubility in the solid, the solution of the diffusion problem becomes more complicated. An analytical solution of the diffusion problem has been derived for diffusion-limited growth of the precipitates. The results show that the maximum radius is determined by the ratio of the interface velocity and the melt diffusion coefficient D_m/v and is given by

$$r_{\max} = \frac{D_m}{v}\left\{\frac{\Delta C_{\alpha m}}{c_\beta}\right\} \tag{5.95}$$

with the supersaturation in the melt given by $\Delta C_{\alpha m} = C_{\alpha m}^0 - C_\alpha^{eq}$. C_α^{eq} is the solubility of the impurity at the melting temperature and C_β is the concentration of the impurity in the precipitate (β-phase).

The melt-interface mechanism has been suggested for the precipitation of carbon in silicon, since high concentrations of carbon are quite common in solar cell silicon, and precipitation reactions in the solid are slow because of the low

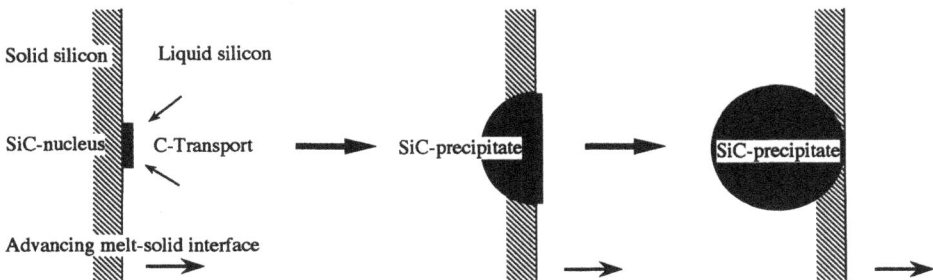

Figure 5.37 Schematic representation of the melt-interface precipitation mechanism, depicted for the case of the SiC precipitation at an advancing melt-solid interface.

diffusion coefficient for carbon in the solid. In particular, the heterogeneous precipitation'at the solidification front may be responsible for the growth of large SiC particles in carbon-rich silicon. According to (5.92), slow interface velocities and high supersaturations of the impurity can lead to large precipitates. Typical values for the directional solidification of silicon melts are interface velocities of $v = 0.001$ cm/s and supersaturations of $\Delta C_{\alpha m} \approx 100$ ppm, which yields precipitate sizes in the micron range. SiC precipitates of this size are actually observed in particular cases, as will be discussed in Chapter 6. Large precipitates are detrimental for the *pn*-junction region of a solar cell because they lead to short-circuit currents and a deterioration of the device performance.

5.4.3 Grain Boundary Contamination

The experimental investigation of impurity segregation at grain boundaries in semiconductors is difficult and hampered by the fact that the impurity concentrations in semiconductors are generally low. Direct methods such as *Auger electron spectroscopy* (AES), *electron energy loss spectroscopy* (EELS), or *secondary ion mass spectroscopy* (SIMS) are frequently not sensitive enough to detect the rather low impurity concentrations at interfaces [102]. Mapping and scanning techniques with a spatial resolution of about 0.5 to 0.8 μm (SIMS) and 300Å to 500Å (AES) give additional information about the distribution of impurities at grain boundaries and in the vicinity if the impurity concentrations are high enough.

Analytical techniques that have a higher lateral resolving power are *transmission electron microscopy* (TEM) and *field ion microscopy* (FIM), but they also require a strong enrichment of impurities at the interface ($>10^{19}$ cm^{-3}). This can be achieved, for instance, in highly doped polycrystalline silicon where segregation at grain boundaries can accumulate enough impurities to be detected by *scanning TEM* (STEM) analysis. Segregation of aluminum, arsenic [103], and phosphorus [104] has been observed in this way, but there are only a few systematic quantitative analyses that give information about the segregation behavior. High-resolution TEM, which can resolve individual atomic columns, is usually not sensitive enough to detect a few segregated impurity atoms. A further improvement can be expected from the *scanning tunneling microscopy* and *spectroscopy* (STM, STS) technique, which has already been applied to show the segregation of boron and hydrogen at a grain boundary in silicon (Figure 5.38) or the enrichment of oxygen at a grain boundary in CuInSe$_2$ [105].

Experimental techniques that have a high chemical sensitivity are radioactive tracer measurements or Mössbauer spectroscopy, but are limited to a few suitable radioisotopes such as ^{14}C, ^{3}H, ^{57}Fe, or ^{57}Co. Quantitative tracer measurements require mechanical or chemical sectioning techniques with a rather low spatial resolution. This technique has been applied to investigate the cobalt segregation

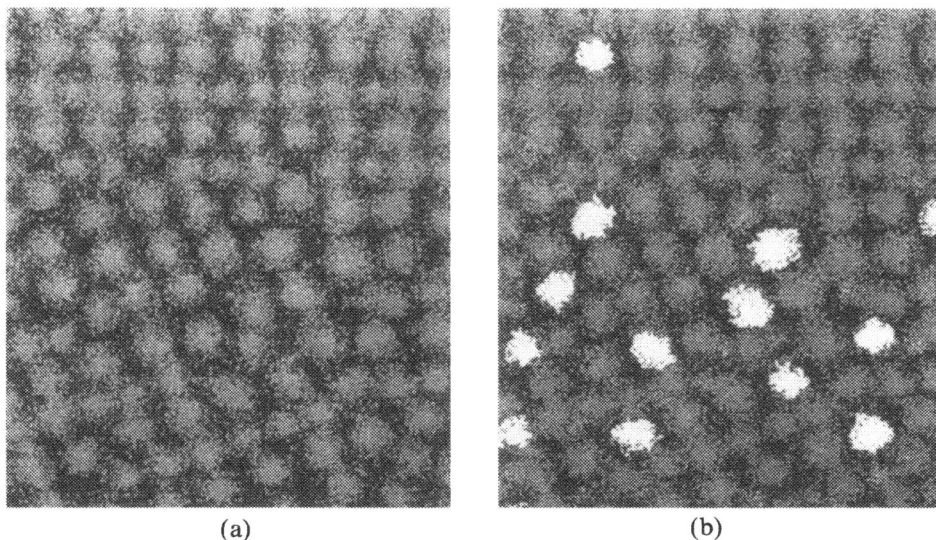

Figure 5.38 *Scanning tunneling microscope* (STM) images of boron-doped silicon, showing a grain boundary [115]: (a) conventional STM image; (b) spectroscopic STM image showing boron (red) and hydrogen (green) atoms in the grain boundary region.

at a single grain boundary in silicon [106]. Despite the low cobalt volume concentration of about 10^{13} cm^3, the quantitative analysis enabled the measurement of an enrichment of about $k_G \approx 50$ to 100 and the depletion of the cobalt concentration in the vicinity of the grain boundary (Figure 5.39(a)). The width of the depleted zone is about 50 to 100 μm and depends on the cooling rate of the specimens. Denuded zones with comparable extensions have also been observed for other impurities, usually by etching or in fourpoint probe resistivity measurements. A typical result is shown in a resistivity topograph of a silicon wafer (Figure 5.39(b)) and demonstrates the ramifications on the impurity distribution near electrically active grain boundaries. These results are particularly important for polycrystalline semiconductors because they show that the interaction of impurities with grain boundaries can change the distribution of electrically active defects in the volume within a range that is comparable to the grain size in many cases.

If impurities precipitate at interfaces or grain boundaries, they can usually be studied by TEM unless the precipitate sizes are very small. A number of investigations have shown, for instance, in silicon that transition metals precipitate at interfaces, forming metal-silicides. An example is given in Figure 5.40 for CoSi$_2$ precipitates that have formed at a grain boundary and extend into the bulk. The distribution and size of the precipitates are rather inhomogeneous and depend

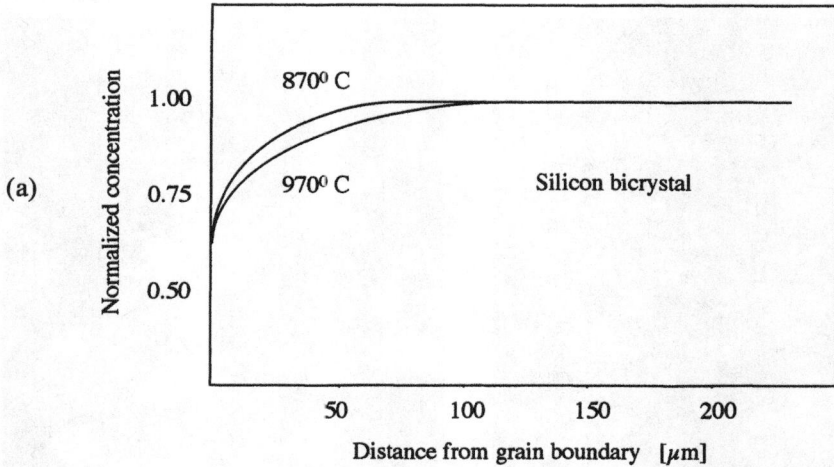

Figure 5.39 Impurity depletion in the vicinity of grain boundaries in silicon: (a) experimentally determined concentration profiles of cobalt near a grain boundary in silicon after diffusion of radioactive tracer atoms. The specimens are quenched after the diffusion of cobalt at 870 and 970°C [106]; (b) four-point probe resistivity mapping of a polycrystalline silicon wafer showing denuded zones of electrically active grain boundaries. The resistivity values are color coded and increase from blue to red (scale marker ≅ 5 mm). (Courtesy of W. Koch, Bayer Co.)

primarily on the cooling rate and to some extent on the structure of the boundary. Random boundaries tend to accumulate more precipitates, partly due to the fact that preferential precipitation occurs at dislocations and steps, which are more frequent in random grain boundaries. The quantitative analysis of the growth of the copper precipitates, for instance, indicates that the rate-limiting step for the growth is not the impurity diffusion. Since copper silicides have a large lattice mismatch with silicon that requires the participation of vacancies or silicon interstitials during precipitate formation, it has been speculated that silicon interstitial injection in the volume may be the rate-limiting step [107].

In most other cases, the analytical techniques cannot distinguish between the accumulation of an impurity due to segregation or precipitation. Therefore, at present, the number of experimental data for semiconductors are not sufficient to develop a consistent picture of the impurity-grain boundary interaction for the different impurities. Qualitatively, the interaction of doping elements (B, P, As), carbon and oxygen, and of the 3d transition elements has been observed in silicon. Whereas the doping elements seem to segregate and remain dissolved at the grain boundary even at room temperatures, the other impurities have only to be observed in the form of precipitates which indicates a low solubility at the interface.

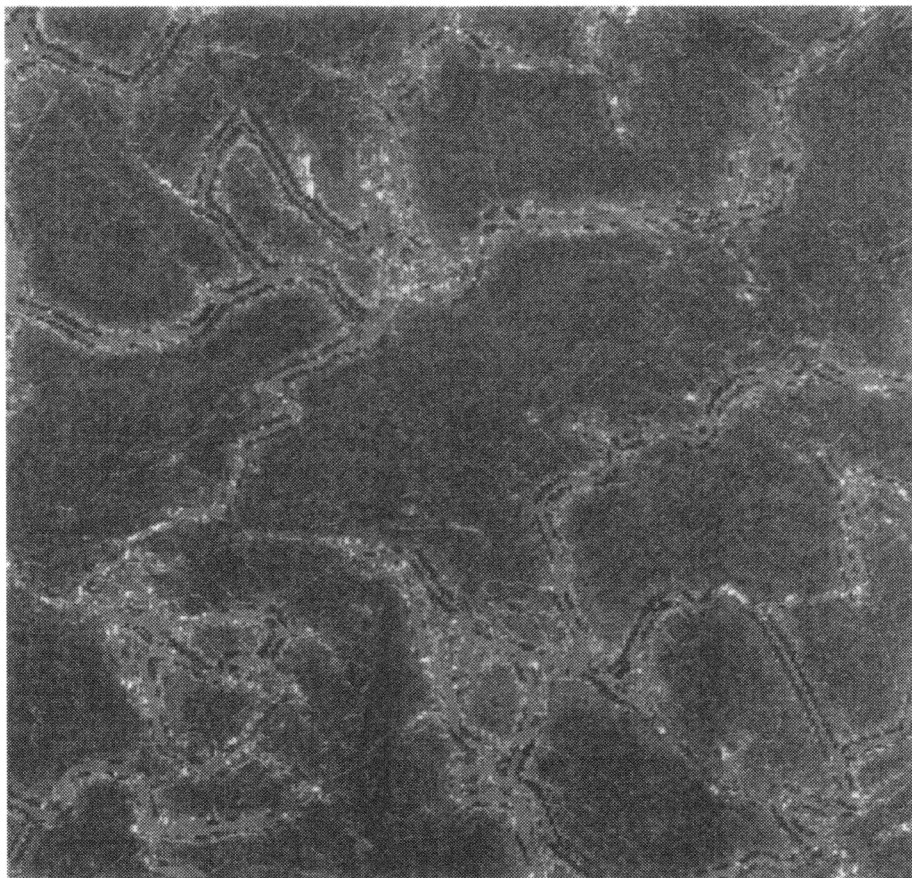

(b)

Figure 5.39 Continued.

Scattered experimental and theoretical investigations have provided some information about the interaction energies of some segregated impurities with grain boundaries in silicon. A summary for a special $\Sigma 9$ grain boundary is given in Table 5.13 for some doping elements and the 3d transition metals (Ti, V, Cr, Mn, Fe, Co, Ni, Cu) [108]. Typically, one obtains interaction energies of about 0.2 to 0.6 eV for the doping elements in silicon and an enrichment corresponding to 0.1 to 0.3 atomic layers. A theoretical calculation for the 3d elements yielded segregation energies of about $\Delta G \approx 0.6$ eV. The results indicate an attractive interaction for

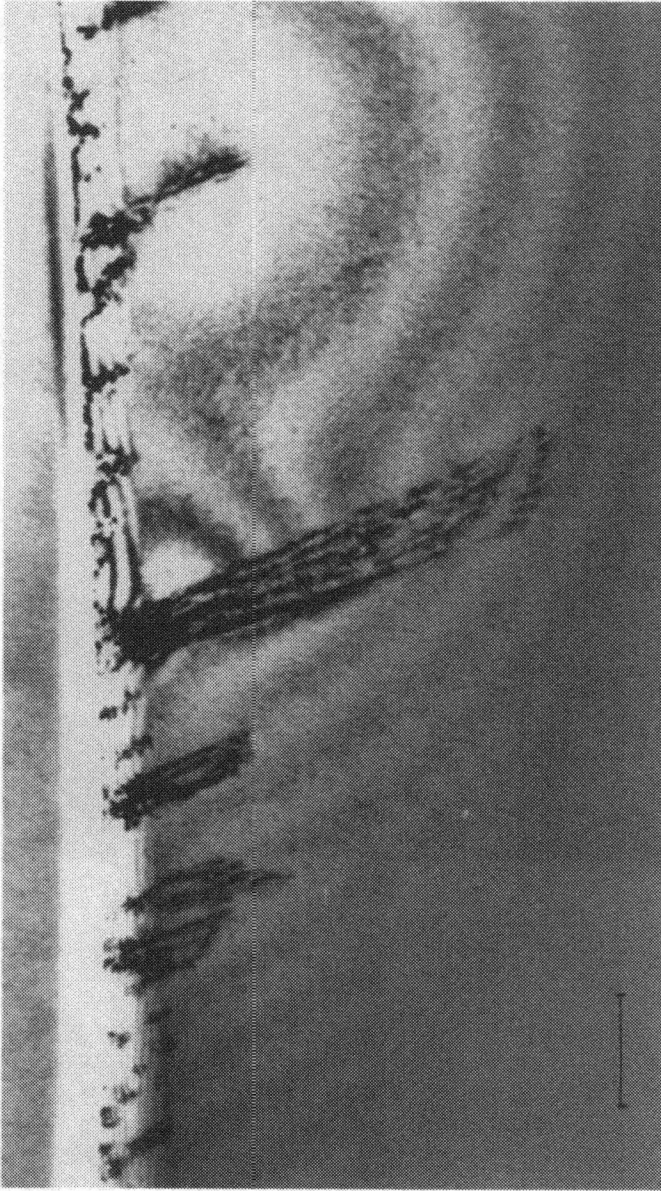

Figure 5.40 TEM image of the distribution of CoSi$_2$ precipitates at a grain boundary in silicon after diffusion of cobalt from a surface source at 1000°C (scale marker ≅ 50 nm).

Table 5.13
Experimental and Numerical Results for the Energy Difference ΔG Between
Bulk and Grain Boundary Sites (Interaction Energies) of Some Doping and
the 3d Transition Elements in Silicon

	B	P	As	3d Elements
ΔG (calc)	0.18	0.27	0.2	0.6
		0.33	0.43	
ΔG (exp)	0	0.44	0.42	

all elements which is smaller for the doping elements and almost negligible for boron, in agreement with the experimental observation.

Much less is known about the interaction of impurities and intrinsic point defects in compound semiconductors. For instance, the segregation of oxygen at grain boundaries has also been observed in GaAs and CuInSe$_2$ [109]. No experimental results are available on the interaction with intrinsic point defects, which can be electrically dominant defects in compound semiconductors.

There are a number of electrical measurements that indicate the influence of segregated impurities on the electrical properties of grain boundaries and interfaces. These techniques include resistivity measurements, admittance spectroscopy, DLTS, and LBIC or EBIC (see Section 4.2.5). In combination with analytical techniques, the segregation of oxygen, carbon, and hydrogen in silicon and various metal impurities in silicon and GaAs has been demonstrated. However, the results are mainly qualitative and have been systematically correlated to the electrical properties of the grain boundary only in a few cases.

5.4.4 Interface Diffusion

Dislocations, grain boundaries, and interfaces affect the diffusion behavior of impurities, since they offer a favorable diffusion path along the defect, which enhances the diffusion coefficient. This aspect is important particularly in polycrystalline semiconductors, where diffusion processes are standard technological steps in the solar cell fabrication and the control of the diffusion profiles are essential.

A relatively simple model has been developed by Fisher [110] and is used in almost all investigations. The grain boundary is considered a thin slab of a finite thickness $d = 2a$, perpendicular to the surface, where the diffusion coefficient D_G differs from the bulk value D_α (Figure 5.41). The concentration profile in the bulk $C_\alpha(x, y, t)$ and in the interface $C_G(y, t)$ is calculated from the standard diffusion equations (Fick's law) for two dimensions (x-axis perpendicular, y-axis parallel to the interface):

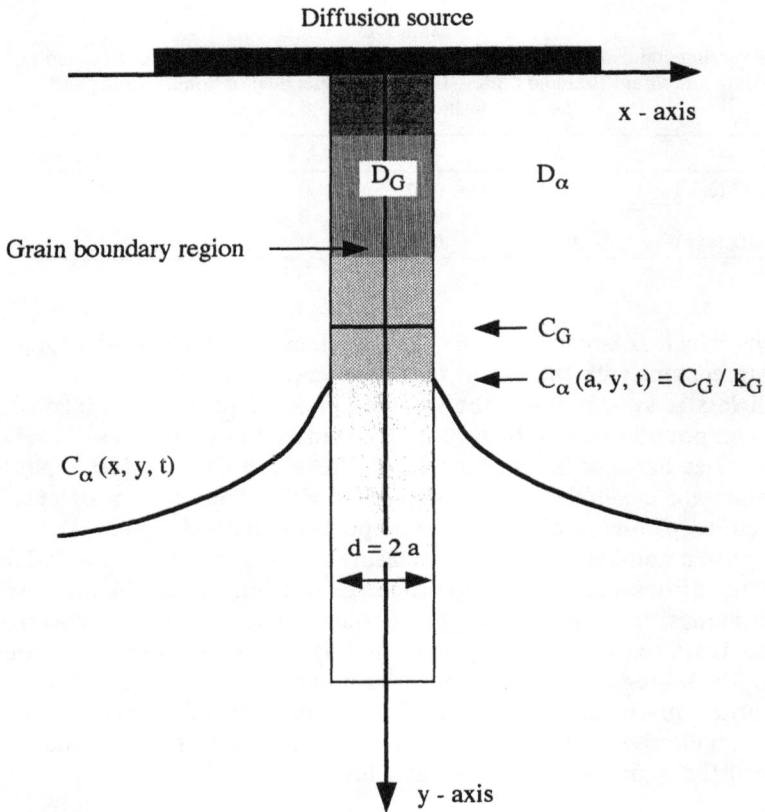

Figure 5.41 Schematic representation of the enhanced diffusion along a grain boundary perpendicular to a surface.

$$\frac{\partial C_\alpha}{\partial t} = D_\alpha \left\{ \frac{\partial^2 C_\alpha}{\partial y^2} + \frac{\partial^2 C_\alpha}{\partial x^2} \right\} \quad \text{for} \quad |x| > a \tag{5.96}$$

$$\frac{\partial C_G}{\partial t} = D_G \frac{\partial^2 C_G}{\partial y^2} + \frac{D_\alpha}{a} \frac{\partial C_\alpha}{\partial x} \quad \text{for} \quad |x| = a \tag{5.97}$$

where it is assumed that $C_G(y, t)$ is constant inside the interface layer in the x-direction. The flux of atoms between the interface and bulk regions is described by the second term in the last equation. For self-diffusion it is assumed that at the edge of the interface region (at $|x| = a$) the concentration profile is continuous; thus $C_G = C_\alpha$. Refinements of the model have assumed that for foreign atom diffusion at $|x| = a$, the ratio of bulk and interface concentration is determined by the equilibrium segregation ratio $k_G = C_G/C_\alpha$. Exact solutions of the diffusion equations have been given for two boundary conditions:

1. The constant source model, where $C_\alpha(a, 0, t) = C_\alpha^0$ [111];
2. The finite source model, where a thin layer of material at the surface is assumed, which can be exhausted and decays with time [112].

The solutions involve non-resolved integrals, which have been tabulated. A simple analytical solution for the constant source case has been obtained by Fisher under the approximation of a very large interface diffusion coefficient D_G:

$$C_\alpha(x, y, t) = C_\alpha^0 \, \text{erfc}\left\{\frac{x - a}{2(D_\alpha t)^{1/2}}\right\} \exp\left\{-\frac{y}{\pi^{1/4}(\beta D_\alpha t)^{1/2}}\right\} \tag{5.98}$$

with

$$\beta = \frac{D_G}{D_\alpha} \frac{k_G a}{(D_\alpha t)^{1/2}}$$

From this solution, it appears that the C_α/C_α^0 ratio varies with $t^{1/4}$, which is different from pure volume diffusion, where the time dependence of the average diffusion length is described by $t^{1/2}$. In many experimental techniques (for instance, sectioning parallel to the surface), an average penetration depth is measured and the concentration profile in x-direction is integrated ($c_\alpha(y, t) = \langle C_\alpha(x, y, t)\rangle$). From the exact solutions, it can be shown that for $\beta > 5$ the average concentration c_α is proportional to the exponential term in (5.98):

$$c_\alpha(y, t) \approx C_\alpha^0 \exp\left\{-\frac{y}{\pi^{1/4}(\beta D_\alpha t)^{1/2}}\right\} \tag{5.99}$$

A plot of $\ln c_\alpha$ versus y should give a straight line then. Provided that the volume diffusion coefficient D_α is known, one can determine the parameter $P_G = D_G k_G a$ from the slope of the plots. The diffusion experiment does not allow one to separate the parameters k_G, a, and D_G, so additional assumptions have to be introduced to determine the interface diffusion coefficient. Considering the atomic structure investigations of grain boundaries and interfaces, it is reasonable to assume that the interface width is of the order of the lattice constant. In fact, the width d is an arbitrary parameter without physical significance. Other models that have been developed to describe the interface diffusion, such as atomic jump models, molecular dynamic studies, or others, do not need this parameter but introduce at least one other structure-related parameter, so the problems remains the same.

It is generally observed that the temperature dependence of the diffusion parameter $P_G(T)$ follows an Arrhenius law:

$$P_G(T) = P_{G0} \exp\left\{-\frac{Q_G}{KT}\right\} \tag{5.100}$$

In the approximations for low interface concentrations given above, the activation energy is $Q_G = \Delta G + H_G^M$, where ΔG is given by (5.85) and H_G^M is the migration enthalpy of the impurity in the interface. More complex diffusion models include the participation of intrinsic defects so that, in general, the activation energy is a sum of the formation and migration energies of these grain boundary defects. Usually, in metals, values for $H_G^M \approx (0.4 \text{ to } 0.7) \times H_V^{SD}$ are found, where H_V^{SD} is the vacancy self-diffusion enthalpy.

In polycrystals, the preceding solutions are only valid if the grain boundaries are perpendicular to the surface and independent from each other. If the grain boundary is inclined by an angle φ, one measures an apparent diffusion coefficient given by $D_G \cos^2\varphi$. If the grain size is smaller than $(D_\alpha t)^{1/2}$, the diffusion profile is typical for volume diffusion with an effective diffusion coefficient $D_{eff} = f D_G$, where f is the fraction of the interface region, calculated on the assumption of the interface width d. In a polycrystalline material with random grain boundaries, one usually considers an average diffusion coefficient, which is determined empirically. The prediction of the diffusion in polycrystalline materials from measurements at single grain boundaries is further complicated, since the structure of the grain boundary has a strong influence on the diffusion coefficient. For instance, it is observed that for twin boundaries in silicon which have a rather perfect atomic structure, no accelerated diffusion occurs, while an enhanced diffusion can be measured along random large-angle grain boundaries. The microstructure of a polycrystal (the distribution and type of the grain boundaries) is thus important, which may explain that diffusion data obtained for various polycrystalline materials are often rather controversial.

In semiconductors, only a few experimental data are available for grain boundary diffusion. Because of the technical importance, these are primarily measurements of the diffusion of doping elements in polycrystalline silicon thin films that have grain sizes in the micron range. Activation energies Q_G (including the segregation energy ΔG) between 1.5 and 1.95 eV have been determined [113] for phosphorus, while antimony seems to diffuse more rapidly, with a lower activation of about 0.83 eV [114].

5.5 POLYCRYSTALLINE SEMICONDUCTORS

For photovoltaic applications, polycrystalline materials are produced by a variety of techniques. Casting and sheet growth techniques, primarily used for silicon, yield polycrystals with grain sizes in the range of about 0.1 to 10 mm, whereas thin-film techniques such as *physical vapor deposition* (PVD) and CVD typically produce grain sizes in the micron range (0.1 to 10 μm). Considering the width of the space charge region, which is typically about 0.01 to 1 μm depending on the doping level, grain boundaries in the first class of materials are electrically separated. The grain boundaries can be considered individual defects embedded in larger regions of monocrystalline material. The bulk properties are usually determined by the doping impurities and other electrically active defects, but may also be changed due to the interaction of grain boundaries with impurities. If one considers the width of the frequently observed denuded zones around grain boundaries of about 10 to 100 μm (Figure 5.39), the ramifications on the electrical properties of sheet-grown semiconductors (for instance, with grain sizes of about 500 to 1000 μm) are not negligible.

Thin-film polycrystalline semiconductors have grain sizes comparable to the space charge width of the grain boundaries, so they interact electrically. This can lead to electrical properties of the material very different from the behavior of single crystals. Contrary to the previous description of the electrical behavior, the grain boundaries cannot be treated as individual defects anymore and the situation requires a new approach to the problem, which will be discussed in this chapter. The high density of grain boundaries in thin-film semiconductors also affects other physical properties, such as light absorption, recombination, and diffusion of impurities, but they are usually explained in terms of the corresponding (averaged) properties of the individual grain boundaries [116–125].

5.5.1 Electrical Transport in Thin Films

Many of the structural and electrical aspects of grain boundaries in elemental and compound semiconductors are well understood. Despite some remaining uncertainties about the origin and the distribution of trap levels for random grain boundaries, the knowledge about the electrical activity of individual interfaces, described in Section 5.3, suffices in many cases to describe the electrical behavior of polycrystalline semiconductors as well.

In photovoltaic and other devices, the conductivity is usually measured parallel to the substrate surfaces. Since the microstructure of thin-film semiconductors often shows a columnar structure, the electrical transport is mainly perpendicular to the grain boundaries. In general, theoretical models are thus based on the assumption that grain boundaries are identical in their electrical behavior and distributed perpendicular to the current direction at an equal distance L corresponding to the average lateral grain size (Figure 5.42). Most of the conduction models are based on the grain boundary barrier model, described in Section 5.3. Refinements of the models include assumptions about additional conduction mech-

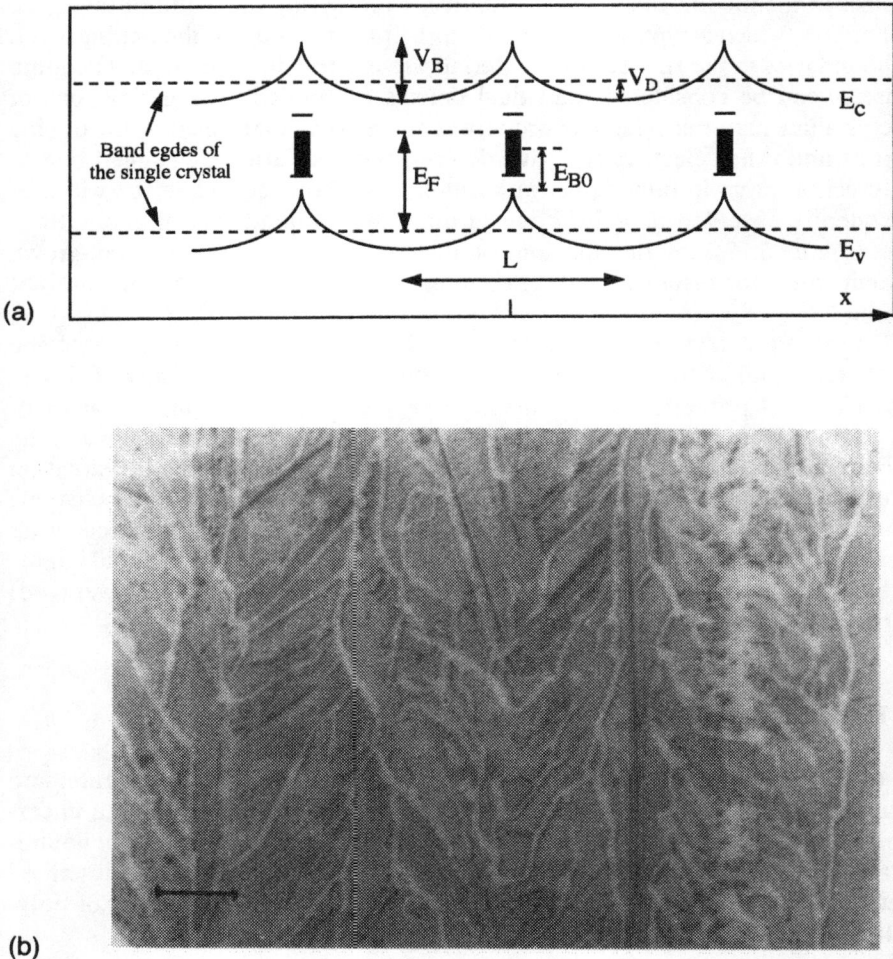

(a)

(b)

Figure 5.42 Schematic band structure of a polycrystalline semiconductor with a periodic array of negatively charged grain boundaries. E_{B0} is the occupation limit of the neutral grain boundaries (a). Etched surface of a polycrystalline silicon thin film showing grain boundaries perpendicular to the surface (b) (scale marker $\cong 50$ nm).

anisms, such as tunneling, or modifications of the shape of the grain boundary potential.

In solar cells, the applied voltages are low; therefore, only the low-voltage (zero-bias) conductance will be considered. The calculation of the conductance requires the determination of the carrier concentration n and p and the mobility μ. The main difficulty for the calculation of the carrier concentration in fine-grained thin films arises when the space charge regions and the potential barriers of adjacent grain boundaries overlap. The band structure in the entire crystal is distorted then and no undisturbed bulk region exists anymore. As for the case of the single grain boundary, the grain boundary charge Q_B, the potential barrier V_B, and the compensating space charge Q_{sc} have to be determined for a given doping concentration N. The calculation requires the determination of the Fermi level E_F, which for a single grain boundary is determined from the doping outside the depletion region (see, for instance, section 5.3.2). In polycrystals with overlapping space charges, no field-free bulk region exists and the Fermi energy is a priori not fixed anymore. In addition, the positions of all energy levels in the entire crystal depend on the distribution of the electrons and holes on grain boundary and bulk states. Thus, the application of the usual Fermi-Dirac statistics is not straightforward.

In many calculations, it is assumed for simplicity that between the grain boundaries at $x = \pm L/2$ in Figure 5.42 the Fermi energy can still be determined from the doping level, analogous to a single crystal. A consequence is that a crystal that is n-doped, for instance, remains n-type for all doping levels. Experimental results show, however, that in some semiconductors (for instance, polycrystalline germanium) an inversion of the conductivity type due to the presence of the grain boundary states occurs, which cannot then be explained.

It is therefore necessary to determine the Fermi energy from a fundamental principle; for instance, the minimization of the free energy of the entire electronic system. Since with the occupation of the grain boundary states and the build-up of the potential barriers all electronic levels are shifted, the positions of all energy levels depend on their occupation [126]. In this case, the usual Fermi-Dirac distribution function is not directly applicable for the calculation of the electron and hole concentrations. A reasonable approximation for the calculation of the occupation of the interface (n_B) or the conduction and valence band states is, however, that a Fermi-Dirac-type function can be used with different quasi-Fermi levels E_{Fi} for each system. The approximation is justified if electrons and holes within the different electronic subsystems (valence or conduction bands, for instance) reach a quasi-thermodynamic equilibrium on a shorter time scale as compared to exchange processes between electronic states of different subsystems.

The first step in further calculation is the determination of the grain boundary potential V_B as a function of the mean electron n and hole concentrations p (averaged over the entire bulk region). These quantities can be measured experimentally, which allows a direct comparison with the theoretical prediction. As with a single grain boundary, the potential barrier is calculated from Poisson's equation, however, taking into account the overlap of adjacent depletion regions. Since the potential barriers in fine-grained polycrystalline semiconductors are usually rather small, $V(x) < KT$, an analytical solution from a linearized Poisson's equation can be obtained, which for $V_B = V(0)$ and $V_D = V(L/2)$, yields

$$V_B = bz \tanh \frac{\pi a}{2} \qquad V_D = bz \frac{\pi a - \sinh(\pi a)}{a^2 \sinh(\pi a)} \qquad (5.101)$$

with

$$b = \frac{e^2 L^2}{4\epsilon\epsilon_0} \qquad a^2 = (n + p) \frac{e^2 L^2}{4\epsilon\epsilon_0 KT}$$

where L corresponds to the average grain size. The other quantities have the same meaning as defined before. $z = n_B/L$ is related to the number of occupied grain boundary states n_B:

$$n_B = \int N_B(E) f_B(E) dE - N_{Bn} \qquad (5.102)$$

and the grain boundary charge $Q_B = e n_B$. The result of the minimization of the total free energy is that the occupation function $f_B(E)$ and the quasi-Fermi energy E_{FB} for the grain boundary states are given by

$$f_B = \frac{1}{1 + \exp\left\{\frac{E - E_{FB}}{KT}\right\}} \qquad F_{FB} = E_F + V_D - V_B \qquad (5.103)$$

The occupation functions for the conduction and valence band states also have the form of the Fermi-Dirac function with a quasi-Fermi energy that is equal for electrons and holes and approximately given by

$$E_F = E_c + KT \ln \frac{n}{N_c} - z \frac{\partial V_D}{\partial n} \qquad (5.104)$$

Together with the neutrality condition

$$N_d - N_a = n - p + z \qquad (5.105)$$

where N_d and N_a are the donor and acceptor concentrations, respectively, this set of equations, (5.101) to (5.105) allows one to calculate the parameters n, p, n_B, and V_B for a given density of states distribution $N_B(E)$.

The main features of the numerical calculations are depicted in Figure 5.43, which show the total free carrier concentration $n + p$ as a function of the n-doping concentration N_d. A single grain boundary level at E_{B0}, with $N_B = 10^{12}$ cm^{-2}, a grain size of 0.1 μm, and the material parameters of germanium (band gap $E_g = 0.66$ eV) have been assumed. For a trap level in the upper half of the band gap ($E_{B0} \geq E_g/2$), the crystal remains n-type, and the average electron concentration decreases rapidly in the doping range $N_d \approx N_B/L$ with decreasing doping level. This behavior is due to the trapping of carriers at the interface and the corresponding depletion of the bulk. At high doping levels, the interface states are filled and the total carrier concentration increases almost linearly with the doping concentration. The rapid change in the carrier concentration depends on the grain size L. For larger grains, the transition region is shifted to lower doping levels and finally does not occur at all.

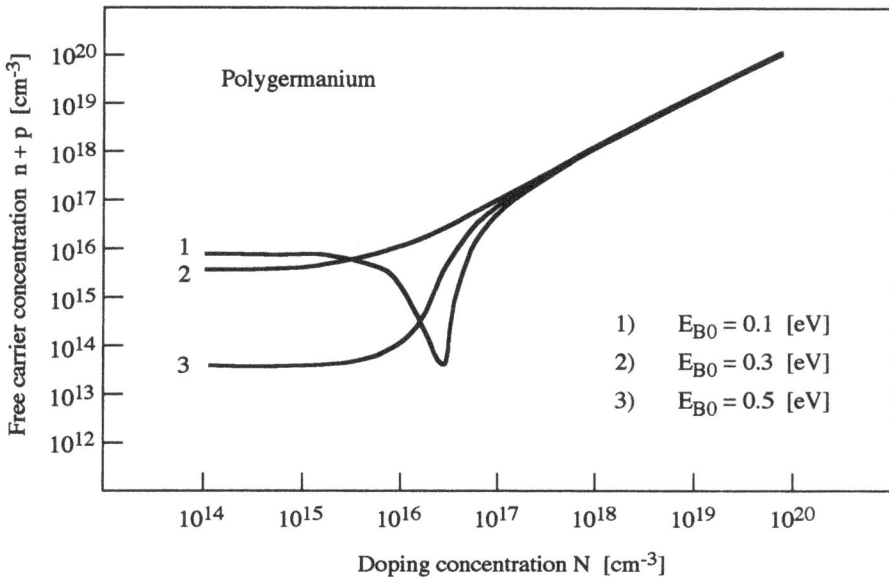

Figure 5.43 Calculated average carrier concentration $n + p$ in polycrystalline germanium as a function of the doping level. A single grain boundary level E_{B0} at different energies is assumed. Grain size $L = 0.1$ μm and $N_B = 10^{12}$ cm^{-2} [127].

A different behavior is observed for a trap level in the lower half of the gap. The electron concentration now decreases even below the intrinsic level so that holes become dominant at low doping levels and the total carrier concentration goes through a minimum. The Fermi level is pinned by the interface states in the lower half of the gap. The implication of the result is that, in this case, a polycrystalline semiconductor can become very conductive again at low doping levels due to the inversion and the increase of the hole concentration. The dependence on the position of the interface level E_{B0} is reversed for p-doping, where an inversion of the majority carrier type occurs for a trap level in the upper half of the band gap.

Figure 5.44 shows the corresponding change of the grain boundary potential, which goes through a maximum when the carrier concentrations decrease rapidly; however, it is quite small at high and low doping levels ($V_B \leq KT$). The results remain qualitatively the same for other density of states distributions $N_B(E)$ and for the same position of the occupation limit of the neutral grain boundary E_{B0}. Since $N_B(E)$ and the position of E_{B0} depend on the material, the behavior of other semiconductors can differ considerably.

The conductivity is usually described within the thermionic emission model. Since the applied voltage drops over a large number of grain boundaries for a polycrystalline semiconductor, the voltage across a single boundary is quite small. Therefore, the conductivity can be described by the zero-bias conductance given in (5.69). Although different models have been developed as well, they result essentially only in modifications of the prefactor so that one can summarize the results, for instance, for n-type material with

$$\sigma = en\mu_n = enA(T)\exp\left(-\frac{V_B}{KT}\right) \tag{5.106}$$

The mobility is determined by the prefactor $A(T)$ and the exponential dependence on the potential barrier. Since the barrier height goes through a maximum in the doping range $N_d \approx N_B/L$, the mobility shows a minimum correspondingly.

5.5.2 Conductivity in Silicon and Germanium

A typical example of the specific resistivity ρ in a polycrystalline thin film as a function of the doping concentration is given in Figure 5.45 for silicon [126] and germanium [127]. For silicon, the main feature is a strong increase of the resistance (over several orders of magnitude) in a narrow range of doping concentrations (usually around 10^{18} cm^{-3}) and a constant value for low doping levels. The increase is observed for both n- and p-type films and is related to the grain size of the films.

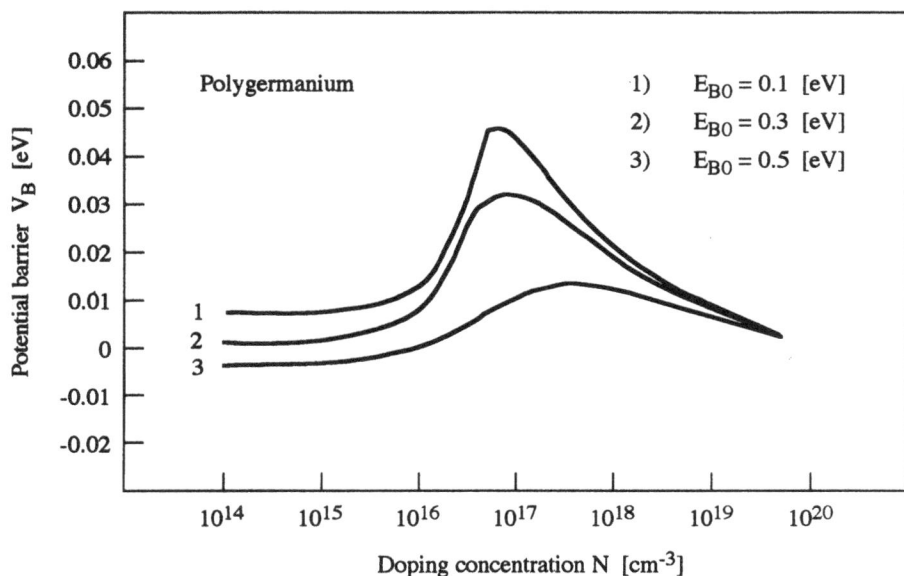

Figure 5.44 Calculated potential barrier height of a grain boundary in polycrystalline germanium. The parameters used in the calculation are the same as those in Figure 5.41 [127].

It disappears for larger grain sizes above about 100 μm, and the conductance approaches the single-crystal behavior. The temperature dependence changes from a positive to a negative slope at high doping levels and is exponential at low doping levels, with an activation energy of about 0.55 eV, as in intrinsic silicon. For germanium, a quite different behavior is observed. The specific resistance decreases for lower n-doping levels and the material becomes even more conductive compared to the single-crystal behavior. For p-doping, the resistivity changes corresponding to the single crystal, but is lower on average.

Hall measurements have confirmed that the change in the specific resistivity ρ or conductivity $\sigma = 1/\rho = e(n\mu_n + p\mu_p)$ is mainly due to a change of the majority carrier concentration (n or p) and is accompanied by an overall decrease in the Hall mobility (Figure 5.46). The results for silicon show that for low doping concentrations and small grain sizes the grains are considerably depleted from majority carriers, whereas the mobility has a minimum at about $N_d \approx 10^{18}$ cm^{-3}. In germanium, the increase in conductivity is due to an inversion in the conductivity from n- to p-type below $N_d \approx 10^{18}$ cm^{-3} and an increase of the hole concentration. These results indicate that a substantial amount of majority carriers are trapped in the grain boundary states. A similar behavior is also observed for non-stoichiometric

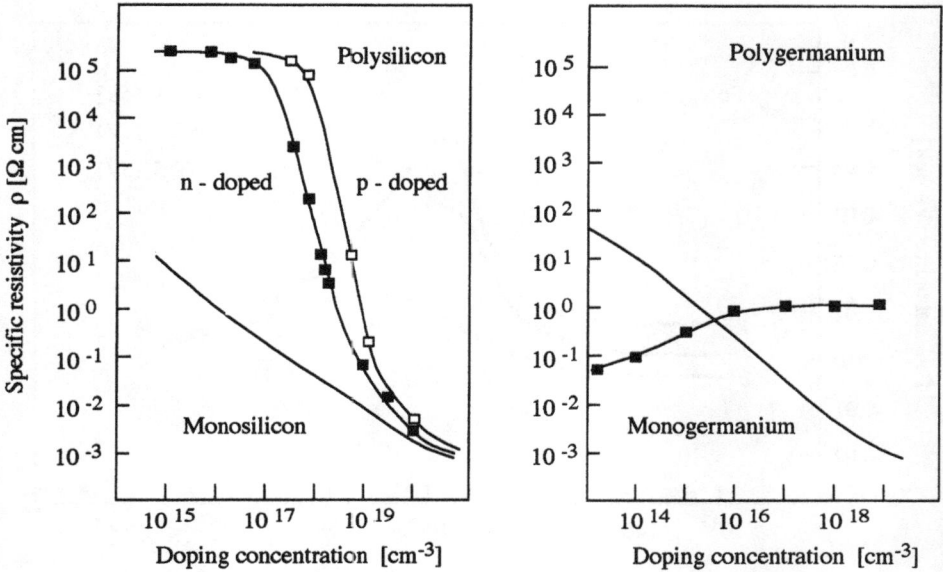

Figure 5.45 Specific resistance of $\rho = 1/\sigma$ as a function of the doping concentration N [cm^{-3}] for monocrystalline and polycrystalline silicon (open squares: B-doped, black squares: P-doped [126], and monocrystalline and polycrystalline germanium, Sb-doped [127]).

polycrystalline CuInSe$_2$, where the doping level is determined by the stoichiometry of the crystals (see Chapter 8) and the conductivity thus changes with the composition of the crystal.

The theoretical model of the previous part can quantitatively explain the behavior of polycrystalline silicon and germanium. The numerical parameters that can be obtained from the comparison with measurements agree with measurements on single grain boundaries. For instance, in silicon, admittance measurements for individual grain boundaries indicate a midgap trap level E_{B0}; therefore, a strong decrease in the carrier concentration is expected for both n- and p-doping. This result is actually observed experimentally. The mobility also shows the predicted minimum in agreement with the model [116,117,126]. The decrease of the carrier concentration and, to a lesser degree, of the mobility is responsible for the sharp rise of the resistance of polycrystalline silicon. The experimental current-voltage results can also be used to determine an averaged density of states $N_B(E)$ of the grain boundaries. The basis for the deconvolution technique is a relationship that can be derived from (5.102) by taking the derivative with respect to the quasi-Fermi energy E_{FB}:

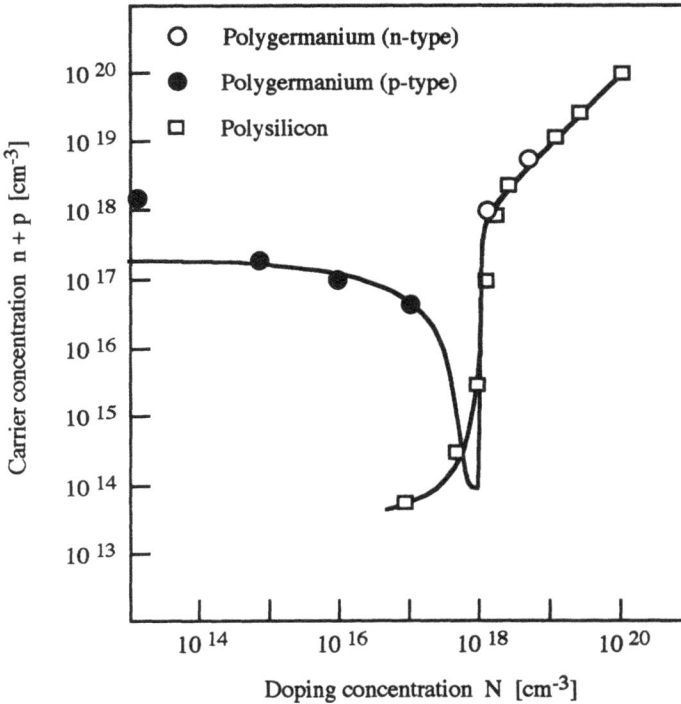

Figure 5.46 Average electron and hole concentration $\langle n + p \rangle$ for n-doped poly-germanium and silicon. Poly-germanium shows an inversion from n- to p-type conductivity below $N \approx 10^{18}$ cm^{-3} [127], whereas polysilicon remains n-type [126]. Solid lines are fitted theoretical curves from (5.101) to (5.105).

$$\frac{\partial n_B}{\partial E_{FB}} = \int N_B(E) \frac{\partial f_B}{\partial E_{FB}} dE \approx N_B(E_{FB}) \qquad (5.107)$$

E_{FB} and n_B are calculated from (5.101) to (5.104) for a wide range of doping levels. The averaged density of states is depicted in Figure 5.47 and compared with measurements of a single grain boundary. The maximum in $N_B(E)$ in the lower half of the band gap is also observed for a single low-energy grain boundary, which may indicate that it is also a characteristic feature of random boundaries. It should be noted that results from other authors [128] have also yielded different density of states distributions; therefore, a consistent picture for the grain boundary states in polysilicon has not emerged yet.

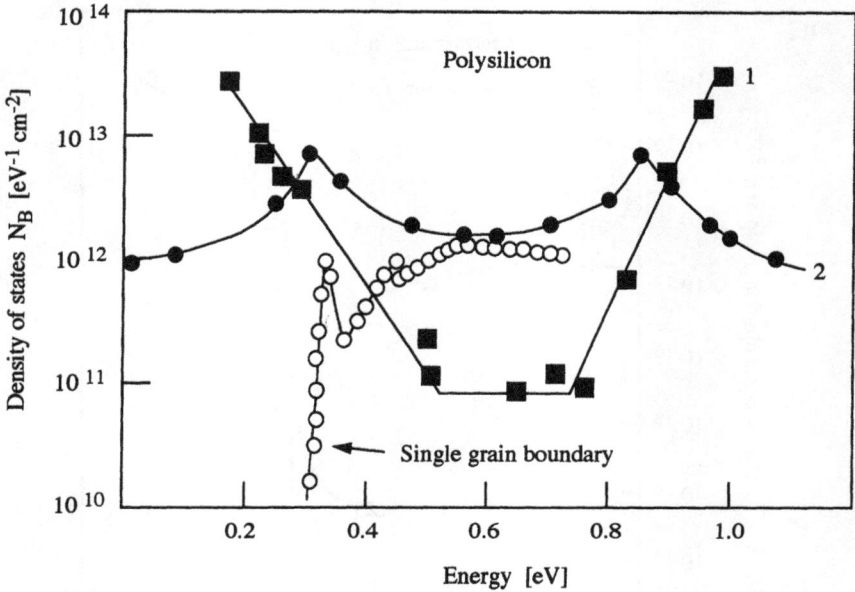

Figure 5.47 Experimental densities of states distribution for grain boundaries in various polysilicon thin films. For comparison, the results for a single grain boundary in silicon as measured by admittance spectroscopy are included (open circles) [129]. (1) from [128]; (2) from [116,126]. The energy is measured from the valence band edge.

The conductivity measurements in poly-germanium [127] are a typical example for a case where the grain boundary levels lie in the lower half of the gap. Since experimental results on single boundaries have shown a trap level at about 0.1 eV above the valence band, the total carrier concentration of poly-germanium should increase again with decreasing n-doping concentrations because of the increase of the hole concentration. This is indeed observed, as can be seen in Figure 5.46. The numerical solutions of (5.101) to (5.104) have been fitted to the data and have yielded the following parameters $E_{B0} = 0.07$ eV and $N_B = 6 \times 10^{12}$ cm^{-2}, for the grain size $L = 75$ nm. The result is in agreement with grain sizes determined by TEM.

Since the results of the conductivity and Hall effect measurements in a polycrystalline semiconductor are determined by the properties of a large number of grain boundaries, the numerical results for the level E_{B0} and the density of states N_B are average values. Only in the case of a rather uniform electrical behavior of grain boundaries with different orientations and structures can one expect to find a correlation with the measurements on individual grain boundaries. Since the results for silicon, germanium, and some compound semiconductors seem to confirm this correlation, one may assume that the same is valid for other semiconductors as well. The result is therefore of particular importance for compound semiconductors, where grain boundaries are difficult to study as a single defect, but measurements on polycrystalline materials are easier to perform or are already available. For instance, for a number of polycrystalline solar cell materials such as $CuInSe_2$ or $CdTe$, conductivity measurements have been carried out (see Chapter 8), but these results have not been discussed from the perspective of the individual grain boundary behavior yet.

REFERENCES

[1] Frank, W., U. Goesele, H. Mehrer, and A. Seeger, *Diffusion in Crystalline Solids II*, New York: Academic Press, 1984, p. 63.
[2] Seeger, A., and K. P. Chik, *Phys. Stat. Sol.*, Vol. 29, 1968, p. 455.
[3] Morehead, F., *Mater. Res. Soc. Symp. Proc.*, Vol. 104, 1988, p. 99.
[4] Watkins, G. D., and J. R. Troxell, *Phys. Rev. Lett.*, Vol. 44, 1980, p. 593.
[5] Lercy, B., *J. Appl. Phys.*, Vol. 50, 1978, p. 7996.
[6] Tan, T. Y., and U. Goesele, *J. Appl. Phys.*, Vol. 53, 1982, p. 4767.
[7] Willoughby, A. F. W., *Impurity Doping Processes in Silicon*, New York: North-Holland, 1982, p. 1.
[8] Antoniadis, D. A., and I. Moskowitz, *J. Appl. Phys.*, Vol. 53, 1982, p. 6788.
[9] Kröger, F. A., *The Chemistry of Imperfect Crystals,* Amsterdam: North Holland, 1973.
[10] *Proc. 4th Int. Conf. on Ternary and Multinary Compounds, Jpn. J. Appl. Phys.*, Vol. 19-3, 1980.
[11] Neumann, I., *Verbindungshalbleiter*, Leibzig: Akadem. Verlag, 1986, p. 392.
[12] Möller, H. J., *Solar Cells*, Vol. 31, 1991, p. 77.
[13] Groenink, J. A., and P. H. Janse, *Z. Phys. Chem. Neue Folge*, Vol. 110, 1978, p. 17.
[14] Mikkelsen, J. C., *Mater. Res. Soc. Symp. Proc.*, Vol. 59, 1986, p. 26.
[15] Newman, R., *Mater. Res. Soc. Symp. Proc.,* Vol. 104, 1988, p. 25.
[16] Corbett, J. W., J. C. Corelli, U. Desnica, and L. Snyder, *Mater. Res. Soc. Symp. Proc.*, Vol. 46, 1986, p. 243.
[17] Weinberg, I., *Current Topics in Photovoltaics*, London: Academic Press, 1990, p. 87.
[18] Ourmazd, A., A. Bourret, and W. Schröter, *J. Appl. Phys.*, Vol. 56, 1984, p. 1670.
[19] Kimerling, L. C., *Mater. Res. Soc. Symp. Proc.*, Vol. 59, 1986, p. 83.
[20] Mathiot, D., *Mater. Res. Soc. Symp. Proc.*, Vol. 104, 1988, p. 189.
[21] Bourret, A., *Microscopy of Semiconducting Materials, Inst. Phys. Conf. Ser.*, Vol. 87, 1987, p. 197.
[22] Ham, F., *J. Phys. Chem. Solids*, Vol. 6, 1958, p. 335.
[23] Leroueille, J., *Phys. Stat. Sol. (a)*, Vol. 67, 1981, p. 177.
[24] Bean, A. R., and R. C. Newman, *J. Phys. Chem. Solids*, Vol. 32, 1971, p. 1211.
[25] Newman, R. C., and J. Wakefield, *J. Phys. Chem. Solids*, Vol. 19, 1961, p. 230.
[26] Kalejs, J. P., L. A. Ladd, and U. Goesele, *Appl. Phys. Lett.*, Vol. 45, 1984, p. 269.
[27] Benton, J. L., M. T. Asom, R. Sauer, and L. Kimerling, *Mater. Res. Soc. Symp. Proc.*, Vol. 104, 1988, p. 85.
[28] Watkins, G. D., and K. L. Brower, *Phys. Rev. Lett.*, Vol. 36, 1976, p. 1329.
[29] Trombetti, J. M., and G. D. Watkins, *Appl. Phys. Lett.*, Vol. 51, 1987, p. 1103.
[30] Kolbesen, B. O., and A. Mühlbauer, *Solid State Electronics*, Vol. 25, 1982, p. 759.

[31] Föll, H., U. Goesele, and B. O. Kolbesen, *J. Crystal. Growth*, Vol. 52, 1981, p. 907.

[32] Goesele, U., *Mater. Res. Soc. Symp. Proc.*, Vol. 59, 1986, p. 419.

[33] Dietl, J., D. Helmreich, and E. Sirtl, *Crystals*, Vol. 5, Berlin: Springer, 1981, p. 43.

[34] Kalejs, J. P., and B. Chalmers, *J. Crystal. Growth*, Vol. 79, 1986, p. 487.

[35] Weber, E., *Appl. Phys.*, Vol. A 30, 1983, p. 1.

[36] Seibt, M., and K. Graff, *J. Appl. Phys.*, Vol. 63, 1988, p. 4444.

[37] Davis, J. R., A. Rohatgi, R. Hopkins, P. Blais, P. Rai-Choudhury, J. McCormick, and H. C. Mollenkopf, *IEEE Trans. Electron. Devices*, Vol. 27 (4), 1980, p. 677.

[38] Milnes, A. G., *Deep Impurities in Semiconductors*, New York: John Wiley & Sons.

[39] Pearton, S. J., J. W. Corbett, and T. S. Shi, *Appl. Phys.*, Vol. A 43, 1987, p. 153.

[40] Picreaux, S. T., F. L. Vook and H. J. Stein, *Def. and Rad. Effects in Semiconductors, Inst. Phys. Conf. Ser.*, Vol. 46, 1979, p. 31.

[41] Pankove, J. I., *Cryst. Latt. Def. Amorph. Mat.*, Vol. 11, 1985, p. 203.

[42] Pankove, J. I., D. J. Zanzucchi, C. W. Magee, and G. Lucovski, *Appl. Phys. Lett.*, Vol. 46, 1985, p. 421.

[43] Van Wieringen, A., and N. Warmholtz, *Physica*, Vol. 22, 1956, p. 849.

[44] Ichimiya, T., and A. Furuichi, *Int. J. Appl. Rad. Isotopes*, Vol. 19, 1968, p. 573.

[45] Corbett, J. W., D. Peak, S. J. Pearton, and A. Sganga, *Hydrogen in Disordered and Amorphous Sol.*, New York: Plenum Press, 1986, p. 61.

[46] Corbett. J. W., and J. C. Bourgoin, *Point Defects in Solids*, New York: Plenum Press, 1975, p. 1.

[47] Tada, H. Y., J. R. Carter, B. E. Anspaugh, and R. G. Downing, *Solar Cell Radiation Handbook, JPl Publication*, Vol. 82, 1982.

[48] Flood, D. J., and H. W. Brandhorst, *Current Topics in Photovoltaics*, Vol. 2, London: Academic Press, 1987, p. 163.

[49] Weinberg, I., C. K. Swartz, R. E. Hart, and R. L. Statler, *Proc. 19th IEEE Photovolt. Specialists Conf.*, New York: IEEE, 1987, p. 548.

[50] Weinberg, I., *Current Topics in Photovoltaics*, London: Academic Press, 1990, p. 87.

[51] Weinberg, I., C. Goradia, J. W. Stupica, and C. K. Swartz, *J. Appl. Phys.*, Vol. 60, 1986, p. 2179.

[52] Hocine, S., and D. Mathiot, *Appl. Phys. Lett.*, Vol. 53, 1988, p. 1269.

[53] Pons, B., *Physica*, Vol. 116B. 1983, p. 388.

[54] Labusch, R., and W. Schröter, *Dislocations in Solids*, F. R. N. Nabarro, ed., New York: North Holland, 1980, p. 129.

[55] Alexander, H., and P. Haasen, *Solid State Physics*, Vol. 22, New York: Academic Press, 1968, p. 27.

[56] Alexander, H., *Dislocations in Solids*, F. R. N. Nabarro, ed., London: Elsevier Science Publ., 1986, p. 114.

[57] Greuter, F., and G. Blatter, *Semicond. Sci. Technol.*, Vol. 8, 1990, p. 111.

[58] Leamy, H. J., G. E. Pike, and C. H. Seager, eds., *Grain Boundaries in Semiconductors, Mat. Res. Soc. Symp. Proc.*, Vol. 5, Amsterdam: North Holland, 1982.

[59] Rocher, A., ed., *Polycrystalline Semiconductors, J. Physique*, Vol. 43-C1, 1982.

[60] Harbeke, G., ed., *Polycrystalline Semiconductors, Springer Series Solid State Science*, Vol. 57, Berlin: Springer, 1985.

[61] Möller, H. J., H. Strunk, and H. Werner, eds., *Polycrystalline Semiconductors, Springer Proceedings in Physics*, Vol. 35, Berlin: Springer, 1989.

[62] Bollmann, W., *Crystal Defects and Crystalline Interfaces*, Berlin: Springer, 1970.

[63] Pond, R. C., *Crystal Defects and Crystalline Interfaces*, W. Bollmann, ed., Berlin: Springer, 1970, p. 27.

[64] Skrotzki, W., H. Wendt, C.B. Carter, and D.L. Kohlstedt, *Phil. Mag.*, Vol. A57, 1988, p. 383.

[65] Papon, A., M. Petit, and J. J. Bacmann, *Phil. Mag.*, Vol. A49, 1984, p. 573.
[66] Vaudin, M. D., B. Cunningham, and D. G. Ast, *Scripta Met.*, Vol. 17, 1983, p. 191.
[67] Bourret, A., and J. L. Rouviere, *Polycrystalline Semiconductors*, H. J. Möller, H. Strunk, and J. Werner, eds., *Springer Proceedings in Physics*, Vol. 35, 1989, p. 8.
[68] Möller, H. J., *Phil. Mag.*, Vol. A43, 1981, p. 1045.
[69] Papon, M., and M. Petit, *Scripta Met.*, Vol. 19, 1985, p. 391.
[70] Möller, H. J., and H. H. Singer, *Grain Boundary Structure and Related Phenomena, Suppl. to Transactions of the Japan Institute of Metals*, Vol. 27, 1986, p. 987.
[71] Wetzel, J. T., A. A. Levi, and D. A. Smith, *Mater. Res. Soc. Symp. Proc.*, Vol. 63, 1986, p. 157.
[72] Phillpot, S., and D. Wolff, *Mater. Res. Soc. Symp. Proc.*, Vol. 122, 1988, p. 103.
[73] Johnson, N. M., D. K. Biegelsen, and M. D. Moyer, *Appl. Phys. Lett.*, Vol. 40, 1982, p. 882.
[74] Pike, G. E., P. L. Gourley, and S. R. Kurtz, *Appl. Phys. Lett.*, Vol. 43, 1983, p. 939.
[75] Spencer, M. G., W. J. Schaff, and D. K. Wagner, *Appl. Phys.*, Vol. 54, 1983, p. 1429.
[76] Ziegler, E., W. Siegel, H. Blumtritt, and O. Breitenstein, *Phys. Stat. Sol. (a)*, Vol. 72, 1982, p. 593.
[77] Hermann, R., G. Nachtwei, and W. Kraak, *Phys. Stat. Sol. (a)*, Vol. 83, 1984, p. K207.
[78] Takeda, Y., *Ceram. Bull.*, Vol. 67, 1988, p. 1961.
[79] Holt, D. B., *J. Phys. Chem. Solids*, Vol. 25, 1985, p. 1385.
[80] Möller, H. J., *Solar Cells*, Vol. 31, 1991, p. 77.
[81] Blatter, G., and F. Greuter, *Polycrystalline Semiconductors, Springer Series Solid State Science*, Vol. 57, G. Harbeke, ed., Berlin: Springer, 1985, p. 118.
[82] Matare, H. F., *J. Appl. Phys.*, Vol. 56, 1984, p. 2605.
[83] Wu, X. J., V. Szkielko, and P. Haasen, *J. Physique*, Vol. 43-C1, 1982, p. 135.
[84] Petermann, G., and P. Haasen, *Polycrystalline Semiconductors, Springer Proceedings in Physics*, H. J. Möller, H. Strunk, and J. Werner, eds., Vol. 35, 1989, p. 332.
[85] Szkielko, W., and G. Petermann, *Poly-Microcrystalline and Amorphous Semiconductors*, P. Pinard, and S. Kalbitzer, Paris: Les Edition De Physique, 1985, p. 379.
[86] Broniatowski, A., *Polycrystalline Semiconductors, Springer Series Solid State Science*, Vol. 57, G. Harbeke, ed., Berlin: Springer, 1985, p. 95.
[87] Blatter, G., and F. Greuter, *Phys. Rev.*, Vol. B 33, 1986, p. 3952; *Phys. Rev.*, Vol. B 34, 1986, p. 8555.
[88] Spicer, W. E., I. Lindau, P. Skeath, and C. Y. Su, *J. Vac. Sci. Technol.*, Vol. 17, 1980, p. 1019.
[89] Ourmazd, A., and W. Schröter, *Appl. Phys. Lett.*, Vol. 45, 1984, p. 781.
[90] Guttmann, M., *Surface Science*, Vol. 53, 1975, p. 213; *Interfacial Segregation*, ASM Seminars, Metal Park, 1979, p. 251.
[91] Hondros, E. D., *J. Physique*, Vol. 36-C4, 1975, p. 117.
[92] Aucouturier, M., *Polycrystalline Semiconductors, Springer Series Solid State Science*, Vol. 57, G. Harbeke, ed., Berlin: Springer, 1985, p. 47.
[93] Bourret, A., J. Thibault, and D. M. Seidman, *J. Appl. Phys.*, Vol. 55, 1984, p. 825.
[94] Ham, F., *J. Phys. Chem. Solids*, Vol. 6, 1958, p. 335.
[95] Lifschitz, I. M., and V. V. Slyozov, *J. Phys. Chem. Solids*, Vol. 19, 1961, p. 35.
[96] Wagner, C. Z,. *Elektrochemie*, Vol. 65, 1961, p. 581.
[97] Langer, J. S., and A. J. Schwartz, *Phys. Rev.*, Vol. A21, 1980, p. 948.
[98] Jendrich, U., and H. J. Möller, *Colloque de Physique*, Vol. 51-C1, 1990, p. 197; *Mater. Res. Soc. Symp. Proc.*, Vol. 163, 1990, p. 579.
[99] Mclean, D., *Grain Boundaries in Metals*, Oxford: Clarendon Press, 1957.
[100] Tyson, W. R., *Acta Met.*, Vol. 26, 1978, p. 1471.
[101] Möller, H. J., *Mater. Res. Soc. Symp. Proc.*, Vol. 205, 1991, p. 476.
[102] Kazmerski, L. L., and P. E. Russell, *J. Physique*, Vol. 43-C1, 1982, p. 171.

224

[103] Grovenor, C. R. M., P. E. Batson, D. A. Smith, and C. Wong, *Phil. Mag.*, Vol. A 50, 1984, p. 409.
[104] Mandurah, M. M., *J. Appl. Phys.*, Vol. 51, 1980, p. 5755.
[105] Kazmerski, L. L., *Polycrystalline Semiconductors*, *Springer Proceedings in Physics*, H. J. Möller, H. Strunk, and J. Werner, Vol. 35, Berlin: Springer, 1989, p. 96.
[106] Tütken, T., W. Schröter, and H. J. Möller, *Mater. Res. Soc. Symp. Proc.*, Vol. 122, 1988, p. 109; *Polycrystalline Semiconductors*, *Springer Proceedings in Physics*, H. J. Möller, H. Strunk, and J. Werner, Vol. 35, Berlin: Springer, 1989, p. 108.
[107] Möller, H. J., U. Jendrich, L. Huang, and A. Foitzik, *Mater. Res. Soc. Symp. Proc.*, 1991, in press; *Mater. Res. Soc. Symp. Proc.*, 1990, p. 579.
[108] Matsuda-Jindo, K., *Polycrystalline Semiconductors*, *Springer Proceedings in Physics*, H. J. Möller, H. Strunk, and J. Werner, eds., Vol. 35, Berlin: Springer, 1989, p. 52.
[109] Matson, R. J., R. Noufi, K. Bachmann, and D. Cahen, *Appl. Phys. Lett.*, Vol. 50, 1987, p. 158.
[110] Fisher, J. C., *J. Appl. Phys.*, Vol. 22, 1951, p. 74.
[111] Whipple, R. T. P., *Phil. Mag.*, Vol. 45, 1954, p. 1225.
[112] Suzuoka, T., *J. Phys. Soc. Jap.*, Vol. 19, 1974, p. 839.
[113] Baumgart, H., H. S. Leamy, G. K. Celler, and L. E. Trimble, *J. Phys.*, Vol. 43-C1, 1982, p. 363.
[114] Liotard, J. L., R. Biberian, and J. Cabane, *J. Phys.*, Vol. 43-C1, 1982, p. 213.
[115] Kazmerski, L. L., A. J. Nelson, and R. G. Dhere, *J. Vac. Soc. Sci. Tech.*, Vol. A5, 1987, p. 1994; *Proc. 19th IEEE Photovolt. Specialists Conf.*, New York: IEEE, 1987, p. 944.
[116] Podbielski, R., and H. J. Möller, *Poly-Micro-Crystalline and Amorphous Semiconductors*, P. Pinard, and S. Kalbitzer, Les Edition de Physique, Les Ulis, 1984, p. 365.
[117] Greuter, F., and G. Blatter, *Semicond. Sci. Technol.*, Vol. 8, 1990, p. 111.
[118] Werner, J. E., *Polycrystalline Semiconductors*, *Springer Series Solid State Science*, Vol. 57, G. Harbeke, ed., Berlin: Springer, 1985, p. 76.
[119] Rocher, A., ed., *Polycrystalline Semiconductors*, *J. Physique*, Vol. 43-C1, 1982.
[120] Harbeke, G., ed., *Polycrystalline Semiconductors*, *Springer Series Solid State Science*, Vol. 57, Berlin: Springer, 1985.
[121] Möller, H. J., H. Strunk, and H. Werner, eds., *Polycrystalline Semiconductors*, *Springer Proceedings in Physics*, Vol. 35, Berlin: Springer, 1989.
[122] Seto, J. Y. W., *J. Appl. Phys.*, Vol. 46 (12), 1975, p. 5247.
[123] Gat, A., L. Gerzberg, and J. F. Gibbons, *Appl Phys. Lett.*, Vol. 33 (8), 1978, p. 775.
[124] Schaber, H., D. Cutter, J. Binder, and E. Obermeier, *J. Appl. Phys.*, Vol. 54 (4), 1983, p. 4633.
[125] Kamins, T. I., *J. Electrochem. Soc.: Solid State Science and Tech.*, Vol. 127, 1980, p. 686.
[126] Podbielski, R., and H. J. Möller, *Proc. 13th Int. Conf. on Defects in Semiconductors*, Metallurgical Society of Aime, L. C. Kimerling and J. M. Parsey, eds., 1985, p. 435.
[127] Möller, H. J., and V. Schlichting, *Polycrystalline Semiconductors*, *Springer Proceedings in Physics*, H. J. Möller, H. Strunk, and J. Werner, eds., Vol. 35, Berlin: Springer, 1989, p. 326.
[128] Werner, J. E., and M. Peisl, *Mat. Res. Soc. Symp. Proc.*, Vol. 46, 1985, p. 575.
[129] Petermann, G., and P. Haasen, *Polycrystalline Semiconductors*, *Springer Proceedings in Physics*, H. J. Möller, H. Strunk, and J. Werner, eds., Vol. 35, 1989, p. 332.

Chapter 6
Monocrystalline and Polycrystalline Silicon

Silicon is the most widely used solar cell material for both high-efficiency cells and low-cost applications. Since it is also the base material for microelectronic devices, silicon is the most thoroughly investigated semiconductor today. Many results are beneficial for solar cell applications; therefore, further improvements of solar cell efficiencies can be expected in the future, particularly for terrestrial solar cells. In this chapter, the two main developments of crystalline silicon will be discussed. In the first section, the most important material problems are addressed which are connected with the growth of single crystals for high-efficiency cells. Although many of these issues are also important for the device technology and have been solved from the perspective of the device performance, there are additional requirements for solar cell silicon, which shall be discussed here. Some of the most promising low-cost growth techniques will be presented in Section 6.2. Since these techniques inherently produce a material with higher concentrations of lattice defects, investigation of them and their relation to the processing is essential for the improvement of the device performance. Silicon is also used in the amorphous state as a thin-film material. Since amorphous semiconductors have quite different properties compared to crystalline semiconductors, this subject shall be discussed separately in Chapter 9.

6.1 HIGH-QUALITY SILICON

Silicon single crystals for high-efficiency solar cells are grown either by the Czochralski (Cz) or float-zone (Fz) technique, which is the same material that is also used for microelectronic devices. The growth methods and technical requirements have been described in numerous papers and shall be discussed here only briefly. In a comparison of the quality of semiconductors, these are the best materials available today. In addition, an enormous wealth of information exists about the behavior of impurities and lattice defects in silicon. These are ideal conditions for the control of all processing steps and the fabrication of high-efficiency cells. Nonetheless, even with the best silicon material, solar cells yield only about 23% efficiency, which is still well below the possible limit for silicon. Further improvements of the material properties and the design of the solar cells are necessary to reach this goal.

6.1.1 Czochralski and Float-Zone Techniques

The processing steps for the fabrication of a silicon wafer, which is the starting material for the fabrication of devices and solar cells, are depicted schematically in Figure 6.1. The primary source for the preparation of silicon is SiO_2 in quartzite rocks or sand. The conventional reduction process is the carbothermic reduction [1]

$$SiO_2 + 2C \leftrightarrow Si + 2\,CO$$

The reduction process yields rather impure silicon which is classified as either *metallurgical-grade* (MG) or *high-purity* (HP 1) silicon, depending on the purity of the starting material. The following refining procedures use the conversion into volatile compounds such as silane (SiH_4), silicon-chloride ($SiCl_4$), or silicon-chlorohydrogen compounds. The volatile compounds are purified by chemical methods such as distillation, which allows a very high degree of purification. The refined gases are reduced to elemental silicon, for instance, by pyrolysis under high purity conditions. The final product is a high-purity polycrystalline rod which is the starting material for the single crystal growth. The chemical composition of the resulting silicon material varies with the applied process, and the terminology given in Table 6.1 is used to characterize the different products according to their impurity content.

Single crystals are grown either by the Cz- or Fz-technique, and both are schematically depicted in Figure 6.2. In the Cz-technique, the crystals are grown from the melt in a crucible. A seed crystal is brought into contact with the melt and is slowly pulled out under rotation. The single crystal grows at the seed crystal with a defined diameter that is determined by the pulling speed and the temperature gradients. The crystals are doped by the addition of the doping element in the melt. Today, about 80% of the single crystals are grown by the seeded Cz-growth method.

Silicon melt is reactive with almost any element and the necessary contact with the crucible in this technique is a crucial problem for maintaining the purity of the single crystal. The number of possible crucible materials is rather limited and based on compounds such as SiO_2 (quartz), Si_3N_4, SiC, or graphite. The elements carbon, oxygen, and nitrogen, which can be incorporated into the melt, are

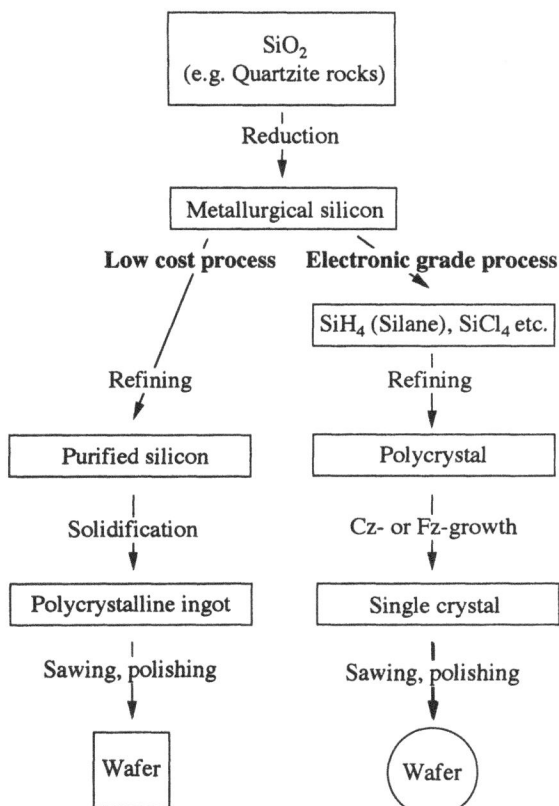

Figure 6.1 Schematic representation of the processing steps for the fabrication of single and polycrystalline silicon solar cells.

the least detrimental ones for the material properties of silicon. For instance, a typical crucible material is high-purity quartz embedded in a graphite mold. However, because of the unavoidable contamination of the melt from the wall of the crucible, the purity of Cz-crystals is limited (see Table 6.2). The maximum resistivity of an uncompensated crystal can hardly exceed 50 Ωcm. Oxygen is particularly present in higher concentrations, usually between 10^{17} and 10^{18} cm^{-3}.

The Fz-technique solves the crucible problem, with the result, however, of smaller crystal diameters and other disadvantages. In the crucible-free method, the crystal is heated locally by an rf heater and the free-floating molten zone is passed through the hanging feed crystal. If the molten zone is brought into contact with a seed crystal, the process can also be used to grow single crystals. Due to the

Table 6.1
Quality Classification of Silicon Materials and
Corresponding Concentrations of Electrically Active
Impurities

Silicon Categories	Electrically Active Impurities (at. %)
Poly hyperpure	1×10^{-7} to 1×10^{-8}
Cz-monocrystals	5×10^{-6} to 5×10^{-7}
Terrestrial solar grade	1×10^{-3} to 1×10^{-6}
High purity (HP2)	8×10^{-3} to 1×10^{-4}
High purity (HP1)	2×10^{-1} to 8×10^{-3}
Metallurgical grade	2×10^{-0} to 5×10^{-1}

molten zone, the weight of the lower part of the crystal is limited and only crystals with smaller diameters (≈ 10 cm) can be grown. The advantage is the higher purity of the crystals. Resistivities of about 300 Ωcm can be easily obtained, and by applying multipass zone melting, resistivities close to the intrinsic level can be reached. Both methods have advantages for particular applications in the device technology. As far as purity is concerned, the Fz-technique is superior. In contrast, the Cz-method is advantageous if crystals with larger diameters and high doping concentrations are required, for instance, as a substrate material for epitaxy. Such a material can hardly be produced by the Fz-method.

After the single-crystal growth, the monocrystalline rods are sliced with diamond or wire saws. About 40% to 50% of the material is lost during the sawing when diamond saws are used, which has led to the development of finer wire saws, reducing the loss of material. The final wafers have a thickness of about 300 to 400 μm, which is also the required thickness for solar cell application. After the sawing, the surfaces of the wafers are mechanically and chemically polished to remove the surface damage.

The application of silicon single crystals for solar cells requires, apart from high quality, low production costs. Improvements of the existing crystallization techniques in terms of lower costs are feasible but certainly restricted. Limitations in the growth rate and other factors do not allow substantial increase of the throughput. For the Cz-method, for instance, reductions in the cost can be achieved by bigger charges, larger diameters, and a better utilization of the expensive crucibles by a continuous melt replenishment. A further reduction of the material loss due to the slicing of the ingots is possible with new slicing techniques, which use thinner blades or wires and yield an increasing number of slices per ingot. The output of total wafer area that can be produced is estimated to be about 200 cm²/min. There are also certain limits on the cost reduction that can probably not be exceeded. Therefore, it is evident that other single-crystal growth techniques have to be developed in the future that are more suitable for the photovoltaic application.

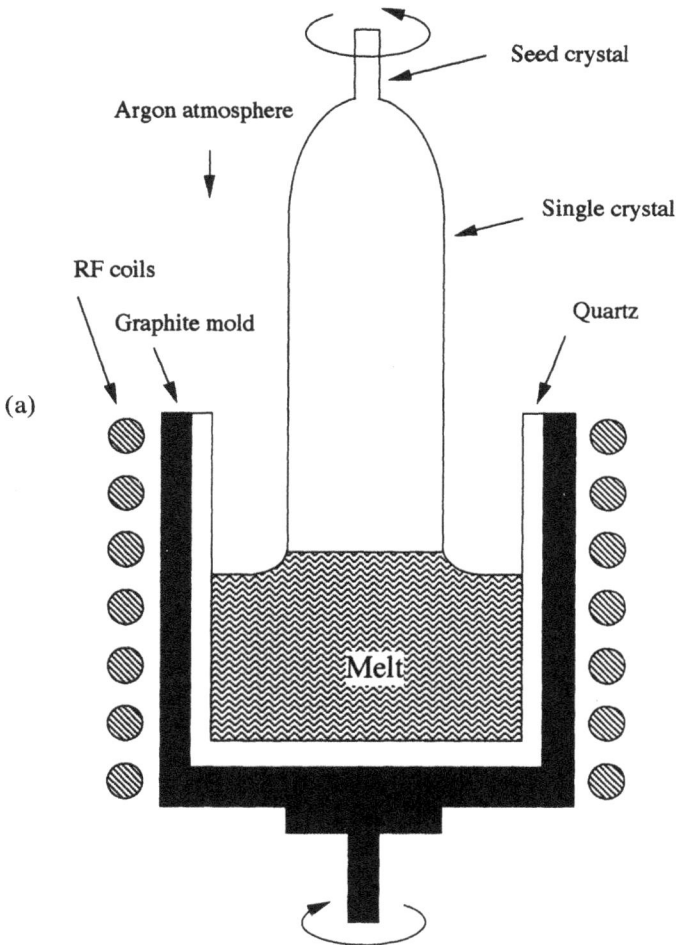

Figure 6.2 Schematic representation of the Czochralski (a) and float-zone (b) single-crystal growth techniques.

6.1.2 Material Properties

Typical material properties of Cz- and Fz-crystals are summarized in Table 6.2. For solar cell applications, the most relevant aspect is the concentration of remaining lattice defects, which reduce the lifetime and the conversion efficiency. One might expect that high-purity, dislocation-free Fz-crystals are particular suitable

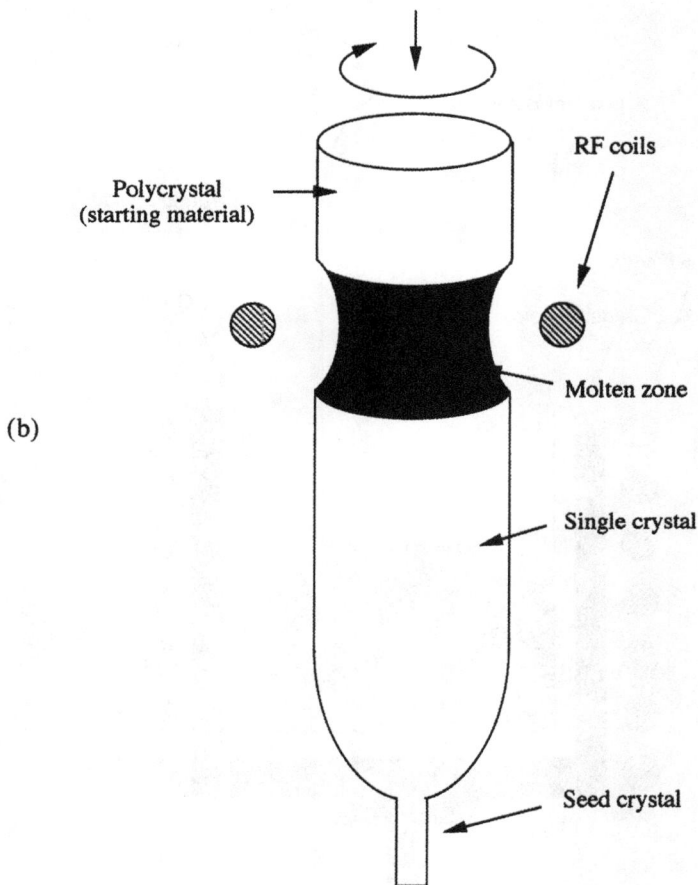

Figure 6.2 Continued.

for the fabrication of high-efficiency solar cells. It contains low concentrations of oxygen and carbon and most transition metals can be kept below 10^{11} cm^{-3}. Some metals such as iron may occur at higher concentration. Despite the relatively low defect concentrations, the lifetime is usually less than 0.5 ms, which is still much lower than the calculated limit determined by the recombination at remaining impurities or by radiative and Auger recombination. The reason is that, in high-purity crystals, a variety of mostly unknown microdefects occur which nucleate during the growth and cooling process.

Table 6.2
Comparison of Typical Material Properties of Cz- and Fz-Grown Silicon

Property	Czochralski	Float-Zone	Units
Output	200	400	cm²/min
Diameter	<20	<10	cm
Resistivity (phosphorus-doped)	1–50	1–300	Ωcm
Resistivity (boron-doped)	0.005–50	1–300	Ωcm
Oxygen	$0.1–2.0 \times 10^{18}$	$0.1–3.0 \times 10^{16}$	cm^{-3}
Carbon	$0.5–2.5 \times 10^{17}$	$0.5–5.0 \times 10^{16}$	cm^{-3}
Transition metals	$10^{10}–10^{14}$	$10^9–10^{13}$	cm^{-3}
Dislocation density	<500	<500	cm^{-3}
Minority lifetime	30–300	50–500	μs
Diffusion length	275–900	350–1100	μm
Efficiency (laboratory)	17–18	20–23	%
Efficiency (production)	13–14	17–18	%

Some of the microdefects can be made visible by etching the surface of the crystals. The types of defects typically observed in Fz-crystals are labeled A-, B-, D-, and I-microdefects [2]. Experimental investigations show that the most important growth parameters for the defect formation are the thermal gradient G and the growth speed V of the crystal. The most thoroughly investigated defects are the A- and B-swirls (because of their swirl patterns in the cross section of a crystal), which form at lower growth rates (typically below about 4 mm/min) and for larger thermal gradients, a parameter mainly determined by the crystal diameter. The condition for a swirl-free crystal is thus a large V/G ratio, and high-quality crystals today are grown under these conditions. D-defects appear at higher growth rates (>4 mm/min), whereas the I-defects are always present to some extent and less affected by these parameters.

Local lifetime measurements on Fz-grown crystals correlate partly with the distribution of some of the microdefects. The results show that the lifetime is clearly related to the formation and distribution of A- and B-swirls, whereas D- and I-defects, which are more homogeneously distributed in the crystal, have only a minor influence.

In addition to the visible defects, the lifetime is also reduced by other factors. This effect is shown in Figure 6.3 and depicts the dependence of the lifetime on the cooling rate $dT/dt = V \times G$. This degradation is a general effect and not related to the distribution and concentration of swirls and D- or I-defects in the crystal. The origin of the recombination defects and especially the dependence of the cooling rate on the lifetime are not understood yet. It has been speculated that fast cooling may produce a supersaturation of vacancy clusters, clusters enclosing impurities, or even impurity complexes which may act as recombination centers.

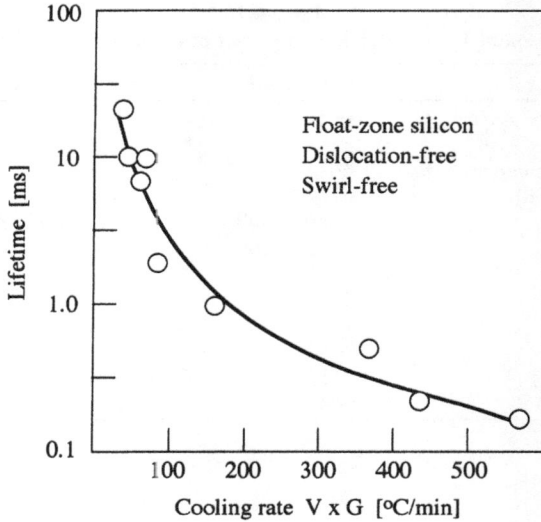

Figure 6.3 Lifetime versus cooling rate for a swirl- and dislocation-free Fz-silicon single crystal [2]. The cooling rate $V \times G$ is determined by the thermal gradient G and the growth rate V.

The identification of the detectable defects is also not yet completely clear. It has also been suggested that D-defects may be vacancy clusters [3], but they are apparently not related to the lifetime degradation. Therefore, D-defects and the nondetectable defects which may form during fast cooling must be of different origin. The I-defects are thought to be microprecipitates of trace oxygen in Fz-silicon and are even smaller compared to D-defects [4].

The more thoroughly investigated defects are the A- and B-swirls, which were discussed in Section 5.1.2. A-swirls have been identified as interstitial-type dislocation loops with diameters of about 1 to 3 μm [5], whereas the nature of the B-swirls, which have a size of about 600Å to 800Å is not quite clear yet. It has been proposed that they are carbon-self-interstitial clusters [6] that eventually transform into A-swirls. EBIC investigations show an effective recombination radius of about 40 μm for A-swirls, so that for an observed density of about 10^6 to 10^7 cm^{-3}, about 5% to 20% of the crystal volume is affected. Since A- and B-swirls can be suppressed by appropriate growth conditions, other defects limit the lifetime, but their nature and origin are basically unknown.

6.2 POLYCRYSTALLINE SILICON

The need to reduce the production costs of silicon for terrestrial solar cells has led to a re-examination of all steps of the fabrication procedure. In Figure 6.4, the

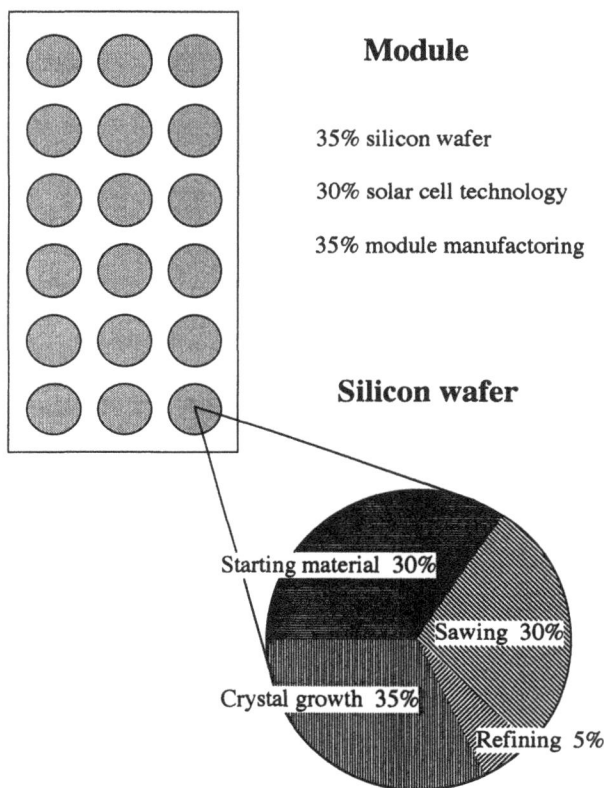

Figure 6.4 Schematic representation of the cost distribution for solar cell and module fabrication.

distribution of the costs is shown schematically for the fabrication of a silicon wafer and a module. The production of the raw silicon and the processing of the single or polycrystals take about 65% of the total wafer cost. Since this is a major fraction of the total cost, alternative processes were looked for, both for the preparation of the starting material from the primary source SiO_2 and for the following refining and crystal growth procedures.

A reduction process that seems to be cheaper compared to the conventional carbothermal reduction is, for instance, by aluminothermal reduction [1]:

$$3\ SiO_2 + 4\ Al \leftrightarrow 3\ Si + 2\ Al_2O_3$$

The raw silicon is characterized by a higher contamination with aluminum, which is also a *p*-doping element. The quality of the silicon can vary considerably depending on the purity of the quartz material and the applied reduction process. In any case, the use of MG or HP 1 silicon (see Table 6.1) requires further refining steps. There are numerous feasible routes to upgrade the purity of silicon, which depend on the desired quality of the final product (Figure 6.1). As discussed in the previous chapter, high-quality silicon for devices and high-efficiency solar cells utilize the conversion into a volatile silicon compound, such as $SiCl_4$, $SiHCl_3$, or SiH_4, which can be refined in the gas phase.

Low-cost processes also start from MG or HP 1 silicon but circumvent the expensive gas phase refining steps and use a number of different refining procedures. Impurities with a low segregation coefficient such as the transition metals can be effectively removed during solidification of the melt (e.g., by *multi-pass zone refining*) below 10^{15} cm^{-3}, whereas this procedure has little effect for the doping elements (e.g., B, P, or As) or oxygen and carbon, which have unfavorable segregation coefficients. Since the crystal growth techniques described below involve the solidification step anyway, this process can be an integral part of the refining procedure. Some impurities like iron and aluminum are concentrated at higher levels at grain boundaries in MG silicon. This offers the possibility of removing them by *acid leaching* after the MG silicon is crushed and milled to a powder size comparable to the grain size. The melt can be purified by *blowing a reactive gas* through the liquid, which can remove aluminum, calcium, carbon, or phosphorus, depending on the gas that is used. Another technique applied is the *liquid-liquid extraction*: using a molten slag that has a higher solubility for the impurities but does not mix with the silicon melt. The final combination of these refining procedures to upgrade silicon will not necessarily involve all the steps discussed and will be selected from the perspective of minimizing the costs and the purity of the silicon that is desired. The terrestrial solar-grade silicon that can be obtained with these procedures has impurity levels ranging from about 10^{16} to 10^{18} cm^{-3} for carbon and oxygen and 10^{11} to 10^{15} cm^{-3} for metal impurities (Table 6.3).

Table 6.3
Comparison of Typical Material Properties of Ingot, Ribbon, and Thin-Film Silicon

Property	Ingot	EFG	RAFT	Thin Films	Units
Output	1,600	160	20,000	100	cm^2/min
Wafer size	10 × 10	10 × 10	10 × 10	10 × 10	cm^2
Thickness	250–350	250–350	250–350	30–50	μm
Grain diameter	>10	>10	≈1	0.05–0.5	mm
Resistivity (*p*-doped)	1	1	1	1	Ωm
Oxygen	10^{16}–10^{17}	<1 × 10^{18}	<1 × 10^{18}		cm^{-3}
Carbon	10^{16}–10^{17}	<1 × 10^{17}	<1 × 10^{17}		cm^{-3}
Transition metals	10^{11}–10^{15}	<10^{15}	<10^{15}		cm^{-3}
Dislocation density	10^5–10^7	10^5–10^7	10^6–10^8		cm^{-2}
Minority lifetime τ_n	0.04–8	1–4	<0.16		μs
Diffusion length L_n	10–140	50–100	<20		μm
Efficiencies (lab)	13–17	10–17	<10	<10	%

Silicon with these impurity levels is usually used for the following low-cost crystal growth processes. There are basically two approaches, both of which result in a polycrystalline silicon material: the ingot technology and the sheet or ribbon growth techniques. They will be discussed in the following sections. A completely different approach is the use of LPE and vapor-deposition techniques (e.g., CVD), which result in either monocrystalline (epitaxial), polycrystalline, or amorphous thin films. So far, only epitaxial and amorphous films seem to be technically feasible solutions, since polycrystalline thin films have never displayed an adequate conversion efficiency. Despite these negative results, however, all possibilities offered by the thin-film technologies are currently being explored. These developments are still at a very early stage and shall be briefly discussed in Section 6.3.

6.2.1 Ingot Technology

Several ways of making mold-based ingots have been explored in the past. They are all based on directional solidification in combination with casting techniques [7]. The main problem with these techniques is the high reactivity of the molten silicon with almost every other material. Therefore, contamination of the melt from the wall of the mold can be quite severe, and the choice of the container material is thus an important aspect of ingot technology. Since the dissolution of the crucible wall in contact with the reactive silicon melt can hardly be suppressed, the material must exhibit a high degree of inertness and purity. There are only a few container materials that fulfill these requirements, such as high-density graphite, SiO_2, Si_3N_4, SiC, mullite $(SiO_2)_x(Al_2O_3)_{1-x}$, or a combination of these. Both SiO_2 and graphite are available today in high-purity grades, so contamination apart from carbon and oxygen is only a minor problem. The solidifying melts adhere to the container walls unless special precautions are taken. In addition, silicon expands during solidification because the melt has a higher density. Since the thermal expansion coefficients of the container material are usually different, either the ingot or the container will crack during the cooling process. Since it is economically unfavorable to use one-way molds, one has to protect the inside of the container with powders of Si_3N_4, SiO_2, Si, or graphite, which prohibit adhesion to the wall during solidification.

The need to cope with the crucible problem has led to different modifications of the directional solidification, which are schematically depicted in Figure 6.5. The silicon is either molten in the crucible or the melt is poured into the mold prior to the subsequent directional solidification. Usually the casting process is operated discontinuously, but continuous casting procedures are under development, since it offers a further cost reduction. The variants of the solidification technique differ in the way the heat is extracted from the melt.

In the *Bridgman-Stockbarger process* (Figure 6.5(a)), the container is moved within a fixed temperature profile which causes the melt to solidify from bottom

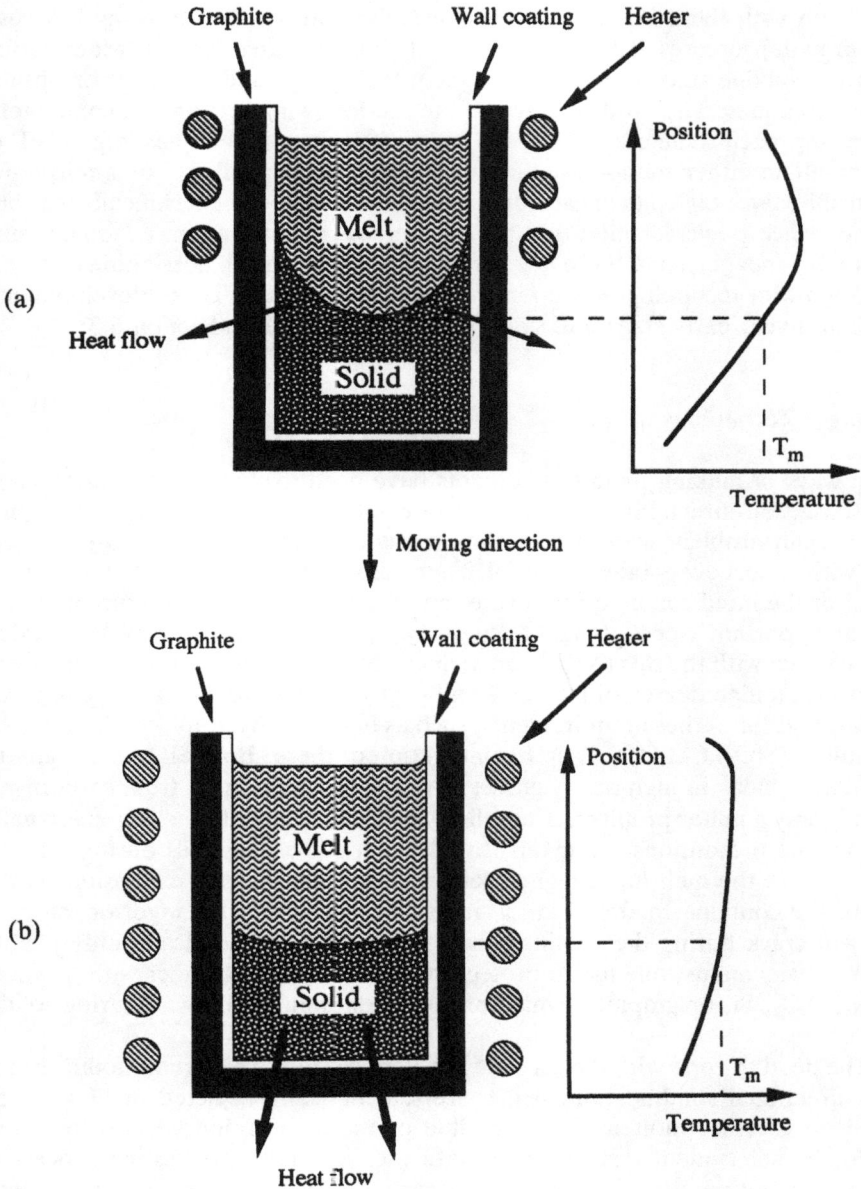

Figure 6.5 Schematic representation of the ingot growth: (a) *Bridgman-Stockbarger* method; (b) *gradient freeze* method. In (a), the crucible is moved in a fixed temperature field, while in (b) the directional solidification is achieved by control of the electric power supply to the heater.

to top [8]. The heat is mainly extracted through the crucible, and crystallization occurs spontaneously at the crucible walls first. The advantage is that silicon solidifies first at the walls and thus protects the melt from further contamination. The drawbacks are that higher thermal stresses occur in the ingot, causing higher dislocation densities, and a rather non-uniform distribution of the grain morphology and size is obtained.

In the *gradient-freeze process* (Figure 6.5(b)), both container and heating system are fixed and the temperature profile is varied by reducing the heat in a controlled way. Most of the heat is extracted over the top and bottom surfaces of the ingot. This yields a very uniform temperature profile over the cross section of the ingot. Consequently, the internal stresses are low and the overall quality of the material is improved. A rather columnar grain structure through the entire ingot and grain sizes of several centimeters can be obtained in this way. Since the crucible walls are kept at elevated temperatures for a longer period of time compared to the first technique, special precautions have to be taken to suppress the adhesion and contamination. The development of a suitable wall material is thus an important requirement for the successful application of this technique.

A particular variant which also yields a very uniform temperature profile in the ingot is the *heat exchanger method* (HEM) [9], where the reduction of the heating is combined with a simultaneous localized heat extraction through a seed crystal at the bottom of the mold by a helium gas stream. This method also results in a rather homogeneous distribution of large grains (several centimeters).

The solidification rates in the ingot technique range from 0.7 to 10 mm/min for ingots with typical cross sections between 5×5 cm^2 to about 20×20 cm^2, and ingot weights between 30 and 120 kg, depending on the techniques. The technology has to be coupled with an efficient and cost-effective slicing method. Recent developments include multiple blade and wire techniques that reduce the cost of the slicing process and the waste of silicon material. The ingots are usually cut into wafers of about 10×10 cm^2 with a thickness of 300 to 400 μm. They are mechanically and chemically etched before the wafers can be processed into solar cells. The typical grain morphology of a polycrystalline silicon wafer is shown in Figure 6.6.

6.2.2 Ribbon Growth Technologies

Despite the promising results for the efficiencies of solar cells produced from multicrystalline ingot silicon, this technology still has the disadvantage that the output (wafer area per time) is only comparable to monocrystalline cells and that the material is still too expensive. For instance, a polycrystalline silicon wafer is only about 30% cheaper than a monocrystalline Cz-grown wafer. One reason is that the technology also requires high-efficiency slicing operations, which waste

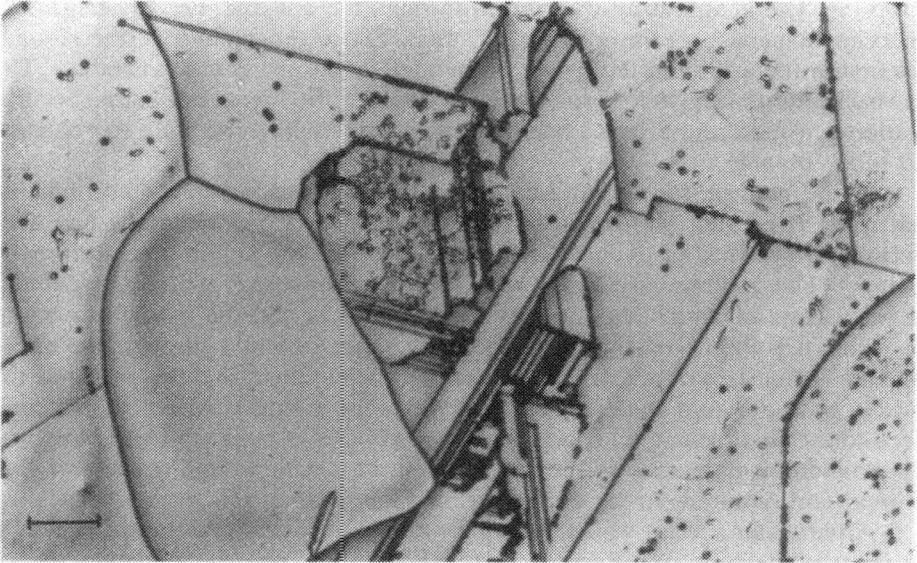

Figure 6.6 Optical micrograph of an etched polycrystalline silicon wafer showing grain boundary and dislocation distribution (scale marker $\cong$ 2 mm).

about 40% to 50% of the material. Many research efforts have therefore been directed towards the development of growth processes where the material is directly grown in the form of sheets or ribbons. A very recent effort is the development of epitaxial and polycrystalline thin films, which shall be discussed in the next section. Numerous sheet growth processes have been developed, but none of these techniques has reached a commercial level yet [7,10–12]. Only a few processes so far have been developed that yield solar cells with more than 10% efficiency under laboratory conditions. Some technical solutions that may have the potential for further improvements are summarized in Table 6.4.

Most of the sheet growth techniques require some guiding system to shape the solidifying melt. The shaping system is either in contact with the melt only during growth or is incorporated in the final ribbon. There are also techniques that avoid a guiding system completely. One way to distinguish the different techniques is by describing them according to the various degrees of interaction between the solidifying melt and the guiding system. Here another aspect shall be emphasized which is related to the direction of the sheet growth and the direction of the proceeding solidification front.

Table 6.4
Maximum Conversion Efficiencies and Output for Different Ribbon Materials

Sheet Growth Technique		η [%]	Output [cm²/min]	Laboratory
Shaping elements incorporated				
Dendritic web	D-Web	<17	10	Westinghouse, USA
Supported web	S-Web	<12	1000	Siemens, FRG
Shaping system				
Edge defined film fed growth	EFG	<15	160	Mobil Solar, USA
Ramp assisted foil technique	RAFT	<10	20000	Wacker, FRG
Ribbon growth on substrate	RGS	<10	6000	Bayer, FRG
Cast ribbon		<10	200	Hoxan, Japan
Spin cast		<12	400	Hoxan, Japan
No shaping system				
Silicon sheet from powder	SSP	<13	20	FhG-ISE, FRG

Solidification Parallel to Sheet Growth Direction

The major techniques described in this section are characterized by the existence of a molten film shaped by a die. The solid-liquid interface proceeds parallel to the sheet pulling direction; therefore, the solidification rate V determines the sheet growth velocity $V_S = V$. In general, the velocity with which the solidification front advances is determined by the thermal gradients and typically varies between 10 and 100 mm/min.

There are several techniques that differ by the degree of wetting between the shaping die and the melt. Processes with little contact between melt and shaping elements are the *dendritic web growth* (D-Web) and the *edge-supported pulling* (ESP) method. In both techniques (Figure 6.7), the contact with the shaping system is limited to the edges of the sheet. The ribbon is stabilized either by dendrites which grow simultaneously with the ribbon (D-Web) or by graphite or quartz filaments which feed through the melt to the crystallization front (ESP). A problem with these growth techniques is that they are difficult to control and require highly skilled workers. To improve the output, the ribbons (in the D-Web process) have been grown as side faces of a hexagon cut into individual ribbons after the crystal growth is completed. Although ribbons of high quality and several meters in length can be grown, the D-Web technique has never reached the production stage.

The contact between melt and shaping system is almost completely avoided in the *silicon sheet from powder* (SSP) method (Figure 6.8). In a first step, a supporting ribbon is formed by partly sintering silicon powder on a high-purity quartz or coated poly-silicon bed. The process is controlled in such a way that the contact between molten powder and quartz bed is avoided. In a second step,

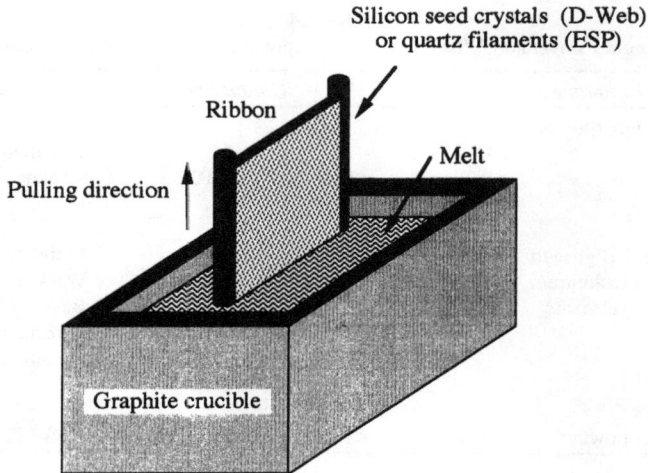

Figure 6.7 Schematic representation of the D-Web. A similar technique is the ESP method, where the ribbon supporting dendrites are replaced by quartz or graphite filaments.

substrate and pre-formed ribbon are separated, and the sheet is recrystallized by pulling a molten zone along the ribbon. Melting is achieved by focusing light with a mirror or from a high-intensity laser on the sheet. The solidification proceeds in the direction of the ribbon, and the rate is determined by the speed with which the molten zone can be pulled along the ribbon [11–13].

Better control of the ribbon growth process can be obtained with a shaping die, though this results in contamination problems. A method whose development is quite advanced is the *edge-defined film fed growth* (EFG) process, which is shown schematically in Figure 6.9. The melt is supplied by capillary forces through grooves or channels inside the die, and the meniscus is anchored at the outer edges of the shaped die. The width of the ribbons can vary between 7 and 10 cm and is determined by the width of the die and the growth speed. In order to illustrate some of the relevant issues for achieving a continuous, stable growth, the EFG method shall be discussed further. The meniscus height adjusts its position with respect to the pulling speed as long as enough melt can be supplied by the capillary action. This adjustment is important, since the process then becomes more stable against fluctuations in temperature or speed. The stability of the meniscus is determined mainly by the temperature gradient near the solid-liquid interface. Heat flow calculations for the meniscus shape as a function of the geometry of the die and the thermal conditions have established the parameters important for the process, especially the thickness of the ribbons d. The maximum growth rate is given by

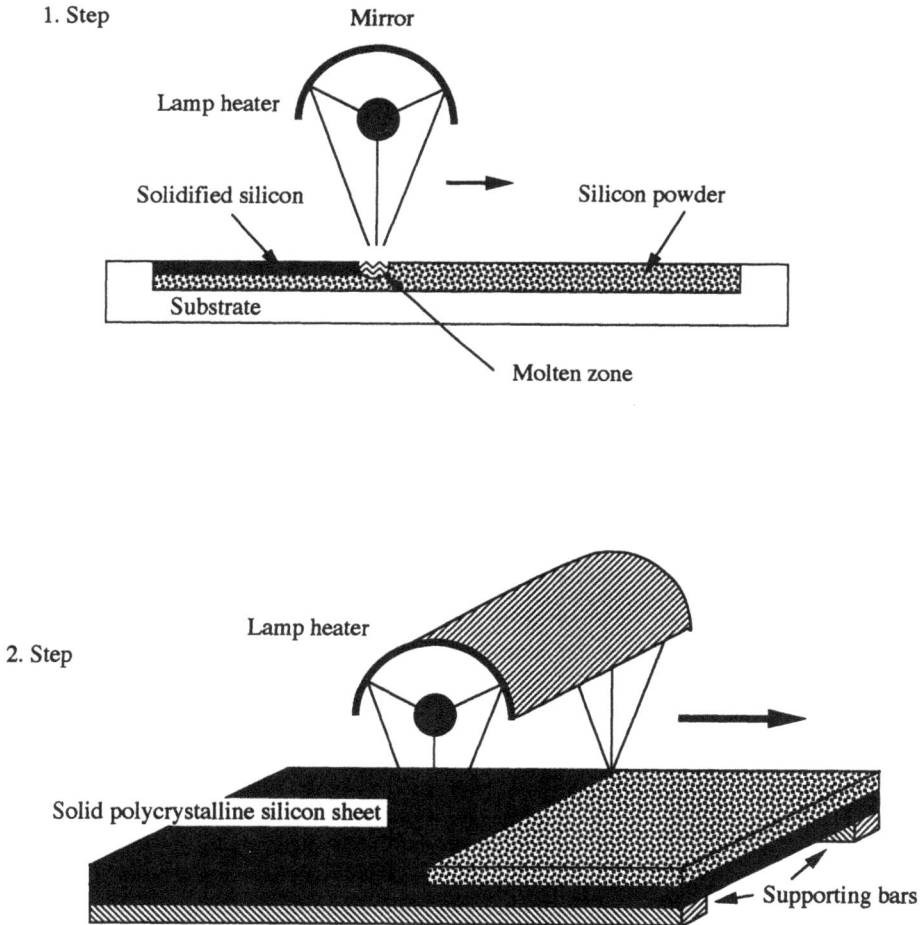

Figure 6.8 Schematic representation of the SSP method. In the first step, the powder is partly melted or sintered to yield a stable sheet which can be separated from the substrate. In the second step, the sheet is inverted and supported only at the edges to reduce the contamination from the supporting structure. The heat for melting and sintering is supplied by lamp heaters or laser light.

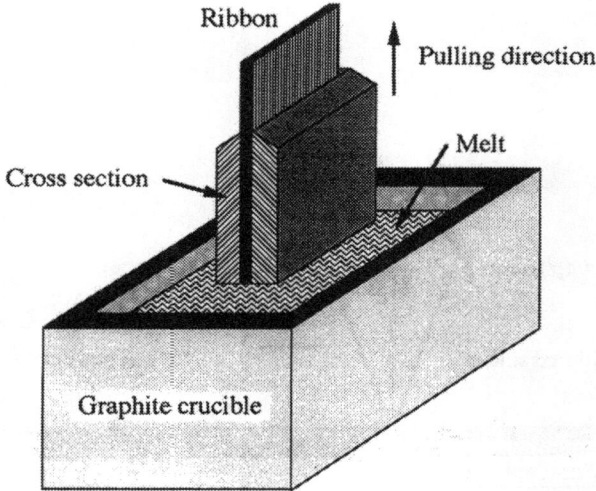

Figure 6.9 Schematic representation of the EFG technique. The melt is fed through small channels and grooves in the shaping die by capillary forces.

$$V_{max} = \left(\frac{\epsilon_m \kappa_S K T_m^5}{d\rho^2 L^2}\right)^{1/2} [1 - \beta^{-4}(1 + 4\ln\beta)]^{1/2} \qquad (6.1)$$

where ϵ_m is the emissivity of the ribbon surface, L the latent heat of fusion, ρ the density of the solid, κ_S the thermal conductivities in the solid, and $\beta = T_m/T_0$. The thermal gradient adjacent to the solid-liquid interface is given by

$$G = (T_t - T_m)\frac{d_t - d}{dh \ln(d_t/d)} \qquad (6.2)$$

where d_t is the thickness of the die top, and T_m, T_t, T_0 are the melting temperature (at the melt interface), the temperature at the die top, and the average temperature of the environment, respectively.

Figure 6.10 depicts the maximum growth rate for silicon as a function of the ribbon thickness d and $T_m \gg T_0$. Typical values for a ribbon thickness of about 430 μm and $T_0 = 1100°C$ yield a growth speed of 4.7 cm/min and temperature gradients between about 20 and 250°C/cm at the ribbon edge and surface, respectively. The growth speeds are larger by about a factor of ten than those for the ingot technology.

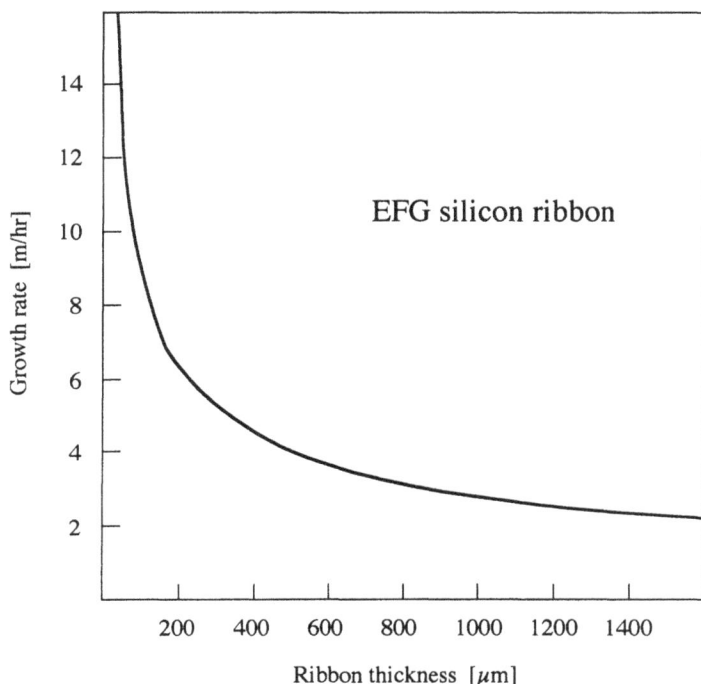

Figure 6.10 Maximum growth rate as a function of the thickness for an EFG silicon ribbon calculated from (6.1) [15].

It is evident that in the EFG technique the die material is of great importance. High-density graphite, vitreous carbon, and silicon carbide are the most promising die materials. Apart from other metal impurities, interaction with the silicon melt introduces mainly carbon into the silicon. Carbon-rich ribbons can be grown with concentrations in the solid of about 5 to 100 ppm. The enrichment of carbon at the melt-solid interface can exceed the maximum solubility limit at the melting point (50 to 100 ppm) and lead to the growth of large SiC precipitates in the melt. The precipitates disturb the meniscus and the smoothness of the resulting ribbon surface, or can even be incorporated into the solid. The problem can be alleviated with specially shaped die tops [14,15].

Solidification Perpendicular to Sheet Growth Direction

Higher ribbon growth velocities can be obtained if the sheet velocity V_S and the solidification velocity V are decoupled to some extent. A number of technical solutions have been proposed, in which the melt solidifies (almost) perpendicular to the sheet growth direction. Several techniques are characterized by a moving support ribbon on which the silicon solidifies. In the *supported web* (S-Web) or *horizontal supported web* (HSW) technique, the substrate is a graphite net which is pulled over the melt (Figure 6.11). The melt is held between the meshes of the net by capillary forces before it solidifies.

Other techniques use a solid substrate and differ basically in the way the substrate is brought into contact with the melt. The melt is cast on the moving substrate through a die which shapes the ribbon and determines the width. In the *ramp-assisted foil transport* (RAFT) method, the substrate ribbon is moved upwards, inclined by an angle of about 45°, whereas in the *ribbon growth on substrate* (RGS) technique, the substrate is moved almost horizontally (Figure 6.12). These techniques allow continuous operation, since the die can be replenished continuously and the substrate ribbon can rotate like a conveyer belt [16]. The requirements for the substrate material are similar to those of other techniques: a high degree of inertness against the silicon melt with no adhesion so that ribbon and substrate can be separated easily. Therefore, the most likely candidates are again graphite materials, usually with an appropriate coating on the surface. The contact between solidifying melt and substrate is a crucial problem in these techniques since it determines to a large extent the resulting microstructure of the ribbon and the contamination with impurities.

In all methods, the solid-liquid interface proceeds almost perpendicularly to the sheet pulling direction; therefore, the solidification rate is generally independent from the sheet growth velocity. In practice, however, there exist other constraints which usually limit the growth rate. For instance, for the RGS technique, it is

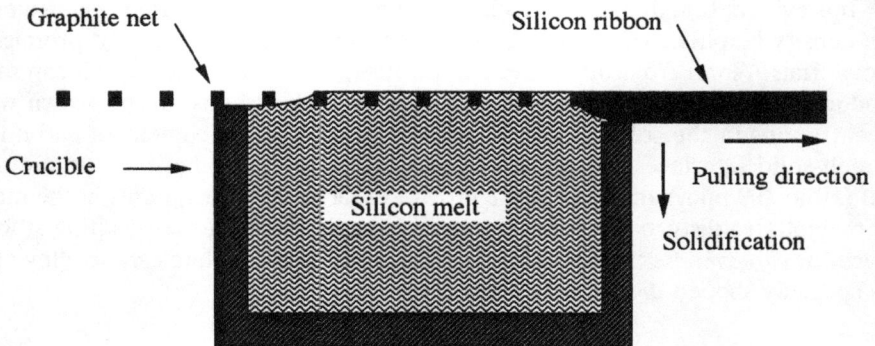

Figure 6.11 Schematic representation of the S-Web technique. A graphite net is pulled over the melt which is held between the meshes of the net by capillary forces before it solidifies. The solidification is perpendicular to the pulling direction.

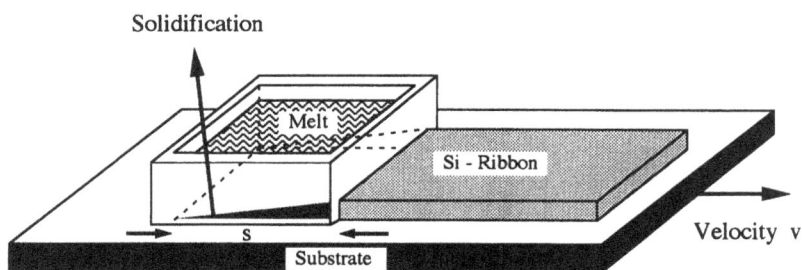

Figure 6.12 Schematic representation of the RGS technique. The solidification is perpendicular to the pulling direction.

desirable to solidify the melt inside the die; therefore, the sheet velocity V_S is to some extent determined by the geometry of the die and the solidification rate V, which depends on the temperature difference between melt and substrate temperature $\Delta T = T_m - T_s$. The sheet velocity has been calculated in a simple heat flow model yielding

$$V_S = \frac{4\alpha\kappa_S s}{\rho L d(2\kappa_S - \alpha d)} \Delta T \tag{6.3}$$

where s is the length of the growth wedge, d is the ribbon thickness, and α is the effective coefficient of heat transfer between silicon and substrate or die material. Figure 6.13 shows the ribbon thickness as a function of the sheet velocity for different ΔT. Typical sheet velocities for a ribbon thickness of about 300 μm are between 0.5 and 3 m/min, which is about a factor of 100 to 400 faster compared to the EFG technique. The growth of 10-cm-wide ribbons with a homogeneous thickness has been reported [16]. From the perspective of a high output of wafer area per time, only the RAFT and RGS techniques show a substantial improvement compared to all other methods (Table 6.4). The disadvantages so far are the problems connected with the quality of the material, which have to be solved before these techniques can offer a reasonable alternative to the existing (mainly ingot) technologies.

6.2.3 Microstructure and Electronic Properties

6.2.3.1 Ingots

Polycrystalline silicon for solar cells produced by ingot techniques has been commercially available for more than a decade. Most of the wafers are produced by

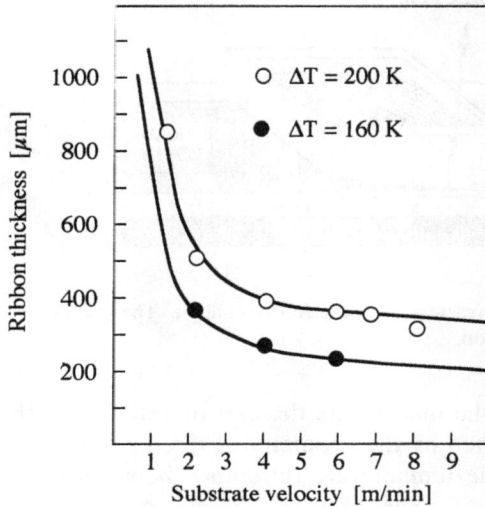

Figure 6.13 The thickness of an RGS ribbon as a function of the substrate velocity for different temperature gradients ΔT perpendicular to the substrate [16].

only a few manufacturers in Germany, France, Japan, and the U.S., and the conversion efficiencies of single solar cells made from these materials currently (1991) ranges between 13% to 13.5%. Applying design concepts developed for high-efficiency single-crystal cells, one can further improve the efficiency to about 15% to 17% (resulting, however, in increased costs).

The material properties of commercial solar silicon wafers are well characterized (see Tables 6.2 and 6.3), but the results are difficult to compare, since the conditions under which the crystals have been produced and treated after crystal growth are usually not published. It has been realized in the meantime that more systematic investigations of the defect behavior are necessary for a further improvement of the solar cell performance, and research programs in the U.S., France, and Germany have initiated support for the research. Contrary to electronic-grade silicon, an important feature of multicrystalline silicon is the inhomogeneous distribution of the electrically relevant defects in a wafer. In addition, the defects contribute differently to the performance of the entire solar cell. The characterization of a polycrystalline wafer, therefore, requires topographical measurement techniques and procedures to evaluate the electrical characteristic of an entire cell (see, for instance, Figure 6.14). These problems have been recognized now, but the development of adequate techniques and their systematic application is still in an early state.

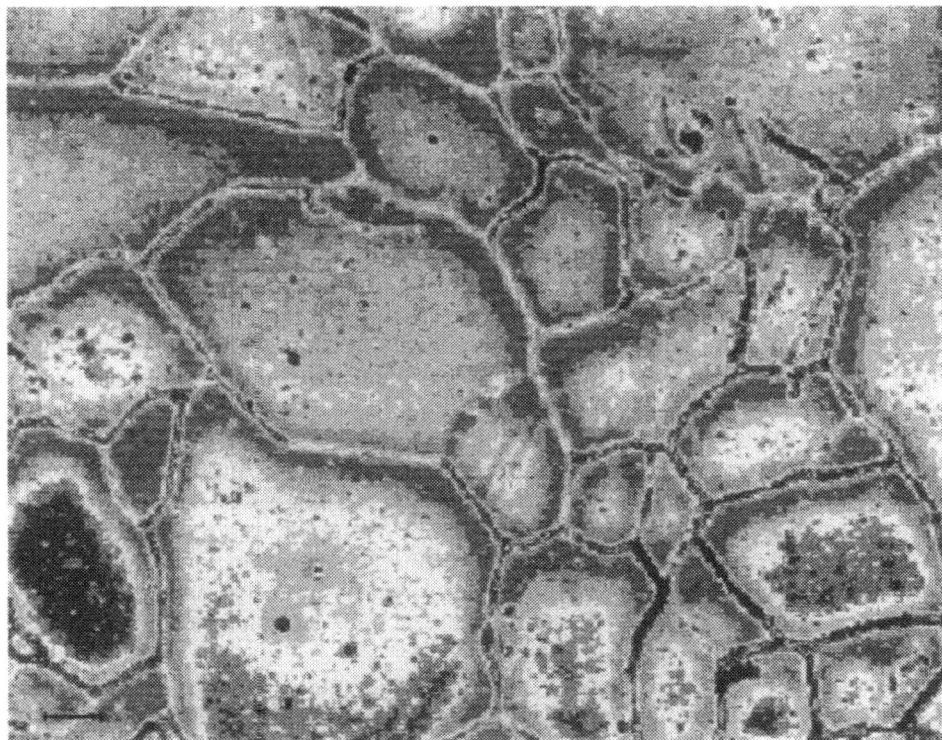

Figure 6.14 Color-coded LBIC mapping of a polycrystalline silicon wafer from cast ingot material showing the increased recombination of grain boundaries. The recombination current increases from red to blue (scale marker $\cong$ 2 mm). (Courtesy of W. Koch, Bayer Co.)

The polycrystalline ingots contain a variety of impurities at various concentration levels, depending on the purity of the starting material and the contamination through the crucible walls. Besides oxygen, carbon, and the transition metals, unconventional main group elements (such as Mg, Ca or S, Se, Te) can be incorporated. Their concentrations are usually low, so they seem to play no significant role. The exponentially decreasing solubility during cooling produces a high driving force for precipitation, complex formation with other point defects, or segregation at grain boundaries and dislocation. Nonetheless, some of the impurities also remain in solution or form only very small microdefects. This is the case for impurities with low diffusion coefficients or if precipitation is prevented during cooling. Small defects in the nanometer range are difficult to observe and therefore have been rarely analyzed. Larger microdefects (>1000Å) have been studied in cases where the concentration of a particular impurity is relatively high and precipitation occurs during cooling. Since it is now widely accepted that the recombination behavior of extended defects in silicon such as dislocations, grain boundaries, or stacking faults owe much of their electrical behavior to the decoration with impurities or precipitates, the investigation of the interaction processes are of particular importance.

Besides the doping elements, carbon and oxygen are generally the impurities with the highest concentrations between about 10^{15} and 10^{18} cm^{-3}. Both impurities have low diffusion coefficients and high nucleation energies, which hampers their precipitation. In addition, oxygen has a higher solubility, so it may not be surprising that the observation of larger SiO_2 precipitates is rarely reported for silicon ingots. The situation is different for carbon, where SiC precipitates may occur in ingots with higher carbon concentrations. They are usually distributed inhomogeneously in the ingots, with densities up to 10^4 cm^{-2} and typical diameters of about 1 μm. TEM and SEM investigations have identified them as either the cubic β-SiC or α-SiC with the hexagonal 15R phase (Figure 6.15) [17]. In ingots with very high carbon concentrations (>200 ppm) grown at low cooling rates, very large precipitates can also occur, with diameters exceeding 10 μm. These precipitates consist of SiC and a second phase, such as AlN or SiO_2N. The occurrence of the second phase is related to a high concentration of a second impurity, such as aluminum or nitrogen, or the presence of small nuclei of the compound incorporated from the crucible walls.

In view of the difficulties of forming SiC precipitates during solidification, as discussed in Section 5.1.2, it has been suggested that the SiC precipitates are already formed in the melt by the melt interface process discussed in Section 5.4.2. The numerical calculation of the growth [18] indeed shows that the observed diameters can be explained quantitatively (Figure 6.16). An important result is that the reduction of the carbon concentrations below the eutectic level (about 50 ppm = 2.5×10^{18} cm^{-3}) is sufficient to suppress the SiC carbide formation from the melt, which is in general agreement with the experimental observation.

Although it is generally possible to prevent the oxygen and carbon precipitation by reducing their concentrations, it is clear from the chemical analysis that solid silicon at room temperature is a supersaturated solution. *Infrared absorption spectroscopy* (FTIR) confirms that the impurities are dissolved, but one cannot rule out completely that very small clusters of a few atoms occur that are hardly detectable by high-resolution electron microscopy (HREM) or other methods. For instance, annealing treatments of as-grown wafers at high temperatures (1200 °C) has produced defects that can be observed by etching but usually not by other methods [7,8]. Their concentration can be reduced in the vicinity of grain boundaries (denuded zone), which indicates a segregation or precipitation at the interface. Although in some cases the SiO_2 and SiC formation after annealing has been confirmed, one must admit that the oxygen- and carbon-related defect formation

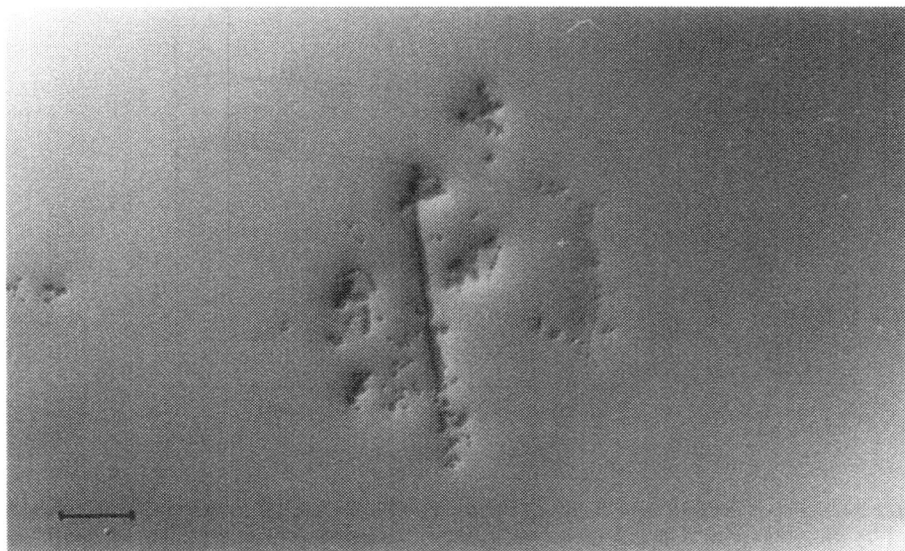

(a)

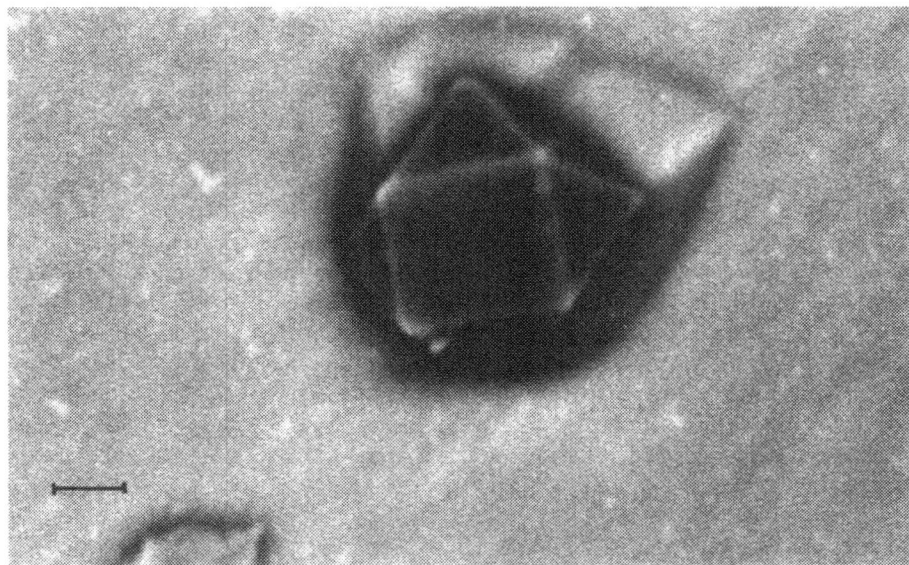

(b)

Figure 6.15 Light optical image of the surface of polycrystalline solar cell silicon showing a colony of β-SiC precipitates (a) (scale marker ≅ 10 μm). SEM image of the same area showing a single SiC precipitate (b) (scale marker ≅ 1 μm).

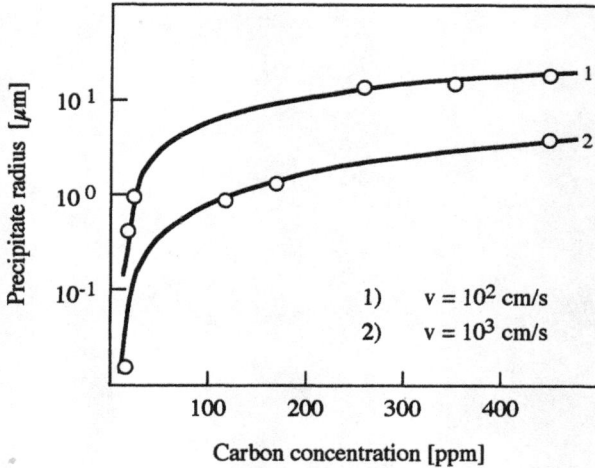

Figure 6.16 Experimental results for the size of SiC precipitates as a function of the carbon concentration in carbon-rich silicon. Solid curves are numerical calculations assuming the melt interface mechanism for the growth of the precipitates [18]. v is the solidification rate.

in as-grown and annealed multicrystalline silicon is almost unknown compared to Cz- and Fz-grown crystals (see Sections 5.1.2 and 6.1). Since annealing treatments are an integral part of the solar cell processing, it is certainly necessary to know more about the formation of these microdefects, which can be electrically active and affect the lifetime of material and the solar cell performance.

Another important class of impurities in polycrystalline silicon are the transition elements, which are present at various concentration levels, very mobile even at low temperatures, and electrically active (see Section 5.1.3). Due to their small segregation coefficients (between about 10^{-3} and 10^{-8}), their concentrations in the solid can usually be kept low (10^{11} to 10^{14} cm^{-3}) depending on the element, but it is still very important to control and reduce the impurity levels in the melt. The deep levels introduced by the transition metals and the concentration of these levels depend on whether the impurity occupies an interstitial or substitutional lattice site or whether they form complexes with other elements (e.g., cluster, precipitates). Measurements on deliberately contaminated solar cells have shown that below certain impurity levels (threshold values) the degradation of the performance of the cells is negligible. Threshold values have been determined for a variety of transition and other elements in p- and n-base solar cells (produced from Cz-grown single crystals) and are summarized in Table 6.5 [19] and Figure 6.17. Although these values may vary with the processing conditions, and thus have to be considered

Table 6.5

Experimentally Determined Threshold Concentrations of Impurities (N_{th}) Above Which the Efficiency of p- and n-Based Silicon Solar Cells Decreases (The solar cells are manufactured from monocrystalline Cz-grown material. Included are the maximum solubility and the diffusion coefficient at 1100°C of the elements if available.)

Impurity	P-base Solar Cell N_{th} [cm^{-3}]	N-base solar cell N_{th} [cm^{-3}]	Max. Solubility N_{eq}^{max} [cm^{-3}]	Diff. Coefficient D [cm^2/s]
Copper	4.1×10^{17}	8.0×10^{16}	2×10^{18}	1×10^{-4}
Nickel	5.0×10^{15}	—	7×10^{17}	4×10^{-5}
Cobalt	1.1×10^{15}	—	3×10^{16}	1×10^{-5}
Iron	2.5×10^{14}	1.8×10^{14}	2×10^{16}	4×10^{-6}
Manganese	1.8×10^{14}	9.5×10^{14}	2×10^{16}	3×10^{-6}
Chromium	1.3×10^{14}	1.2×10^{14}	1×10^{16}	2×10^{-6}
Vanadium	2.5×10^{12}	4.1×10^{13}	2×10^{16}	1×10^{-7}
Titanium	2.6×10^{12}	3.7×10^{13}	1×10^{15}	4×10^{-9}
Silver	4.0×10^{15}	—	5×10^{17}	3×10^{-9}
Palladium	2.0×10^{15}	—	3×10^{16}	—
Molybdinum	6.0×10^{11}	1.3×10^{12}	2×10^{14}	—
Niobium	1.6×10^{12}	—	5×10^{14}	—
Zirconium	3.6×10^{11}	—	5×10^{14}	—
Gold	1.0×10^{13}	—	1×10^{17}	1×10^{-6}
Tungsten	1.2×10^{12}	—	9×10^{13}	—
Tantalum	2.3×10^{11}	—	1×10^{14}	4×10^{-14}
Carbon	1×10^{18}	—	3.5×10^{17}	8×10^{-12}
Oxygen	1×10^{19}	—	2.0×10^{18}	9×10^{-11}
Aluminum	4.4×10^{15}	8.5×10^{16}	1.5×10^{21}	2×10^{-12}
Phosphorus	1.7×10^{18}	—	1.8×10^{22}	3×10^{-13}

with some caution, they give some indication about the relative influence of the transition elements.

The results indicate that the highly mobile elements (e.g., Cu, Ni, Co), which tend to precipitate or form complexes, introduce electrically active defects in concentrations much below the total impurity concentration, whereas slower diffusing elements (e.g., Ti, V, Mo), which remain dissolved to a larger extent, introduce trap levels in concentrations closer to the total impurity concentration (see also Section 5.1.3). In some cases, the ratio of electrically active centers N_T with respect to the total impurity concentration N_T/N_{tot} has been measured and the following values, 0.23 (Cr), 0.28 (V), 0.4 (Ti), and 1.0 (Mo) [19] which were obtained show indeed the almost total electrical activation of these elements. Although in general the ratio N_T/N_{tot} should depend on the thermal history (cooling conditions during

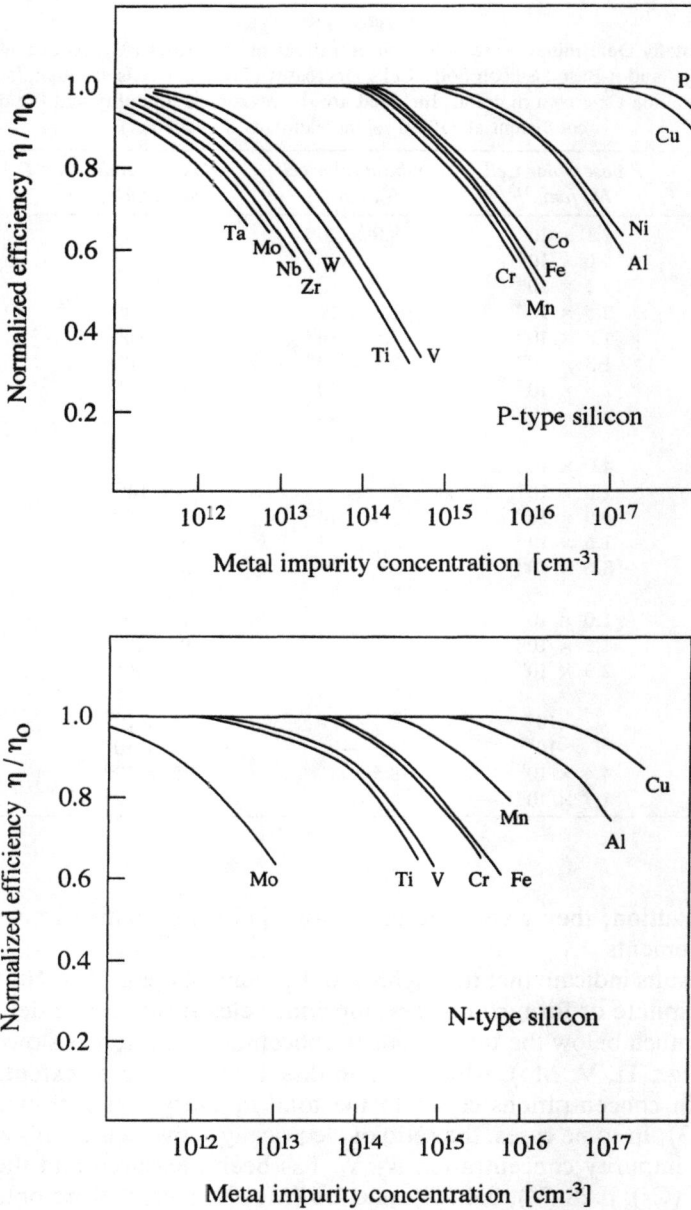

Figure 6.17 Experimental solar cell efficiencies versus impurity concentration for a 4-Ωcm p-base (a) and 1.5-Ωcm n-base device (b) [19].

solidification, annealing treatments, etc.) of the solar cells, these values can serve as guidelines for the impurity concentrations that can be tolerated without loss in the efficiency.

Rather detrimental are the slow-diffusing elements (e.g., Ti, V, Mo, Nb), which degrade the lifetime already in concentrations of about 10^{11} to 10^{12} cm^{-3} because of their high-capture cross sections. Most of these elements also have a low solubility, so their total concentrations can be kept low, but vanadium and titanium, for instance, can occur in concentrations of up to 10^{15} to 10^{16} cm^{-3}, so they can introduce rather high concentrations of trap centers. In practice, one has to monitor all of these impurities and eliminate their source if they occur in higher concentrations.

The threshold values have been determined in single-crystal cells containing only one impurity. On the other hand, a polycrystalline material contains a variety of impurities, and their interaction with other point and lattice defects can lead to new electrical defects or even a passivation of other impurity levels. For instance, it has been observed that additions of copper can reduce the recombination centers, which are introduced by titanium, vanadium, and zirconium in Cz-silicon. The limited amount of experimental data available for threshold values for polycrystalline cells indicates that the results do not differ too much from single-crystal cells, but one should be careful so far to use threshold values for polysilicon.

A further important aspect of the electrical behavior of polycrystalline ingot silicon is the influence of dislocations and grain boundaries. The ingots have typical grain sizes between 5 and 20 mm, and in some cases (e.g., HEM—silicon) even larger. This means that, in a final wafer, grain boundaries extend from the top to bottom and a columnar structure results, which reduces the influence of the grain boundaries in the solar cell. The grain boundaries are mainly large-angle grain boundaries, but small-angle boundaries also occur in areas with higher dislocation densities. In addition, coherent $\Sigma 3$ twin boundaries and very thin (<1 μm) twin lamellae are frequently observed. Dislocation densities vary in the range between 10^4 and 10^6 cm^{-2}, but local densities above 10^8 cm^2 can be observed, which indicates the presence of local thermal stresses during solidification. As discussed in Section 5.2, dislocations are very mobile in silicon above about 500 to 600°C, which facilitates their nucleation, but a high contamination with impurities during cooling can effectively reduce their mobility. In as-grown silicon, many dislocations are thus effectively pinned and the reduction of their density by further annealing treatment is only partly successful. It is therefore important to minimize thermal stresses during growth and thus keep the dislocation density low in the ingots. Dislocations have been shown to be very active recombination centers. Measurements of their influence on the lifetime, shown in Figure 4.7, indicate that the dislocation density has to be kept below about 10^5 cm^{-2} for a high-quality ingot material.

A large number of investigations on single grain boundaries in bicrystals and in polycrystalline silicon have demonstrated their electrical activity as recombination centers. The main results of these investigations have been discussed in Section 5.3. Topographic evaluations of the grain boundary activity in ingot silicon, for instance, by four-point probe resistivity mapping (e.g., Figure 5.36) or the EBIC and LBIC techniques (Figure 6.14) show a nonuniform behavior:

- Small-angle grain boundaries ($\theta < 10°$) which consist of an array of dislocations show an efficient current degradation (20% to 30%).
- Random, large-angle grain boundaries are usually also strong recombination centers, but their electrical activity can vary considerably at the same boundary and for different boundaries.
- Coherent twins which occur rather frequently and other low-energy (near-) CSL boundaries are usually not active or are only weakly active.

Despite the rather sophisticated description of the electrical activity of grain boundaries which has emerged in the past few years, the fundamental question about the origin of the electrical activity have not been completely answered. Although there is general agreement now that the recombination behavior of grain boundaries is mainly of extrinsic origin, the results in a polycrystalline ingot indicate, however, that the grain boundary structure is also important. Two examples shall be mentioned. Electrically inactive coherent twin boundaries can become electrically active if their structure contains dislocations or steps. This is depicted in the TEM images in Figure 6.18, where the structure of coherent, electrically inactive twin boundaries is compared with the structure of an electrically active boundary. It is evident that the electrical behavior is related to the presence of steps and dislocations in the boundary. Although some of the defects in the TEM image are decorated with impurity precipitates, other investigations show that, in general, no visible impurity contamination is necessary to cause electrical activity of the grain boundary. Nonetheless, one cannot rule out that even in these cases low concentrations of segregated impurities are still present in the core of the steps and dislocations and are responsible for the electrical activity.

A second case is the precipitation of copper at grain boundaries, which has been shown to enhance the recombination activity. In the TEM image in Figure 6.19(a), the formation of copper Cu_3Si precipitates at a twin boundary is shown. Due to the large lattice mismatch of the copper silicide phase with the silicon lattice, dislocations are punched out to relieve the strain. A random grain boundary with a less perfect structure can accommodate the strain more easily, and no dislocation generation is necessary during precipitation (Figure 6.19(b)). The TEM image also indicates that the distribution of the precipitates is related to structural features, such as dislocations in the boundary plane [20–24]. The morphology of precipitates at different grain boundaries thus varies considerably and one can expect that the electrical activity also changes correspondingly.

The precipitation behavior at grain boundaries is also different for other impurities. For comparison, see Figure 5.40 for the behavior of cobalt which forms

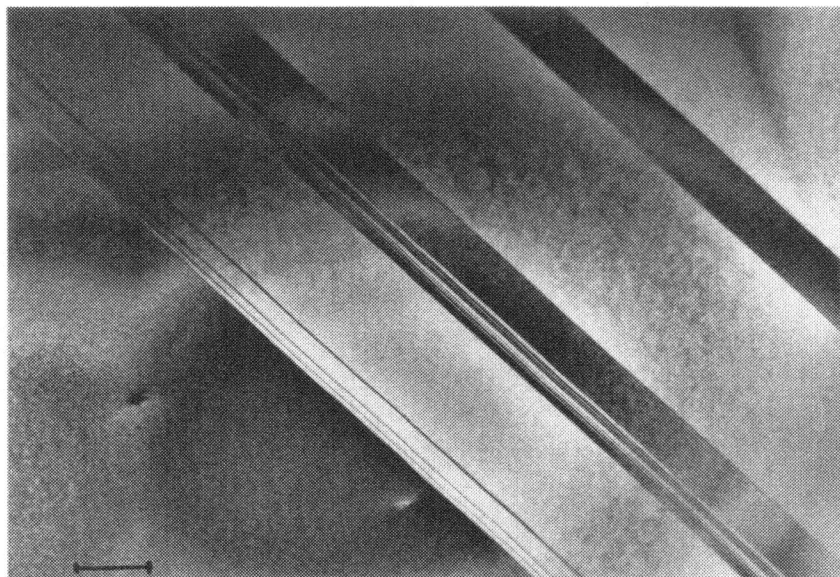

(a)

Figure 6.18 TEM images of twin boundaries in polycrystalline solar silicon: (a) twin lamellae consisting of coherent, electrically inactive twin boundaries (scale marker $\cong$ 0.5 μm); (b) HREM showing the perfect atomic structure of the twin boundaries (scale marker $\cong$ Å); (c) TEM micrograph of an electrically active twin boundary in silicon containing steps and dislocations which are partly decorated with copper-silicide precipitates (scale marker $\cong$ 0.1 μm).

$CoSi_2$ precipitates. The precipitates nucleate at the boundary plane and extend into the bulk. Since $CoSi_2$ has a good lattice match with the (111) planes in silicon, it is energetically favorable to form platelike precipitates along these planes. Other elements such as nickel have only a weak tendency to precipitate at grain boundaries, and the precipitates are more or less uniformly distributed in the crystal (Figure 6.20).

A difficulty for the experimental investigation of electrically active grain boundaries by TEM is that as-grown polycrystalline ingots with low impurity concentrations rarely show precipitates at grain boundaries. Only in cases of a rather high concentration of impurities do precipitates begin to nucleate at grain boundaries as shown in Figures 6.18 to 6.19. Therefore, one has to assume that the electrically active impurities are rather dissolved in the grain boundary. Because of the low concentrations, it is experimentally almost impossible to identify the impurities that cause the electrical activity.

(b)

Figure 6.18 Continued.

Although the resistivity topographies show an enhanced resistance for a few grain boundaries only, EBIC and LBIC images indicate that almost all grain boundaries show an enhanced recombination. This indicates that most grain boundaries introduce trap levels but remain neutral or only weakly charged, so they can hardly be detected by local resistivity measurements. A strong indication of the interaction of impurities with grain boundaries is the observation that in some cases the lifetime is not only enhanced at the grain boundary but also decreased in the vicinity. This is demonstrated by an LBIC scan across a grain boundary, shown in Figure 6.21. The change in lifetime is due to the enrichment of impurities at the grain boundary and a corresponding reduction in the vicinity. The decrease of the impurity concentration near grain boundaries has been measured quantitatively for the case of cobalt (Figure 5.39) and some other transition elements, and is also in agreement

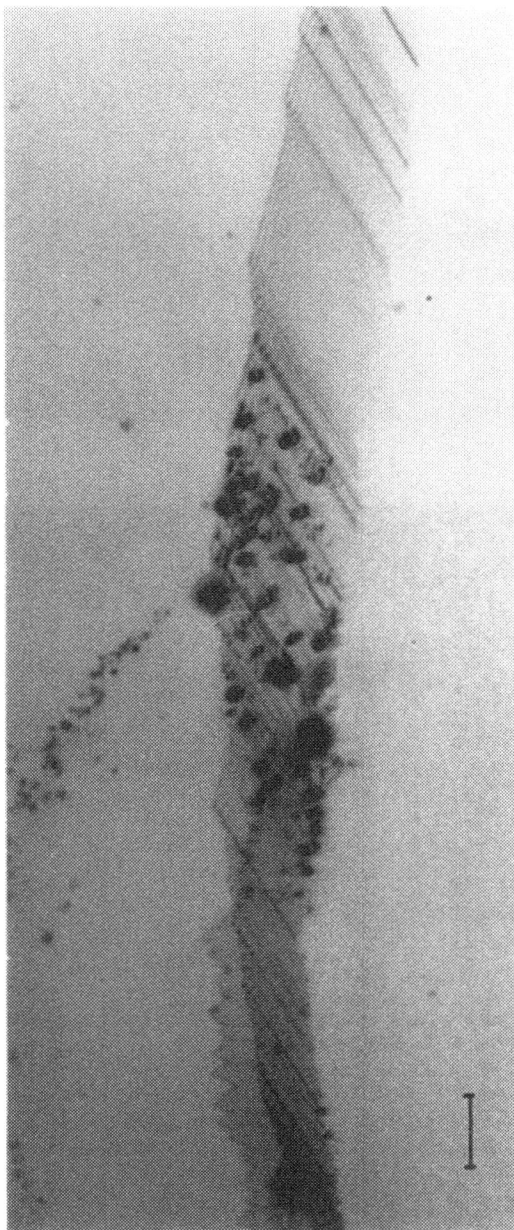

Figure 6.18 Continued.

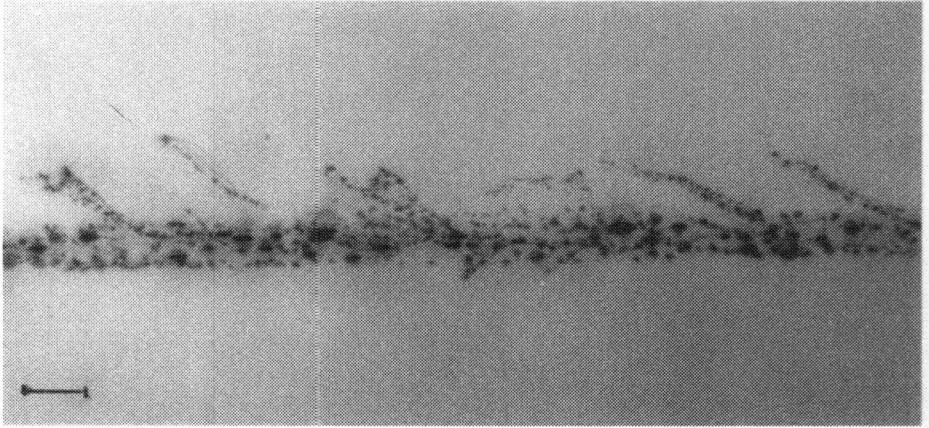

(a)

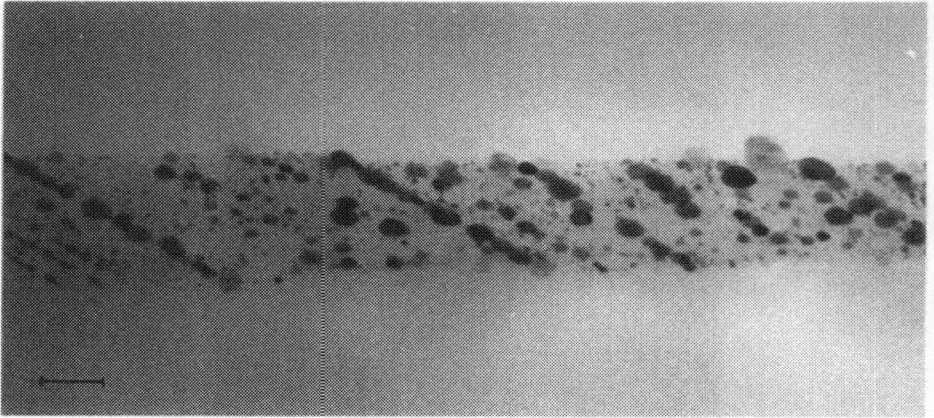

(b)

Figure 6.19 TEM micrographs of copper precipitates at a coherent twin (a) and a random grain boundary (b) in polycrystalline silicon (scale marker $\cong 0.1$ μm). (c) Enlargement of the precipitates in part (b) (scale marker $\cong 10$ nm). Because of large mismatch stresses between the Cu_3Si particle and the silicon matrix, dislocations are punched out during precipitation at the twin boundary. Random grain boundaries relieve the strain without dislocation nucleation. The precipitates nucleate preferentially along dislocations and steps in the boundary plane [21].

with experimentally observed denuded zones of about 10 to 100 μm measured by other methods.

The grain boundary recombination contributes to the total lifetime reduction in a polycrystalline material. Taking into account the width of the denuded zone,

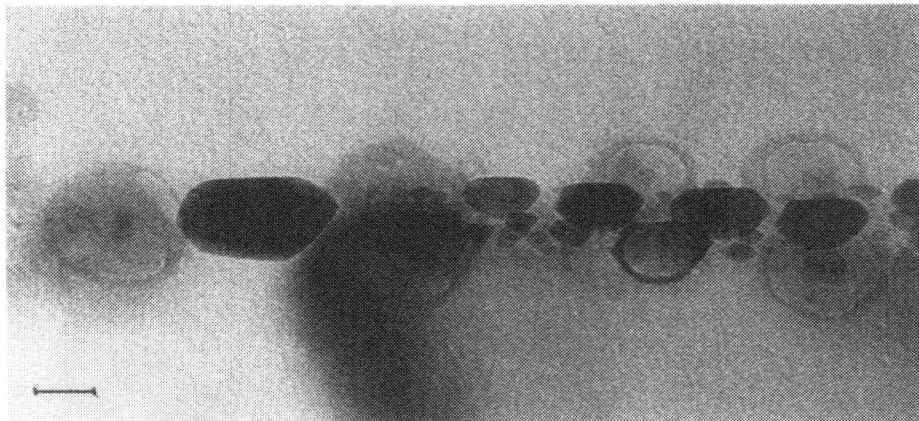

(c)

Figure 6.19 Continued.

a considerable fraction of the bulk material is affected by the presence of electrically active grain boundaries. For the current impurity levels which can be achieved in a low-cost ingot process, grain sizes above a few millimeters appear, however, to be sufficient to reduce the grain boundary activity to an acceptable level.

Several ways have been proposed to reduce the influence of the grain boundary recombination after the ingot growth in the processed wafers. One possibility is to use the enhanced diffusion of doping impurities along grain boundaries. For instance, in a p-base material, phosphorus is diffused into the crystal, preferentially along the grain boundaries. The resulting pn junction around each grain boundary acts as a barrier for minority carriers before they can recombine at the interface. This process has been shown to reduce the electrical activity of the grain boundaries.

Another possibility is the passivation of grain boundaries with hydrogen, which was described in Section 5.1.4. The hydrogenation is carried out in a plasma or by ion implantation (Kaufmann source) and is a standard process in solar cell technology. Hydrogen passivation reduces the recombination at grain boundaries (and other defects such as dislocations or precipitates) and thus improves the efficiency. A problem with the hydrogen passivation is that the improvement of the performance varies considerably with the material. For instance, a polycrystalline silicon with already a rather high efficiency (e.g., about 13%) shows only a marginal increase (0.1% to 0.5%, depending on the material), which in many cases does not justify the increased processing costs.

One problem with the hydrogen passivation technique is the low penetration depth of hydrogen of only a few microns. Responsible for this behavior is the complex diffusion mechanisms which involve trapping at defects and the formation

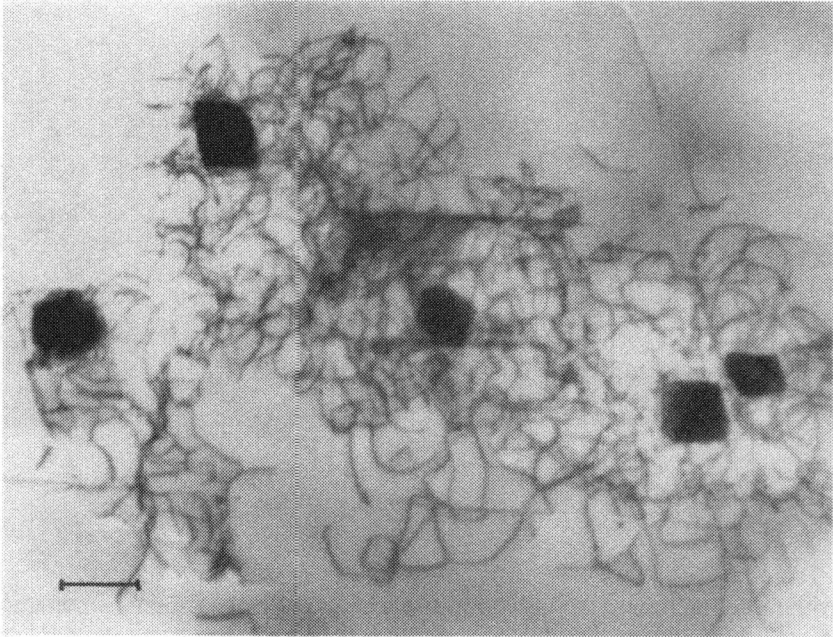

Figure 6.20 TEM micrograph of NiSi₂ precipitates in the bulk of polycrystalline silicon wafer. No preferential precipitation at grain boundaries is observed. Large precipitates punch out dislocations during cooling of the wafers due to lattice mismatch stresses between the precipitates and the matrix (scale marker $\cong 0.5~\mu m$).

of the rather immobile H_2 molecule. Although in a polycrystalline material the penetration is probably enhanced because of the accelerated diffusion along grain boundaries, the distribution of hydrogen in the bulk is difficult to predict. It is quite possible that, in some cases, only near surface recombination centers are passivated and most of the defects in a 300- to 400-μm-thick solar cell are not affected. Since at the same time shallow acceptor impurities are also passivated which may change the properties of the near-surface pn junction, the influence of the hydrogen on the performance of a solar cell is complex and probably not yet completely understood.

6.2.3.2 Ribbons

Solidification Parallel to Sheet Growth Direction

EFG, D-Web, ESP, SSP and related techniques generally produce high-quality material. Some of the properties are summarized in Table 6.4. The ribbons are

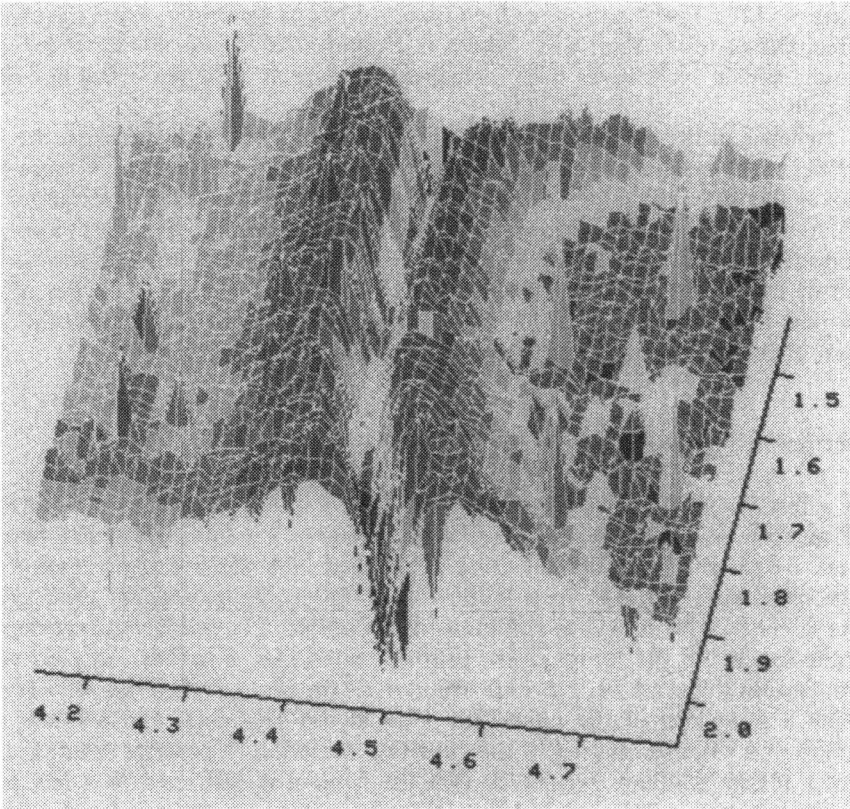

Figure 6.21 Color-coded LBIC scan across a grain boundary in cast polycrystalline silicon showing an enhanced recombination at the grain boundary and a reduced recombination in the vicinity. The recombination current increases from red to blue (scale ≅ 100 μm). (Courtesy of W. Koch, Bayer Co.)

either single crystalline or characterized by a very low density of grain boundaries, most of them being coherent twin boundaries with (111) planes. For instance, in EFG ribbons, the twins are arranged in thin lamellae which run parallel to the growth direction. The other main crystallographic defects in this material are dislocations with average densities of about 10^4 cm^2 and local densities up to 10^6 cm^2, stacking faults and SiC precipitates. The precipitates (mostly near the surface) can become rather large between about 30 to 300 μm; however, their density can be reduced by elevating one side of the die over the other, which causes a considerable reduction of the SiC precipitates on both sides of the ribbon. All of these defects are electrically active to some extent, which is partly due to the contamination with impurities. TEM investigations of high-quality ribbons generally reveal the absence of precipitates other than SiC, which is comparable to the situation for ingot material, and in most cases it is unclear which defects limit the efficiency of the solar cells.

Because of the high quality of the ribbons grown by these methods, the solar cell efficiencies range between about 14% to 17%. A disadvantage of these techniques is the still rather low growth velocity. Although the total throughput in a production facility can certainly be increased, for instance, by continuous melt replenishment and growth of longer ribbons, there are inherent limitations. Estimates of the total solar cell area that can be produced in a given period of time show that these sheet growth techniques are probably not superior to the ingot growth in connection with improved slicing techniques (Table 6.4).

Solidification Perpendicular to Sheet Growth Direction

Only a few results are available for the quality of sheets grown by the RAFT or RGS method. A typical feature of these techniques is a small grain size between 100 and 1000 μm. Since the substrate provides a high density of nucleation sites for the solidifying melt and the solidification is rapidly completed, the grain growth due to selective growth of larger grains is rather limited. The microstructure depends to some extent on the temperature profile, which can, however, only be varied within certain limits. A further improvement of the grain size is thus only feasible through a reduction of the nucleation sites at the substrate, for instance, by a selection of a suitable coating of the substrate material. A possible solution is the use of a textured silicon surface so that the contact is restricted to a few points where the growth is initiated and the silicon provides seed crystals for the nucleation. The drawback is that the substrate coating can only be used a few times and then has to be replaced, which hampers the continous operation of the process.

The processes are characterized by rather high thermal gradients and stresses. Therefore, besides the high grain boundary density, the ribbons also contain high dislocation densities. The average densities range from 10^5 to 10^7 cm^2, with local density up to 10^8 cm^2. An optimization of the temperature profiles and postannealing treatments may reduce the dislocation densities in the future to acceptable levels.

SiC precipitates are also observed in these materials due to the intensive contact of the melt with the graphite die and substrate. Empirical observations yield precipitate diameters of about 1μm. Due to the rapid cooling of the ribbon, it is unlikely that these precipitates have formed in the solid. Instead, it is more likely to assume that precipitation occurs by the melt interface process described in Section 5.4.2.3. Considering the solidification velocity (≈ 0.005 cm/s) and typical carbon concentrations $\Delta C = 100$ ppm that can be expected in the melt in contact with a graphite die, about 0.5 to 2.0 μm is obtained (see Figure 6.16), which is in agreement with experimental observation.

No systematic results for the impurity contamination and electrical properties of any of the sheet grown materials have been published. Scattered results for the various techniques and reported efficiencies are given in Table 6.4. A comparison of some properties of ingot material and ribbons grown by a slow (EFG) and fast process (RAFT) are given in Table 6.3. Considering the difficulties of controlling the microstructure and impurity levels of materials grown by the ribbon techniques, it is probably too early to evaluate the future of any of these methods.

6.3 THIN-FILM SILICON

One of the major difficulties with the silicon technology for solar cells is the low absorption coefficient and the requirement for rather thick crystals to absorb the light. This has the consequence of producing a high-quality material with diffusion lengths larger than 100 μm to obtain sufficient efficiencies above 10%. So far, all efforts have been directed towards the improvement of the defect structure and reduction of impurity levels while keeping the production costs low. With the advancement of semiconductor and solar cell technology, new approaches become feasible and are beginning to be explored. It is evident that a lower quality of material can be tolerated if the absorbing layer becomes thinner. This can be achieved, for instance, if the light is reflected and scattered at the back and front surfaces and thus kept inside the layer by multiple reflection. Theoretical estimations show that the effective absorption thickness can be reduced by a factor of 50 for efficient light trapping techniques. Assuming Auger and radiative recombination in the bulk and no surface recombination, numerical calculation for an ideal *pn* junction yields efficiencies above 28% for cells with a thickness between 10 to 60 μm [25]. Since considerable improvements of light trapping techniques have been obtained in the past, the realization of thin-film solar cells below 60 μm becomes feasible for silicon.

A prerequisite for the preparation of a thin-film cell is a processing technique allowing the preparation of a thin silicon layer on a substrate while maintaining a low defect density. Mainly two approaches are currently being explored: the growth of high-quality layers by LPE and the deposition of polycrystalline thin films by CVD. Both techniques allow the growth of layers with a thickness between about 1 and 50 μm. The microstructure and quality of the films depends on the material used for the substrate. In general, monocrystalline substrates with a good lattice match to silicon are necessary for the epitaxial growth of a layer with a low defect density, while polycrystalline silicon films can be grown on a variety of materials with a polycrystalline or amorphous microstructure.

In the LPE technique, a silicon film is grown from a metallic solution at temperatures close to the solubility temperature of silicon in the solvent. In practice, solvents of indium, gallium, tin, or bismuth are used [26]. If a monocrystalline silicon substrate is used, the films grow epitaxially as a monocrystalline film. Recent results have shown that a high-quality silicon film with a thickness of about 20 μm could be obtained with an open-circuit voltage of about 0.66V. This is comparable to a high-quality Fz-crystal with a diffusion length above 300 μm [27].

In general, epitaxial growth may be prohibited because of the necessity of incorporating a metallic reflector at the back surface between the substrate and the silicon layer. Although in this case a polycrystalline film with a higher defect density is obtained, it may be possible in the future to improve the microstructure by selecting an appropriate reflector material. The application of the LPE on low-cost substrates such as polycrystalline silicon or ceramic materials is therefore a promising new approach.

An alternative to LPE is the deposition of a polycrystalline thin film by CVD. The CVD films have grain sizes below 1 μm if they are deposited at lower temperatures (600 to 800°C). This grain size is too small for photovoltaic applications. An improvement of the microstructure can be obtained if the films are either deposited at higher temperatures (>1000°C) or if they are recrystallized by remelting. Several techniques are available, such as laser or electron beam recrystallization or lamp annealing. Grain sizes between 300 and 500 μm are reported, which is sufficient for a thin-film cell. CVD techniques are a standard technology in the semiconductor industry and it can be expect that the porcesses can be further optimized for the solar cell fabrication. CVD methods have the advantage that films can be deposited on a variety of substrates, which offers a great flexibility in the design of the solar cell.

Silicon thin-film solar cells are still at a very early stage of development. Only a few scattered results are available today, and these are summarized in Table 6.6. The efficiencies still vary between 4 and 10%, but the techniques have the potential for considerable improvements and thus offer a promising alternative to other thin-film solar cell materials.

Table 6.6
Maximum Conversion Efficiencies and Properties of Thin-Film Silicon (Wafer size below 4 cm^2 unless otherwise stated; *wafer size 10 × 10 cm^2)

Technique	Substrate	η [%]	Laboratory
LPE from tin	Ceramic + metal film	15(10*)	Astropower, USA
LPE from indium	monosilicon	>10	MPI-Stuttgart, FRG
CVD + remelting	monosilicon + SiO$_2$	7	Mitsubishi, Japan
CVD + remelting	monosilicon + SiO$_2$	4	FhG-ISE, FRG
CVD	MG-silicon + SiO$_2$	<10	IMEC, Belgium

REFERENCES

[1] Dietze, W., W. Keller, and A. Mühlbauer, *Crystals*, Vol. 5, Berlin: Springer, 1981, p. 2.

[2] Wang, T. H., T. F. Ciszek, and T. Schuyler, *Solar Cells*, Vol. 24, 1986, p. 136.

[3] Roksnoer, P. J., and M. M. B. Van Den Boom, *J. Crystal. Growth*, Vol. 53, 1981, p. 563.

[4] Dekock, A. J. R., *Appl. Phys. Lett.*, Vol. 16, 1970, p. 100.

[5] Petroff, P. M., and A. J. R. Dekock, *J. Crystal. Growth*, Vol. 30, 1975, p. 117.

[6] Föll, H., U. Goesele, and B. O. Kolbesen, *J. Crystal. Growth*, Vol. 52, 1981, p. 907.

[7] Helmreich, D., *Silicon Processing for Photovoltaics II*, C. P. Khattak and K. V. Ravi, eds., London: Elsevier Science Publ., 1987, p. 97.

[8] Helmreich, D., *Proc. Symp. on Electr. and Opt. Properties of Polycrystalline or Impure Semiconductors and Novel Silicon Growth Methods*, K. V. Ravi and B. O'Mara, eds., Electrochemical Soc., Princeton, 1977, p. 184.

[9] Schmid, F., and D. Viechnicki, *J. Am. Ceram. Soc.*, Vol. 53, 1970, p. 528.

[10] Surek, T., *Proc. Symp. on Electr. and Opt. Properties of Polycrystalline or Impure Semiconductors and Novel Silicon Growth Methods*, Princeton: Electrochemical Soc., 1977, p. 173.

[11] Goetzberger, A., and A. Räuber, *Tech. Digest of Photovoltaic Solar Energy Conf.*, Vol. 4, 1989.

[12] *Proc. 1st Int. Conf. on Shaped Crystal Growth*, *J. Crystal. Growth*, Vol. 87, 1987.

[13] Eyer, A., A. Räuber, and J. G. Grabmeier, *Proc. 9th E.C. Photovolt. Solar Energy Conf.*, Dordrecht: Kluwer Academic Publ., 1989, p. 17.

[14] Wald, F. V., *Crystals*, Vol. 5, Berlin: Springer, 1981, p. 147.

[15] Ciszek, T. F., *Crystals*, Vol. 5, Berlin: Springer, 1981.

[16] Lange, H., and I. Schwirtlich, *J. Crystal. Growth*, Vol. 104, 1990, p. 108.

[17] Möller, S., and H. J. Möller, *Proc. 9th E.C. Photovolt. Solar Energy Conf.*, Dordrecht: Kluwer Academic Publ., 1989, p. 439.

[18] Möller, H. J., *Mater. Res. Soc. Symp. Proc.*, Vol. 205, 1991, p. 476.

[19] Davis, J. R., A. Rohatgi, R. Hopkins, P. Blais, P. Rai-Choudhury, J. McCormick, and H. C. Mollenkopf, *IEEE Trans. Electron. Devices*, Vol. 27 (4), 1980, p. 677.

[20] Kazmerski, L. L., *Polycrystalline Semiconductors, Springer Proceedings in Physics*, H. J. Möller, H. Strunk, and J. Werner, eds., Vol. 35, Berlin: Springer, 1989, p. 64.

[21] Möller, H. J., U. Jendrich, L. Huang, and A. Foitzik, *Mater. Res. Soc. Symp. Proc.*, 1991, in press; *Mater. Res. Soc. Symp. Proc.*, 1990, p. 579.

[22] Kazmerski, L. L., P. J. Ireland, and T. F. Ciszek, *Appl. Phys. Lett.*, Vol. 36, 1980, p. 323.

[23] Jendrich, U., and H. J. Möller, *Colloque de Physique*, Vol. 51-C1, 1990, p. 197; *Mater. Res. Soc. Symp. Proc.*, Vol. 163, 1990, p. 579.

[24] Aucouturier, M., A. Broniatowski, A. Chari, and J. L. Maurice, *Polycrystalline Semiconductors, Springer Proceedings in Physics*, H. J. Möller, H. Strunk, and J. Werner, eds., Vol. 35, Berlin: Springer, 1989, p. 64.

[25] Blakers, A. W., *Advances in Solid State Physics*, Vol. 30, R. Rössler, ed., Vieweg, Braunschweig, 1990, p. 403.

[26] Bauser, E., *Advances in Solid State Physics*, Vol. 23, R. Rössler, ed., Vieweg, Braunschweig, 1983, p. 141.

[27] Blakers, A. W., J. Werner, E. Bauser, and H. J. Queisser, *Proc. 10th E.C. Photovolt. Solar Energy Conf.*, Dordrecht: Kluwer Academic Publ., 1991, p. 692.

[28] Dietl, J., D. Helmreich, and E. Sirtl, *Crystals*, Vol. 5, Berlin: Springer, 1981, p. 43.

[24] Aucouturier, M., A. Broniatowski, A. Chari, and J. L. Maurice, *Polycrystalline Semiconductors*, *Springer Proceedings in Physics*, H. J. Möller, H. Strunk, and J. Werner, eds., Vol. 35, Berlin: Springer, 1989, p. 64.

[25] Blakers, A. W., *Advances in Solid State Physics*, Vol. 30, R. Rössler, ed., Vieweg, Braunschweig, 1990, p. 403.

[26] Bauser, E., *Advances in Solid State Physics*, Vol. 23, R. Rössler, ed., Vieweg, Braunschweig, 1983, p. 141.

[27] Blakers, A. W., J. Werner, E. Bauser, and H. J. Queisser, *Proc. 10th E.C. Photovolt. Solar Energy Conf.*, Dordrecht: Kluwer Academic Publ., 1991, p. 692.

[28] Dietl, J., D. Helmreich, and E. Sirtl, *Crystals*, Vol. 5, Berlin: Springer, 1981, p. 43.

Chapter 7
Single-Crystal and Epitaxial Compound Semiconductors

Monocrystalline and polycrystalline silicon for solar cells has the disadvantage that silicon is an indirect semiconductor requiring rather thick crystals to absorb the light and therefore a high crystal quality. It is quite evident that a direct-gap semiconductor can be made much thinner with the advantage of a greater tolerance to the material quality, or made with a higher efficiency if a high-quality material is used. Most of the compound semiconductors have a direct band gap, and there are a number of binary compounds with an optimum band gap between 1.4 and 1.6 eV (see Figure 2.11). Not surprisingly, they have been considered and investigated for use in solar cells. Since only thin layers of about a few microns are required due to the high absorption, there are several routes that can be taken for the fabrication of the absorbing layer.

First, one can start from a single crystal as a substrate and fabricate the monocrystalline solar cell only in the top region of the crystal. In practice, this can be achieved by epitaxial growth techniques such as MOCVD, LPE, or MBE. Although from the perspective of the photovoltaic conversion a good material is only required in the active top region, the quality of the substrate is equally important. The best epitaxial layers are usually obtained if the semiconductor is deposited on substrates of materials that have similar crystal properties. The important parameters are the lattice constants and the band gap energy, which are summarized for some semiconductor compounds in Figure 7.1. Although monocrystalline layers have also been obtained for compound semiconductors with a rather large lattice mismatch between the substrate and the deposited material, the layers suffer from a high density of lattice defects, mainly dislocations, stacking faults, and twins, which are introduced to accommodate the mismatch stresses during growth. Therefore, the conversion efficiencies of heteroepitaxial cells are rather unsatisfactory so far because of the high defect densities in the active region.

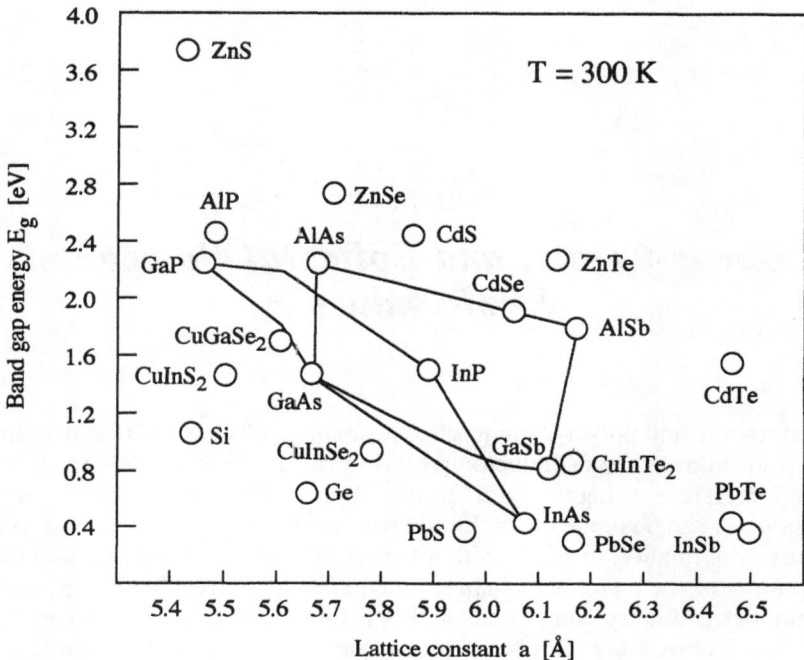

Figure 7.1 Energy gap E_g and lattice constants for some semiconductor elements and compounds. Binary compounds used to form ternary compounds are connected by solid lines. For copper ternary compounds with non-cubic chalcopyrite structure, the smaller lattice constant is taken.

The epitaxial growth of layers with a low density of defects thus requires a monocrystalline substrate of the same material, or at least a semiconductor with a similar lattice constant. However, even for a system with a good lattice match, the defect density of the epitaxial layer is to a large extent determined by the defect density of the substrate material. Experimental results show that, for instance, dislocations from the substrate grow into the epilayer or give rise to the nucleation of other defects. Therefore, for high efficiencies, one has to start from a substrate crystal with a controlled and high material quality, which in most cases is a requirement that at present can only be fulfilled by a few compound semiconductors. Epitaxial techniques are therefore rather expensive at present and only useful for high-efficiency solar cells or if cost considerations are less important (for instance, spacecraft applications).

The alternative to monocrystalline or epitaxially grown compound semiconductors is the application of low-cost deposition techniques with the consequence of a polycrystalline or amorphous thin film with a high density of crystal defects. Although lower efficiencies are generally obtained then, the advantage is that these processes offer a wider flexibility in the substrate material or the design of the solar cell, for instance, as multijunction structure. Low-cost deposition techniques are thus preferred for most of the compound semiconductors and will be discussed in Chapters 8 and 9.

7.1 III-V COMPOUNDS

Many binary semiconductors with a suitable band gap have been explored for photovoltaic applications, but for most of them the development is still at a rather early stage and shall not be discussed here. The situation is different for the III-V semiconductors GaAs, InP, and their multinary compounds, such as AlGaAs, InAsP, or GaInAsP. Since these materials have also received considerable attention for high-speed and opto-electronic applications, their technology is far more advanced. Because of the high cost of single crystals, solar cells using III-V semiconductors are usually only considered for concentrator and space solar cells. As discussed in Section 3.6, solar cells perform with higher efficiencies if they operate under higher irradiance conditions. The drawback is that concentrator cells require focusing lenses or mirrors and direct sunlight and therefore can only compete with their flat panel counterparts in very sunny locations. Nonetheless, attempts are made to reduce the cost for III-V semiconductors also so that they may offer alternative solutions, even for low-cost cells in the future.

GaAs and InP cells have also been developed for space applications because they show a higher resistance to the increased radiation in space compared to high-efficiency silicon cells. In particular, the development of InP cells is mainly driven by the demand in spacecraft, where efficient cells with a long lifetime are required. Although it is clear that the degradation of the efficiency is caused by radiation-induced defects, the details of the physical processes in each case are not completely understood yet (Section 5.1.5).

Gallium arsenide has a much higher electron ($8500 \ \text{cm}^2/\text{Vs}$ in undoped GaAs) than hole mobility ($400 \ \text{cm}^2/\text{Vs}$), which makes it very suitable for the p-base of a pn-junction solar cell. Contrary to silicon, the emitter region is usually not fabricated by diffusion, but both the p- and n-doped regions are grown epitaxially on GaAs single-crystal substrates (Figure 7.2). Although the substrate is not important for the photovoltaic conversion process, a good quality is important for the successful growth of epitaxial layers with a low defect density, as mentioned above.

The growth of single-crystal GaAs and the investigation of the defect structure has been an area of intense research for years now and has provided comprehensive insight into the basic properties of compound semiconductors. Therefore, in the first part of the chapter, the main results for the growth of single crystals and the

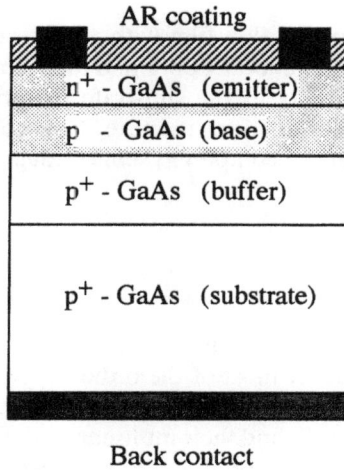

Figure 7.2 Schematic representation of a common structure for a high-efficiency GaAs homojunction cell. The structure is configured to minimize the high surface recombination.

structural and electronic properties shall be discussed. In the second part, the properties of epitaxially grown GaAs and related compounds such as AlGaAs are presented.

7.1.1 Single-Crystal Growth Techniques

Single-crystal GaAs today is mostly either used for opto-electronic applications or as a substrate material for *integrated circuit* (IC) devices. Because of the different requirements for the material in each case, basically two crystal growth techniques are employed, which are based on the method of directional solidification. Doped crystals with a low defect density for the opto-electronic application are usually grown by the Bridgman technique. The GaAs crystal is grown in a boat or crucible, which is passed through the hot zone of a furnace (Figure 7.3). Both a *horizontal Bridgman* (HB) and *vertical Bridgman* (VB) setup for the boat and furnace are possible. Vertical methods generally cause more problems because of the volume expansion of III-V compounds during solidification (between about 5% and 15%). GaAs expands by 10% upon freezing, so crystal and the crucible material may have to withstand quite high stresses. The stresses in the growing crystal may lead to the nucleation of dislocations, stacking faults, twins, or even cracks. The advantage of the Bridgman technique is that it uses low temperature gradients, which reduces the thermal stresses in the growing single crystal. Therefore, one can keep the dislocation density low, which is essential for the stability of the optical per-

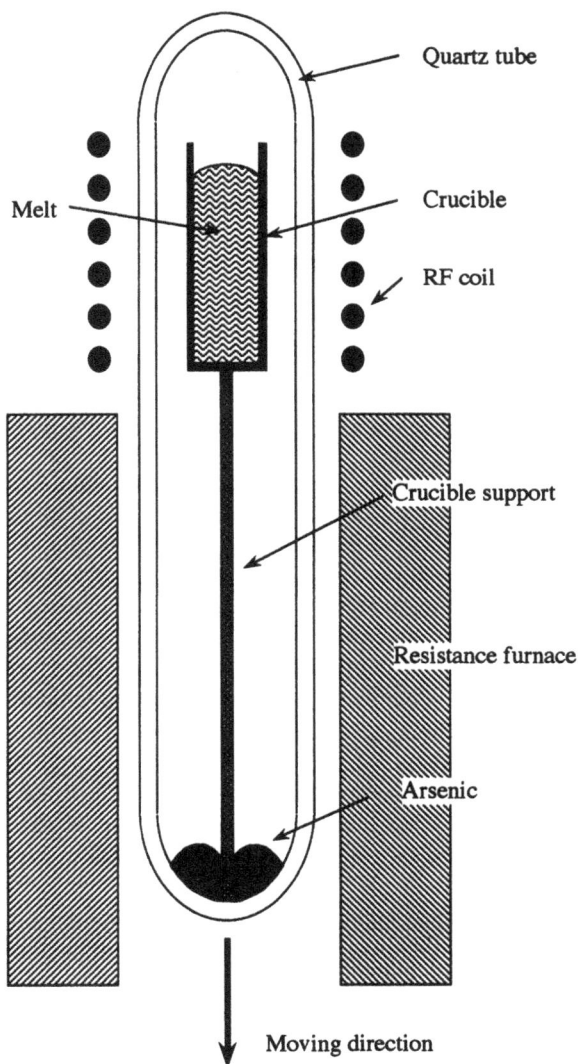

Quartz tube

Melt

Crucible

RF coil

Crucible support

Resistance furnace

Arsenic

Moving direction

Figure 7.3 Schematic representation of the VB technique for the growth of GaAs single crystals.

formance of laser diodes. However, since the GaAs crystals are grown in a boat or crucible, some kind of contamination from the crucible walls is unavoidable in this technique.

A general problem of the growth of many binary compound semiconductors is that the melt decomposes at the melting point. The dissociation pressure at the melting point is equal to the vapor pressure p_m of the group V elements, which is usually several orders of magnitudes higher compared to the vapor pressure of the group III elements. For instance, for GaAs it is $p_m = 0.976$ bar for arsenic compared to $p_m = 10^{-4}$ bar for gallium. Typical vapor pressures for phosphides are about $p_m \approx 30$ bar, for arsenides $p_m \approx 1$ bar, and for antimonides $p_m \approx 10^{-5}$ bar. In order to avoid the decomposition of the melt, the equilibrium pressure of the volatile component has to be maintained inside the growth chamber. Because of the high vapor pressures for the antimonides and phosphides, the growth has to be carried out in a pressure chamber. The equilibrium pressure of the volatile component in the gas phase can be formed by (1) a tolerable decomposition of the melt, (2) a calculated amount of the volatile component added into the chamber, or (3) a supply of the volatile element kept at the corresponding temperature (see Figure 7.3).

For IC-technology, undoped *semi-insulating* (SI) GaAs is required, which is difficult to grow by the Bridgman technique because of the contamination problem. For this purpose, the *liquid encapsulated Czochralski* (LEC) technique has been developed which does not require a boat or crucible and allows the growth of purer crystals (Figure 7.4). Since the decomposition of the melt leading to the high partial pressures causes construction problems with the pulling mechanism inherent to the Cz-technique (see Figure 6.2), the liquid encapsulation method has been developed. In this technique, the melt is covered by an inert liquid on which a counterpressure of an inert gas is exerted which overcompensates the dissociation pressure of the melt. Some of the requirements for the selection of a material suitable as a liquid encapsulation are (1) no mixing and reaction with the melt of the III-V compound, (2) density lower than that of the melt, and (3) wetting of the crucible and the grown crystal. Boron oxide (B_2O_3) meets these and other requirements and is mostly used.

The LEC technique needs higher temperature gradients for the crystal growth compared to the Bridgman technique, which results in higher dislocation densities. Typically, the dislocation densities in LEC material are an order of magnitude larger compared to HB GaAs. Since dislocations are less important for IC fabrication, most of the commercial GaAs substrates are grown by this technique. So far, semi-insulating GaAs serves mainly as a substrate material for the following epitaxial growth of the GaAs layers. Nonetheless, the semi-insulating properties of undoped GaAs are decisive for the fabrication of ICs and thus have been studied intensely during recent years.

The requirements for GaAs that can be used for solar cells differ from those for device application. If one considers the design of a high-efficiency homojunction solar cell, for instance, in Figure 7.2 (see also Chapter 3), where the two contacts are placed on opposite sides of the structure and the photo-generated current passes

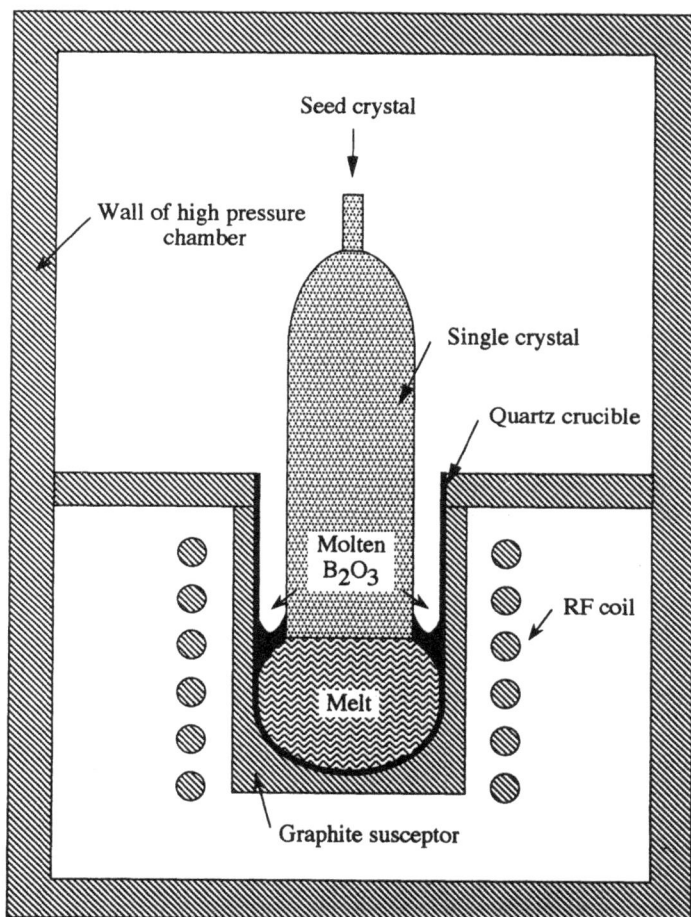

Figure 7.4 Schematic representation of the LEC technique for the growth of GaAs single crystals.

current passes through the entire cell, it is evident that a good conductivity in all layers is required. Since only doped GaAs substrates with a low resistance can be employed, the semi-insulating material itself has no importance for photovoltaic applications. Nonetheless, the GaAs solar cell technology benefits from the developments in the GaAs device technology and applies the materials and techniques that are most appropriate for the fabrication of the cells.

7.1.2 Defects in GaAs

The material problems for GaAs and other III-V compounds are more complex compared to elemental semiconductors for several reasons. Besides the usual lattice defects such as impurities or dislocations, compound semiconductors also contain numerous intrinsic point defects, such as vacancies, on the different sublattices and antisites. These defects are mainly introduced when deviations from the stoichiometry occur. Since in many cases the growth of compound semiconductors is difficult because one of the constituents is rather volatile (for instance, arsenic in GaAs), the control of the stoichiometry is a complex task.

Since semi-insulating GaAs is the compound semiconductor that is commercially important, it is also the most thoroughly investigated material. Although for solar cells it is not directly relevant, these investigations have provided, however, enormous insight into the structural and electronic properties of this material and of compound semiconductors in general.

Most of the investigations focused on a single defect which dominates the electronic properties of undoped, mainly meltgrown GaAs. This defect, referred to as EL2, introduces a donor level in the middle of the band gap and occurs with a rather constant concentration of about 1 to 5×10^{16} cm^{-3} in single crystals. The characteristics of the EL2 defect are two charge states at $E_V + 0.52$ eV and $E_V + 0.75$ eV (Table 7.1) [1,2]. EL2 occurs in arsenic-rich crystals and epitaxial layers independent of the growth technique. In semi-insulating material it compensates residual acceptor impurities. Since EL2 appears to be an intrinsic defect of the semiconductor, its presence determines the electrical properties of semi-insulating GaAs considerably. Because of the importance of controlling the defect from the perspective of the production of high-quality GaAs material, the nature and properties of the EL2 defect have been studied intensively and have received greater attention in the last 10 years than probably any other aspect of semiconductor materials.

Some of the conclusions that point towards the intrinsic origin of the EL2 defect are based on the observation that the concentration of the defect is linked

Table 7.1
Energy Levels of the EL2 and As$_{Ga}$
Antisite Defects in GaAs

Defect	Donor Level $E_V + E_T$ [eV]	
EL2	0.52	0.75
As$_{Ga}$ antisite	0.52	0.70

to the stoichiometry of the compound. In particular, it has been found that the EL2 concentrations depend on the ratio $\Delta s = $ [As]/[Ga], which increases linearly with the excess of arsenic [3]. The concentrations are also higher if the GaAs crystals are prepared at the elevated temperatures necessary for melt growth, which evidently promotes atomic disorder. On the other hand, GaAs prepared by thin-film deposition techniques at lower temperatures contains lower concentrations. Table 7.2 shows typical concentrations for the different growth techniques.

It is now commonly assumed that the EL2 defect is related to the arsenic antisite, since a number of different experimental techniques have demonstrated similarities between the structural and electronic properties of the EL2 and the As_{Ga} defect. For instance, the antisite defect has two extra electrons that are not used for the covalent bond formation and is thus a double donor with the energy levels given in Table 7.1. The first donor level $(0/+)$ is close to the level observed in crystals containing the EL2 defect. Despite the similarities, there are experimental results indicating that the EL2 defect is not completely identical to the arsenic antisite but rather a complex $As_{Ga} - X$. Several suggestions for the unknown component X have been made, and the most widely accepted assumption now seems to be that an arsenic interstitial $(X = As_i)$ is bound to the antisite defect.

One of the characteristic features of the EL2 defect is that it can be transformed into a metastable state by optical excitation (with $h\nu \approx 1.1$ eV) at low temperatures <130K (persistent photoquenching effect). In this configuration, the

Table 7.2

Typical Concentrations of the EL2 Defect for GaAs Crystals Prepared by Various Growth Techniques and Conditions (For comparison, the concentrations and energy levels of some other electrically detectable defect levels are also included)

Growth Method	Defect	Energy Level $E_C - E_T$ [eV]	Concentration $[cm^{-3}]$	Growth Conditions	Composition
Horizontal					
Bridgman (HB)	EL2	0.77	5×10^{16}	Slow cooling, low	As-rich
	EL6	0.28	5×10^{15}	thermal gradients	
	EL3	0.53	5×10^{14}		
Czrochralski (LEC)	EL2	0.82	1×10^{16}	Fast cooling, high	As-rich
	EL6	0.28	7×10^{15}	thermal gradients	
	EL5	0.59	1×10^{14}		
Vapor phase	EL2	0.82	2×10^{14}	Low temperatures	$\Delta_{As}/\Delta_{Ga} \approx$
epitaxy (VPE, MOVPE)	EL16	0.39	1×10^{13}	$(\approx 700°C)$	10^{-1} to 10^2
Liquid phase epitaxy (LPE)	EL2			Low temperatures $(\approx 700°C)$	Ga-rich
Molecular beam epitaxy (MBE)	EL2		$<10^{14}$	Low temperatures $(<600°C)$	

trap level disappears completely from the bandgap but can be reactivated by annealing above >130K. It was suggested that the metastable state is formed by a slight shift of the atomic positions of the arsenic in the As_{Ga}-As_i complex towards a split interstitial configuration (Figure 7.5). Although there is still a controversy about the exact nature of the As_{Ga}-X complex and the metastable configuration, the main aspects seem to be understood now.

EL2 is not the only electrically active defect that appears after crystal growth in GaAs. Other deep levels have been observed, but the concentrations of these defects is usually less then 10% of the EL2 concentration (see Table 7.2). In most cases, the defects have not been yet identified and their origin and nature are still under discussion. Besides the intrinsic point defects, there are also impurities that can become relevant for the electrical properties of GaAs. Of the shallow donor atoms that may be unintentionally present, silicon poses the greatest problems. This contaminant may be introduced if the GaAs ingots are sealed in quartz ampoules or tubes. At the melting temperature of GaAs of about 1238°C, quartz decomposes rapidly enough to release silicon and contaminate the melt.

Carbon on substitutional arsenic sites is the most common unintentional acceptor impurity. This contaminant is introduced from heated graphite parts in the LEC pullers. The presence of carbon is usually beneficial, since it compensates for some of the EL2 defects and whatever donors may be present. This raises the resistivity, but at the expense of the carrier mobility, which is lowered by high impurity concentrations. Thus, it is generally desirable to keep both the carbon and EL2 concentration low [4].

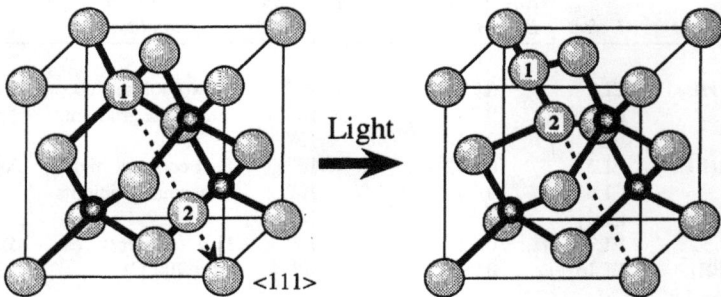

Figure 7.5 Structural model for the As_{Ga}-As_i (EL2) defect in GaAs and the proposed metastable configuration after illumination. Arsenic atoms (open circles), gallium atoms (dark circles), As_{Ga} (atom 1), and As_i (atom 2) on a tetrahedral interstitial site (T_{d1}), and after illumination in a split interstitial configuration [15].

7.2 EPITAXY OF GaAs AND RELATED TERNARY COMPOUNDS

The conventional doping techniques for silicon such as diffusion and ion implantation are less suitable for GaAs. The main problem is the evaporation of arsenic during the necessary heat treatments for either the diffusion or removal of the lattice damage after implantation. This phenomenon leads to a change of the stoichiometry near the surface and the introduction of native defects as discussed in the previous section. Although several techniques such as the capping of the surface with dielectric films (Si_3N_4, SiO_2), rapid thermal annealing, or annealing in arsenic overpressure can be applied to suppress the out-diffusion of arsenic, the current GaAs technology uses to a large extent epitaxial growth techniques for the fabrication of opto-electronic devices and integrated circuits.

The most widely used techniques are the deposition from the gas phase by vapor phase epitaxy (VPE, MOVPE) or from the liquid phase (LPE), and by MBE. The typical processing temperatures and conditions are shown in Table 7.2. These techniques not only allow the deposition of very thin and doped layers (<10 nm), but also a variety of compositions and thus the formation of heterostructures. The most widely used configuration involves the deposition of AlGaAs layers on GaAs. These heterostructure configurations are extensively used for devices where the confinement of charge carriers or light in a two-dimensional layer is required. MBE-grown layers have the advantage of a rather precisely controlled impurity profile, particularly of the aluminum across the AlGaAs/GaAs interface. Nonetheless, MBE material still suffers from the drawback that it contains a higher density of so-called oval defects compared to MOVPE material, which has prevented its commercial use so far [5]. Generally, the epitaxially grown layers have, however, superior qualities compared to the single-crystal bulk material. For instance, the concentrations of EL2 and other deep trap centers are at least an order of magnitude lower, as can be seen from Table 7.2. This is partly due to the lower process temperatures or growth under excess gallium concentrations (e.g., for LPE), which also suppresses EL2 formation.

From the perspective of photovoltaic application, GaAs heterostructures are the natural choice for the fabrication of solar cells. Since the absorption of light requires only a thickness of a few microns, solar cells can be easily fabricated entirely by a suitable sequence of epitaxial layers. Typical device concepts are depicted schematically in Figures 7.2 and 3.7. The heterostructure also uses the optical properties of ternary AlGaAs compounds. The replacement of gallium by aluminum atoms in the compound increases the band gap proportionately to the aluminum content. In addition, $Al_xGa_{1-x}As$ becomes an indirect semiconductor for $x > 0.4$, with a considerable decrease in the absorption. Therefore, these layers are very suitable as a window materials for solar cells, which passivate the surface and reduce the surface recombination at the AlGaAs/GaAs interface usually inherent to direct band gap semiconductors.

7.3 GROWTH DEFECTS IN EPITAXIAL LAYERS

Epitaxially grown layers can contain a variety of lattice defects, such as microtwins, stacking faults, dislocations, or antiphase domain boundaries, which limit their utilization for device applications. These defects arise during the growth process and originate at the substrate interface or the surface. Further processes that degrade the film properties are the interdiffusion of atoms at the interfaces, for instance, of aluminum across the AlGaAs/GaAs interface, which is of particular importance for superlattice structures where the thickness of the epilayers is very low (<100Å), and atomically sharp transitions are required. The composition of epitaxial layers can also be changed by phase separation and long-range ordering, particularly in ternary or quaternary compounds. These phenomena have been observed, for instance, in InGaAs epilayers [6].

The interface between substrate and the epilayer plays an important role in the formation of the lattice defects. The structure of the substrate surface before the deposition determines to a large extent the quality of the deposited layer. The surface orientation, the roughness or the occupation by foreign atoms (adsorption, chemisorption), are relevant for the quality of the deposited films. On the atomic scale, the bond reconstruction, the atomic distances, the occurrence of small (atomic) steps, or the intersection of lattice defects from the substrate (dislocations, grain boundaries) are factors that determine the occurrence of lattice defects in the growing layer. In order to illustrate some of the problems, the following examples shall be discussed.

A particular lattice defect that can occur during heteroepitaxial growth of compound semiconductors is the *inversion-domain boundary* (IDB) (also called *anti-phase boundaries*). It separates two neighboring domains that are related to each other by an inversion operation. This is equivalent to the exchange of atomic species occupying different sublattices. The IDBs in AB compounds are characterized by a high density of incorrect bonds of the type A-A or B-B. In a layer-by-layer growth process, inversion-domain boundaries can occur at atomic steps on the substrate surface, since the stacking sequence of the atomic layers in the film is shifted by the height of the steps (Figure 7.6). Since for particular step heights the correct stacking sequence is maintained, one can reduce or eliminate the IDBs by growing on slightly inclined surfaces which contain atomic steps of the correct height if the inclination angle is chosen correctly.

It is evident that the substrate surface is important for the initial stages of growth. Analysis and study of the nucleation process are therefore necessary for the understanding of the structure and morphology of the growing layer. Dislocation and microtwin generation are of particular concern for heteroepitaxial growth of semiconductors with different lattice constants and thermal expansion coefficients. A brief description of the nucleation models that have been developed sheds some light on the process of dislocation generation.

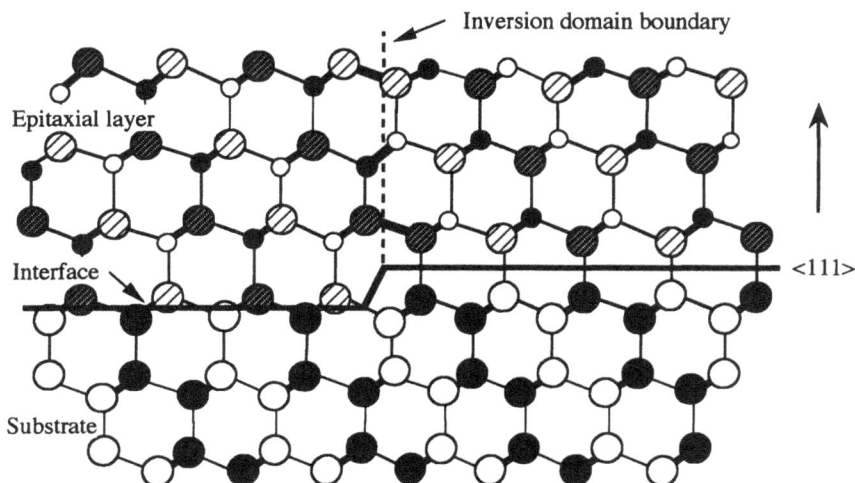

Figure 7.6 (011) projection of a structural model for an IDB in a binary *AB* compound with sphalerite structure. The IDB in the epitaxial layer is formed at a step in the surface of the substrate. *A* atoms (small circles), *B* atoms (large dotted atoms). Light and dark atoms occupy positions in adjacent planes at a distance *a*/2, respectively (*a*: lattice constant).

The theoretical treatments [7] of the nucleation modes show that they can be differentiated by the following factors: the bond energies between two atoms in the epitaxial film E_{AA} and across the interface to the substrate E_{AS}, respectively, and the relative (absolute) lattice mismatch between film and substrate $\Delta a/a$. Three different modes of growth are usually considered and are illustrated in Figure 7.7.

1. The *Frank-van der Merwe growth mode* is the coherent two-dimensional layer-by-layer growth. It is favored when the lattice mismatch is very small, $\Delta a/a \approx 0$, and the bond energy between substrate and layer atoms is higher compared to the bond energy between layer atoms, or $E_{AS}/E_{AA} > 1$.
2. The *Stranski-Kristanov growth mode* is an intermediate case and is favored by the condition $\Delta a/a \approx 0$ and $E_{AS}/E_{AA} < 1$. The growth begins two-dimensionally, but only a few continuous atomic layers are formed. Then the growth mode changes, and separate islands begin to form. The islands grow in three dimensions until they coalesce and a continuous epitaxial layer forms.
3. The *Vollmer-Weber growth mode* is characterized by the nucleation and growth of three-dimensional crystallites from the beginning, and occurs for $|\Delta a|/a \gg 0$ and $E_{AS}/E_{AA} < 1$. Because of the larger lattice mismatch, the growth of the crystallites is accompanied by the formation of misfit dislocations, stacking faults, and grain boundaries.

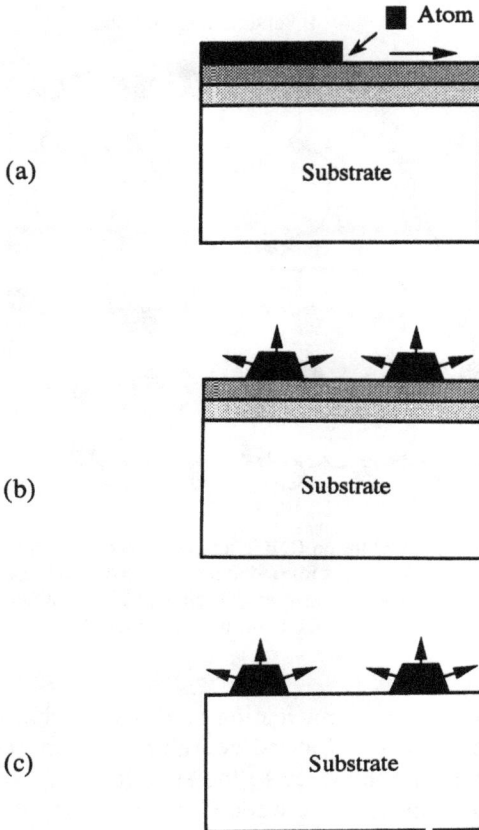

Figure 7.7 Schematic representation of three different modes of epitaxial growth: (a) coherent two-dimensional layer-by-layer growth (*Frank-van der Merwe mode*); (b) an intermediate case where initially a strained layer forms before three-dimensional nucleation begins (*Stranski-Kristanov mode*); (c) the nucleation and growth of three-dimensional crystallites (*Vollmer-Weber mode*).

In the first two cases, which occur for epitaxial growth of systems with a good lattice match, it is observed that the thickness of the layers or islands reaches a certain thickness before the first misfit dislocations are nucleated. Below the critical thickness, the misfit strain is accommodated elastically and a strained layer forms. Since strained-layer heterostructures are important because they are dislocation-free, investigation of them has attracted much attention. Experimental results show that the critical thickness depends on the lattice mismatch and the temperature at which the layers are formed. For instance, experimental data for the system Si_xGe_{1-x}/Si show that at low temperatures of about 550°C, values up to about 100 nm can be obtained (Figure 7.8). Several models, which are more or less in agreement with experimental results, have been developed to calculate the critical thickness [8–10]. The results of a kinetic model, based on the description of the dislocation motion and multiplication given in Section 5.2 (equations (5.45) and(5.49) to (5.51)), are also given in Figure 7.8 and show good agreement with the experimental data. The nucleation of misfit dislocations in this model is thus due to the misfit strains and is a result of the plastic flow of the strained layer.

It is quite evident that in lattice-matched systems such as the growth of AlGaAs on GaAs, the nucleation of defects is less severe. The experimental approach to reducing the defect generation in a system with a large lattice mismatch between substrate and layer material is to grow an intermediate lattice-matched buffer layer of a different material or composition. The buffer layer is kept thin

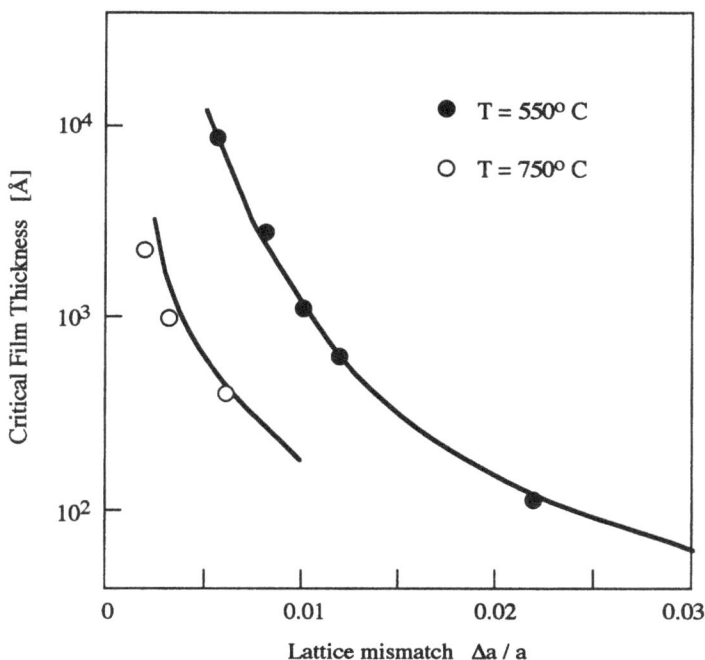

Figure 7.8 Critical thickness of an epitaxially grown Si_xGe_{1-x} layer on a silicon substrate as a function of the lattice mismatch between Si_xGe_{1-x} and silicon. Open circles: experimental results; solid line: calculated result of the kinetic model [9]. The lattice mismatch is related to the composition by $\Delta a/a = (1 - x) (a_{Ge} - a_{Si})/a_{Si}$.

enough to accommodate the strain elastically and provides a better match to the next epitaxial layer above. A technologically important example is the growth of GaAs layers on silicon substrates, because silicon substrates are much cheaper compared to a high-quality GaAs substrate. According to Figure 7.1, the lattice match between GaAs and silicon is poor, but rather favorable for GaAs on germanium. Therefore, the lattice match to the GaAs layer can be improved if a strained Si_xGe_{1-x}/Si buffer layer is grown first. If the composition, thickness, and deposition temperature of the buffer layer are optimized, a rather defect free layer with a better lattice match to the GaAs layer can be obtained. Although GaAs solar cells grown on germanium substrates have shown good conversion efficiencies of 21% at AM1 [11], no satisfactory results have been obtained yet for the Si/SiGe/GaAs system [12,13]. A summary of recent results for III-V compound solar cells is given in Table 7.3.

A further attempt to reduce the dislocation nucleation during epitaxial growth is the doping with isoelectronic elements which are electrically not relevant. For instance, in GaAs it has been found that doping with indium at high concentrations of about 2×10^{19} cm^{-3} can efficiently reduce the dislocation density in LPE-grown layers. The doping concentrations are probably high enough so that the dislocation motion is impeded by elastic interaction [14].

Besides the occurrence of extended lattice defects, the quality of MBE- or MOVPE-grown epilayers is also degraded by the presence of deep-level point defects. High-temperature growth usually results in a lower density of traps but has major drawbacks as far as achieving abrupt interfaces on the atomic scale, in minimizing interdiffusion in the heterostructure, and in maintaining narrow diffusion profiles. One growth scheme to achieve better quality at lower temperatures is also the doping with isoelectronic elements such as indium or antimony. This has been shown to reduce the density of deep traps and enhance the mobility of the charge carriers of the material (see Table 7.4).

Table 7.3
Comparison of Conversion Efficiencies of Different Cell Designs for Single-Crystal III-V Compounds (Results are for *pn*-homojunction cells unless otherwise stated)

System	Efficiency [%]	V_{oc} [V]	I_{sc} [mAcm^{-2}]	Fill Factor [%]
GaAs	24.3	1.035	27.57	85.3
AlGaAs/GaAs	24.8	1.029	27.89	86.4
AlGaAs	16.0	1.200	16.20	82.6
GaAsP	17.7	1.213	17.60	83.0
GaAs/Si	17.6	0.891	25.50	77.7
GaAs/Ge Tandem	24.1	1.190	23.80	84.9
InP	20.4	0.860	27.80	84.4
GaInP/GaAs tandem	25.1	2.293	12.50	87.5

Table 7.4
Mobility of Isoelectronically Doped and Undoped MBE-Grown
GaAs and GaAlAs

	Doping Impurity	Mobility $[cm^2/Vs]$
MBE-GaAs (550°C)	Sb (isoelectronic)	2880
	Undoped	2020
MBE-AlGaAs (620°C)	Sb (isoelectronic)	2230
	Undoped	1770

Because of the technological importance of epitaxial GaAs layers for device applications, there is a strong driving force for further investigation of the growth processes and the defect generation linked to it. From the perspective of the photovoltaic application of GaAs, these developments can be directly incorporated into solar cell technology and, basically, no additional efforts are necessary that are relevant only for this technology. It is, however, evident, considering the complexity of the problems encountered for GaAs, that the utilization and investigation of other binary or multinary compounds as solar cell materials will be a formidable task for the future.

REFERENCES

[1] Gatos, H. C., and J. Lagowski, *Mater. Res. Soc. Symp. Proc.*, Vol. 46, 1985, p. 155.
[2] Meyer, B. K., D. M. Hoffmann, and J. M. Spaeth, *Proc. 14th Int. Conf. Defects in Semiconductors*, Switzerland: Transtech, 1986, p. 311.
[3] Samuelson, L., P. Omling, H. Titze, and H. G. Grimmeiss, *J. Crystal. Growth*, Vol. 55, 1981, p. 164.
[4] Kremer, R. E., D. Francomano, G. H. Beckhart, K. M. Burke, and T. Miller, *Mater. Res. Soc. Symp. Proc.*, Vol. 144, 1989, p. 15.
[5] Mattson, S., and H. D. Shih, *Appl. Phys. Lett.*, Vol. 48, 1986, p. 47.
[6] Mahajan, S., and M. A. Shahid, *Mater. Res. Soc. Symp. Proc.*, Vol. 144, 1989, p. 169.
[7] Al-Jassim, M. M., J. P. Goral, P. Sheldon, and K. M. Jones, *Mater. Res. Soc. Symp. Proc.*, Vol. 144, 1989, p. 183.
[8] Van Der Merwe, J. H., and N. G. Van Der Berg, *Surface Sci.*, Vol. 32, 1972, p. 1.
[9] Dodson, B. W., and J. Y. Tsao, *Appl. Phys. Lett.*, Vol. 51, 1985, p. 1325.
[10] Matthews and A.E. Blakeslee, *J. Crystal. Growth*, Vol. 27, 1974, p. 118.
[11] Fan, J., C. Bozler, and B. Palm, *Appl. Phys. Lett.*, Vol. 35, 1979, p. 875.
[12] Gale, R., J. Fan, B. Tsaur, G. Turner, and F. Davis, *IEEE Electron. Dev. Lett.*, Vol. 2, 1981, p. 169.
[13] Katoda, T., and Kishi, *J. Electron. Mater.*, Vol. 9, 1980, p. 738.
[14] Blom, G. M., and J. M. Woodall, *J. Electron. Mater.*, Vol. 17, 1988, p. 391.
[15] Von Bardeleben, H. J., and D. Stievenard, *Mater. Res. Soc. Symp. Proc.*, Vol. 104, 1988, p. 353.

Chapter 8
Thin-Film Compound Semiconductors

Compound semiconductors usually have high absorption coefficients so that most of the light can be absorbed in a layer of about 1-μm thickness or less. This is the typical range of the thin-film technology, which is a standard technology in the microelectronics industry. A variety of thin-film deposition techniques are thus available, offering great flexibility for the thin-film preparation. Other advantages of thin-film solar cells are that less material is required and that the thin layers can be deposited on many different substrates. Apart from epitaxial growth techniques, thin films are deposited either as polycrystalline, nanocrystalline, or amorphous layers. In this chapter, the properties and material issues of the major polycrystalline thin-film materials will be discussed.

The choice of semiconductors for photovoltaic conversion is based on a number of requirements which have been discussed in previous chapters. They include the following conditions for a solar cell material:

1. A direct band gap with nearly optimum values for either homojunction or heterojunction devices;
2. A high optical absorption coefficient, which minimizes the requirement for high minority carrier lengths.
3. The possibility of producing n- and p-type material, so that the formation of homojunction as well as heterojunction devices is feasible. Generally, p-type material is preferred because electrons in many cases have a higher mobility, and the materials therefore exhibit a higher minority carrier length. Another reason is that most suitable window materials have an n-type character, and a p-type absorber is needed in a heterojunction device.
4. A good lattice and electron affinity match with large band gap (window) materials such as CdS or ZnO so that heterojunctions with low interface state densities can be formed and deleterious conduction band spikes can be avoided.

These requirements are fulfilled by a number of II-VI compounds (which will be discussed in the first part of this chapter) and a wide range of multinary semiconductors mainly based on copper ternary compounds with the chalcopyrite structure. This group of ternary semiconductors includes promising candidates for solar cell applications; however, many basic properties are still not well known. Therefore, it is justifiable to discuss some fundamental problems of these semiconductors in more detail.

8.1 II-VI COMPOUNDS

The major representatives of binary II-VI compounds are based on the elements cadmium, zinc, and mercury in combination with sulfur, selenium, and tellurium. Some of the compounds and their relevant electrical properties are summarized in Table 8.1. Compounds with smaller band gaps have attracted considerable interest for optical applications such as photodetectors and lasers. For photovoltaic applications, only the cadmium and zinc compounds are directly suitable. Since they are direct band gap semiconductors, they have high absorption coefficients and

Table 8.1
Structural and Electronic Properties of Some Monocrystalline and Polycrystalline II-VI Compounds
(The carrier concentrations are for unintentionally doped crystals)

Single Crystals	a, c [Å]	E_g [eV]	μ_n [cm^2/Vs]	μ_p [cm^2/Vs]	Carrier Concentration [cm^{-3}]	
ZnS		2.8				
CdS		2.3				n
ZnSe	5.669	2.7	0.53×10^3	28	$(0.3 - 4.7) \times 10^{16}$	n, p
CdSe	4.299	1.8	0.65×10^3	50	$(1.0 - 9.6) \times 10^{15}$	n
	7.015					
HgSe	6.084	-0.15	2.20×10^4		$(0.8 - 3.1) \times 10^{17}$	n
ZnTe	6.104	2.26	0.34×10^3	110	$(1.0 - 6.1) \times 10^{16}$	p
CdTe	6.481	1.50	1.05×10^3	80	$(0.7 - 4.2) \times 10^{15}$	n, p
HgTe	6.461	-0.14	3.20×10^4		$(2.2 - 6.9) \times 10^{17}$	n, p
Polycrystals						
ZnSe			0.01–0.1		10^9–10^{12}	
ZnTe			0.1–10		10^{10}–10^{20}	
CdSe			20–380		10^{11}–10^{16}	
CdTe			20–400		10^{12}–10^{14}	i
CdTe			200–400		10^{20}–10^{21}	n
CdTe			30–40		1.5×10^{22}	p
$Hg_{1-x}Cd_xTe$						
$x = 0.10$			20×10^3		2×10^{17}	n
$x = 0.15$			12×10^3		—	n
$x = 0.20$			8×10^3		1×10^{17}	n

can therefore be used as thin-film materials. CdS was among the first semiconductors used for solar cell applications, and several attempts have been made to develop a commercial CdS/Cu₂S solar cell. Although frequently described as the "CdS-cell," the optically active layer is Cu_2S, whereas CdS with a rather large band gap of about 2.3 to 2.5 eV serves mainly as a substrate material for the deposition of the Cu_2S layer (see Figure 3.10). Because of its wide band gap and the ease with which CdS can be fabricated, it is still frequently used as a window material, particularly in a variety of thin-film solar cells.

The most suitable II-VI semiconductor for solar cells is CdTe with a band gap of about 1.5 eV at room temperature, which matches the solar spectrum perfectly. This semiconductor has been receiving variable attention as a candidate for photovoltaic applications since the early 1960s. Despite its potential, the utilization of this semiconductor has been hampered by the difficulty of controlling the material properties. Much of the pioneering and early work with this semiconductor was accomplished on single crystals. As-grown single crystals tend to be semi-insulating and doping is therefore required. Metallic dopants such as copper that do lower the resistivity have very high diffusion coefficients, which may be one reason for the frequently observed instability of the material and the solar cells. It is also difficult to make electrical contacts to CdTe, especially to the *p*-type material. Therefore, the development of a *pn*-junction solar cell was not very successful or promising in the past.

The interest in CdTe increased enormously when it was reported that efficiencies of greater than 10% were achieved for sintered polycrystalline thin-film cells [1]. This observation stimulated the further research on polycrystalline thin films, and several techniques were explored to produce larger area cells and higher efficiencies. In fact, it turns out that a variety of methods, such as electrodeposition and chemical vapor deposition, exist for making good device-quality thin films, which makes this material particularly attractive from the commercial point of view.

8.1.1 Thin-Film Preparation and Microstructure

Most of the II-VI compounds crystallize in the cubic zincblende structure; only CdSe prefers the hexagonal wurtzite structure at room temperature. There seems to be, however, only small energy differences in the lattice structures of the cubic and hexagonal phases of CdTe and ZnSe, since the wurtzite structure can easily occur during thin-film growth. With the exception of the mercury compounds, the semiconductors have a direct band gap which varies considerably within this group of compounds. HgTe and HgSe are already semimetals, indicated by their negative band gap energy (Table 8.1). HgTe can easily form a ternary compound with CdTe which offers the possibility of varying the band gap continuously by changing the composition. Experimental results for $Hg_{1-x}Cd_xTe$ show that the band gap changes linearly with the composition x according to the relation $E_g = E_{HgTe} + x(E_{CdTe} - E_{HgTe})$. For small x, the band gap width is close to zero, which can be used to develop a semiconductor with an adjustable small band gap for infrared detectors. For photovoltaic applications, larger band gaps are required, and $Hg_{1-x}Cd_xTe$ with small additions of mercury ($x > 0.85$) have been used because the ternary compound offers a lower resistivity and easier contacting. The wide band gap II-VI semiconductors have absorption coefficient α, usually exceeding 10^4 cm^{-1} above the absorption edge so that films with a thickness of about 1 μm are typically sufficient to absorb the light.

Many of the crystal growth and thin-film deposition techniques are carried out at high temperatures. It is therefore important for the determination of the stoichiometry of the compounds to control the vapor pressures of the components at these temperatures. This becomes difficult or even impossible if one component of the compound is very volatile like, for instance, arsenic in GaAs or antimony in InSb. The vapor pressures of the components of the cadmium and zinc tellurides and selenides in equilibrium with the compound differ only slightly. Therefore, one can generate stable vapor pressure conditions and obtain crystals with a homogeneous stoichiometry rather easily. In the case of compounds containing mercury, which is a rather volatile element, some deposition techniques cannot be employed because of this difficulty. In the field of vacuum technology, methods with one, two, or three (for ternary compounds) sources for the thermal evaporation are used [2]. None of these techniques seems to have a distinct advantage for the preparation and quality of the II-VI compounds, and their application is thus mainly determined by other (practical) considerations.

The microstructure of the films is mainly determined by the substrate temperature and to some degree by the lattice match of the compound. In general, substrate temperatures above about 100°C are necessary for the zinc and cadmium compounds to obtain a polycrystalline film, whereas lower temperatures can be used for the mercury compounds. Selenides have a greater tendency to form the amorphous phase, probably because of the occurrence of Se$_n$ chains ($n = 2$ to 8) in the gas phase, which makes it more difficult to form the compound during deposition. The grain sizes of the polycrystalline films are typically in the range below about 2 μm. For oriented substrates and a lattice structure that matches the lattice of the compound to some extent, epitaxial growth can be achieved in a particular temperature range.

The deposition of CdTe can result in the formation of the cubic β- and hexagonal α-CdTe modification. For instance, thin films prepared by the MBE technique with neutral beams mainly consisted of α-CdTe when deposited at low substrate temperatures ($T \approx 100$°C), whereas the β-modification dominates at temperatures above 300°C. It has also been observed that silver, which is an acceptor dopant, also promotes film recrystallization, and under optimal conditions grain sizes of about 1 to 2 cm can be obtained. In addition, the hexagonal phase forms predominantly during recrystallization.

8.1.2 Electrical Properties

Some of the fundamental electronic properties of the II-VI compounds are listed in Table 8.1. Since some of the compound semiconductors have the tendency to occur both in the sphalerite and wurtzite structure if deposited as a thin film, and since both structures also have different electronic properties, this causes problems concerning the homogeneity of the films. The electronic behavior may thus vary considerably and unpredictably with the deposition conditions.

In general, the mobilities of electrons are considerably larger compared to the hole mobility; therefore, the *p*-doped base material is preferred. Undoped crystals have a conductivity type determined by the stoichiometry, and changing the composition often converts the conductivity type. In most cases, it is not clear which intrinsic defects determine the conductivity in undoped crystals. For instance, for CdSe, it is assumed that the *n*-type conductivity is caused by either the cadmium interstitials or the selenium vacancies. The carrier concentrations of as-grown crystals are in the range of 10^{17} to 10^{18} cm^{-3}, and particular precautions have to be taken to obtain a stoichiometry of the crystal with a lower carrier concentration.

The situation is even less clear for CdTe, which is also *n*-type in the as-grown state, with a carrier concentration of about 10^{16} cm^{-3}. Annealing in cadmium vapor leads to an increase of the conductivity similar to CdSe, although it is not clear which intrinsic defects cause it. Correspondingly, the conductivity of *p*-type material heated in tellurium vapor increases for high resistivity material. A shallow level acceptor with a level at 0.05 eV above the valence band edge is observed and has been related to the cadmium vacancy, which may occur in excess in tellurium-rich CdTe. For low-resistivity CdTe, however, the opposite effect is observed due to the occurrence of a number of defect levels in the range 0.1 to 0.28 eV above the valence, which compensate the semiconductor. The nature of these defects has not been identified yet. The situation is also complicated by the fact that the crystals and particularly the thin films are not homogeneous and may contain impurities, precipitations, or dislocations in an uncontrolled way, which makes the comparison and interpretation of the results of different authors difficult. Generally, the identification of the defect structure and the corresponding trap levels is certainly still in an infant state for the II-VI compounds compared to the elemental and some of the III-V compounds.

The electrical properties of the homogeneous polycrystalline thin films generally differ from the properties of the single crystals. This is due to the high density of the grain boundaries and their influence on the carrier concentration and impurity distribution, as discussed in Section 5.5. The main effects are the compensation of the doping elements due to the grain boundary states and the change of the mobility because of the grain boundary barriers. The experimental data for the zinc and cadmium compounds given in Table 8.1 show the much lower carrier concentrations compared to the single crystals [2]. Although there are no systematic data on the doping dependence available, the results for CdTe show that for high doping levels, both for indium-doped n-type CdTe and silver-doped p-type CdTe, carrier concentrations can be reached that are only slightly smaller that the single crystal values for the same doping level. This is in general agreement with the model for grain boundary-controlled conductivity in polycrystalline materials, which shows that, in a narrow range of doping levels (usually about 10^{18} cm^{-3} for this grain size), the conductivity increases over several orders of magnitude and saturates almost at the single-crystal values. Another way to reduce the resistivity at lower doping levels is the formation of $Hg_{1-x}Cd_xTe$ films. As can be seen from the data in Table 8.1, the addition of smaller amounts of mercury ($x > 0.85$) still provides a large band gap but reduces the resistivity.

The electrical properties of CdSe and CdTe are influenced by the adsorption of impurities at the surface, particularly oxygen and water vapor. This causes instabilities of the material which is not desired for solar cell applications. The doping and stability problems and to some extent the difficulty of fabricating good ohmic contacts on CdTe are partly overcome by the development of suitable cell designs. Most of the cell concepts are based on heterostructure designs employing a variety of materials (Figure 8.1). High efficiencies have been reported for CdS/CdTe (13% to 15%) and SnO_2/CdTe (10.5%) solar cells. Other concepts use the ternary compound $Hg_xCd_{1-x}Te$, which provides the lower resistivity and easier contacting. Efficiencies of about 10% have been achieved for thin-film CdS/$Hg_{1-x}Cd_xTe$ solar cells. Some results are compiled in [1].

The full potential of polycrystalline CdTe-based solar cells has certainly not been fully explored yet, but this semiconductor is considered by some photovoltaic experts to be one of the most promising materials for thin-film solar cells. The variety of methods of making device quality CdTe thin films in large areas is one strength of the technology. Another is that CdTe and its alloys have other applications besides photovoltaics, which permits the coemergence of technologies. However, the material and contacting problems, the instability, and the low commercial interest for solar cells must be offset. It should also be mentioned that environmental concerns about the use of cadmium and other heavy metals may entirely prohibit the utilization of cadmium-based solar cells in the future.

8.2 CHALCOPYRITE SEMICONDUCTORS

A large number of ternary and quaternary semiconductors are currently being investigated for their potential as high-performance, inexpensive solar cells, which

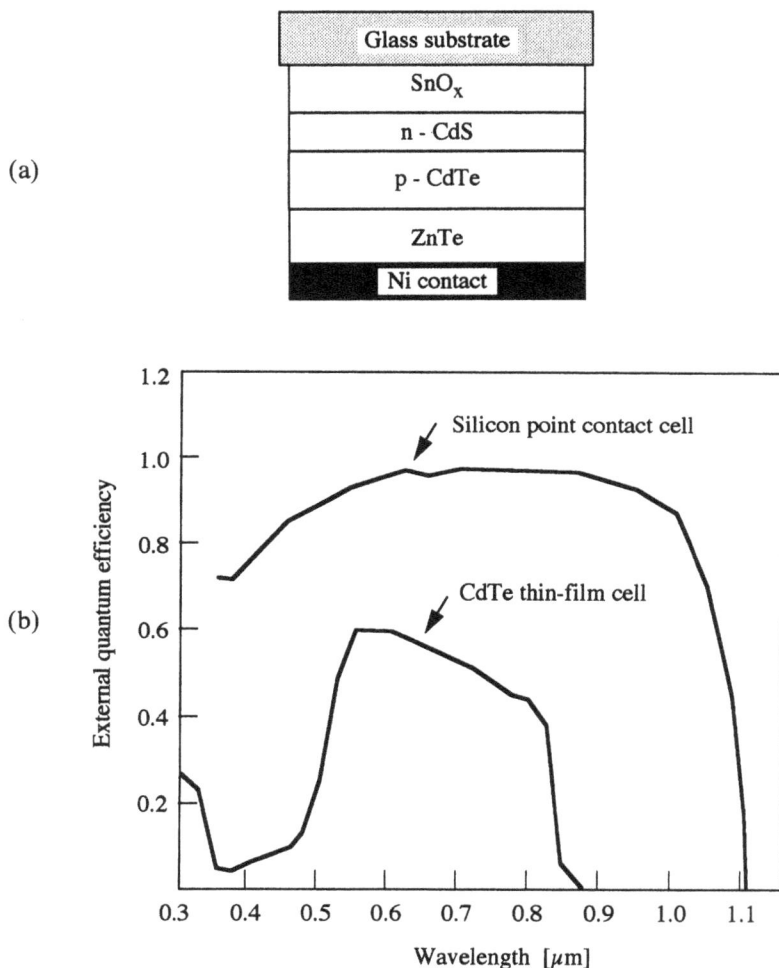

(a)

(b)

Figure 8.1 Cross section of a high-efficiency thin film CdS/CdTe cell (a) and performance characteristics (b) of a cell fabricated by electrodeposition. The cell parameters are $V_{oc} = 0.555$ V, $I_{sc} = 33.71$ mA, $FF = 65.67\%$, $\eta = 12.5\%$. For comparison, the quantum efficiency versus wavelength of a silicon point contact cell with $\eta = 22.3\%$ is included [1].

can serve as an alternative to the monocrystalline and polycrystalline silicon technology. Several groups of ternary semiconductors with the composition $A^I B^{III} X_2^{VI}$ and $A^{II} B^{IV} X_2^V$ (chalcopyrites) have attracted interest because of their diverse optical, electrical and structural properties [3,4]. These compounds offer a broad range of optical band gaps and carrier mobilities, as well as the ability to accommodate different dopants, which has led to their emergence as technologically significant device materials, including applications in photovoltaic solar cells, light-emitting diodes, and various nonlinear optical devices. Quaternary compounds (e.g., $AB_xC_{1-x}X_2$) can be formed by substituting isoelectronic atoms and offer the possibility to change the band gap and other electronic properties continuously. Other quaternary compounds with chalcopyrite structure, which can be derived from ternary $A^IB^{III}X_2^{VI}$ compounds, are of the type $A^IB^{III}C^{IV}X_4^{VI}$. In this structure, vacant lattice sites on the cation sublattice are included. These compounds seem to have band gaps similar to the corresponding ternary compound [5], but their potential as photovoltaic materials has not been systematically explored so far.

Some properties of chalcopyrite compounds with band gap energies in the range 1.0 to 2.5 eV are summarized in Table 8.2. The ternary compounds that

Table 8.2

Structural and Electronic Properties of Ternary ABX_2 Compounds With Chalcopyrite Structure: (a) A = Ag, Cd, Zn; (b) copper ternaries; (c) $CuInSe_2$. (u is the displacement of the cations given in Figure 4.2, χ the electron affinity, E_g the band gap energy, μ the mobility, $T_{\gamma\delta}$ the transition temperature between ordered and disordered phase, N_c and N_v the density of states of conduction and valence band, respectively, m_e and m_h the electron and hole mass, respectively (m_0 free electron mass))

Ternary	a (Å)	c/a	u	E_g (eV)	χ (eV)	μ (cm²/Vs)
$AgInS_2$	5.87	1.910	0.25	1.87		
$AgGaSe_2$	5.985	1.793	0.278	1.83		
$AgInSe_2$	6.09	1.916	0.25	1.24		600 (n)
$AgGaTe_2$	6.283	1.897	0.26	1.2		
$AgInTe_2$	6.406	1.962	0.25	1.0		
$CdGeP_2$	5.74	1.8772	0.283	1.72	3.98	100 (n)
$CdSnP_2$	5.901	1.951	0.265	1.17	4.41	2000 (n)
$ZnSnP_2$	5.651	2.00	0.239	1.66	4.25	55 (n)
$CdSiAs_2$	5.885	1.849	0.298	1.55	3.85	500 (n)
$ZnGeAs_2$	5.672	1.966	0.250	1.15	4.20	23 (n)

Ternary	a (Å)	c/a	u	E_g [eV]	μ [cm²/Vs]	$T_{\gamma\delta}$ [°C]
$CuGaS_2$	5.35	1.959	0.272	2.43		
$CuInS_2$	5.52	2.016	0.2295	1.43	300 (n)	975
$CuGaSe_2$	5.596	1.966	0.243	1.68	20 (p)	
$CuInSe_2$	5.782	2.0097	0.235	1.04	320 (n)	810
$CuAlTe_2$	5.964	1.975	0.25	2.06		
$CuGaTe_2$	5.994	1.987	0.25	1.23	40 (p)	
$CuInTe_2$	6.167	2.00	0.225	1.04	200 (n)	672

E_g [eV]	χ [eV]	μ_n [cm²/Vs]	μ_p [cm²/Vs]	N_c [cm^{-3}]	N_v [cm^{-3}]	m_e/m_0	m_h/m_0	Mismatch with CdS [%]
1.04	4.48	900	10	1.1×10^{18}	2.6×10^{19}	0.09	0.73	1.16

have been most thoroughly investigated so far are the chalcopyrites with the composition $A^IB^{III}X_2^{VI}$ and based on the transition elements A = copper or silver, B = aluminum, gallium, indium, thallium, and X = sulfur, selenium, tellurium. Historically, the copper ternary appeared to be the most attractive candidates because of the previous developments of a similar Cu_2S-CdS solar cell, and the research focused on three of the copper ternary compounds: $CuInS_2$, $CuInSe_2$, and $CuInTe_2$, of which $CuInSe_2$ seems to be the most promising semiconductor so far.

The extensive progress that has been made in experimental studies of the fundamental electrical and structural properties of ternary and quaternary copper chalcopyrites will be presented first. Since $CuInSe_2$ is the leading candidate for photovoltaic application, most of the experimental data have been accumulated for this material. The emphasis will be on the critical discussion of the defect chemistry from the perspective of the photovoltaic properties. In the following section, various processing techniques for bulk and thin-film materials will be presented and the problems that are connected with the design of a solar cell.

The more complex chalcopyrite crystal structure compared to elemental and binary semiconductors, which crystallize in the diamond cubic, sphalerite, or wurtzite structure, and the larger number of atomic species leads to a greater variety of intrinsic point defects and possible atomic structures of lattice defects like grain boundaries and dislocations. It has been established that the electrical and optical properties are dominated by the presence of structural defects, and their control and investigation is, besides the technical design of the solar cells, the greatest challenge for the further development of these materials.

8.2.1 Crystal and Film Growth Techniques

The chalcopyrite lattice can be developed from the sphalerite structure by duplicating the unit cell and arranging the cations A,B on the cation sublattice, as seen in Figure 5.4. The chalcopyrite unit cell is characterized by a tetragonal distortion along the c-axis $\eta = c/2a$ and the anion displacement u. Table 8.2 gives a compilation of experimental data for a, η, u, and the band gap energies for some ternary chalcopyrites [6,7]. The closed packed planes in the chalcopyrite lattice are the {112} planes (corresponding to the {111} planes in the sphalerite structure), which consist of a double layer of anions (X) and cations (A,B). The usual growth direction of thin films is perpendicular to these planes along the ⟨221⟩ direction. Together with the {100} planes, they are also the common cleavage planes of single crystals.

The ternary phase diagrams for most of the compounds have not been completely determined yet. For the Cu-In-Se and Cu-In-S systems, the most complete data on the crystal structure and composition are available. The binary phase relations that border the ternary composition field [8,9] show a variety of intermetallic compounds (Figure 8.2) and several pseudobinary relations between some of these compounds have been established. The best known pseudobinary phase diagrams [10] are published for the phases $Cu_2S-In_2S_3$ and $Cu_2Se-In_2Se_3$, one of which is shown in Figure 8.3. Several other pseudobinary relationships have been determined; for instance, the $CuInS_2$ phase can also coexist with InS and indium.

It appears to be a common feature of many copper ternaries that a phase transition from the chalcopyrite (γ) to the cubic sphalerite structure (δ) occurs at elevated temperatures due to the lowering of the free energy by the entropy contribution of the random distribution of A and B atoms in the cation sublattice (some ternaries exist in the cubic phase already at room temperature). The transition temperature $T_{\gamma\delta}$ for the three compounds $CuInSe_2$, $CuInS_2$, and $CuInTe_2$ are listed in Table 8.2.

It appears that the chalcopyrite structure can be maintained over a rather wide range of compositions. Results for polycrystalline $CuInSe_2$ thin films show a composition window of about 10% for all three components. The deviations from the composition and valence stoichiometry of a compound ABX_2 are usually char-

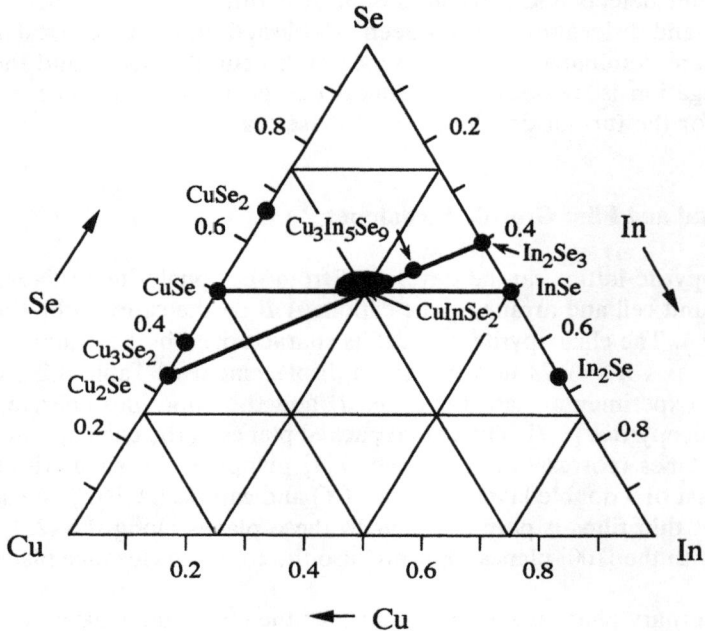

Figure 8.2 Schematic ternary phase diagram for $CuInSe_2$, derived from crystal structure considerations [9].

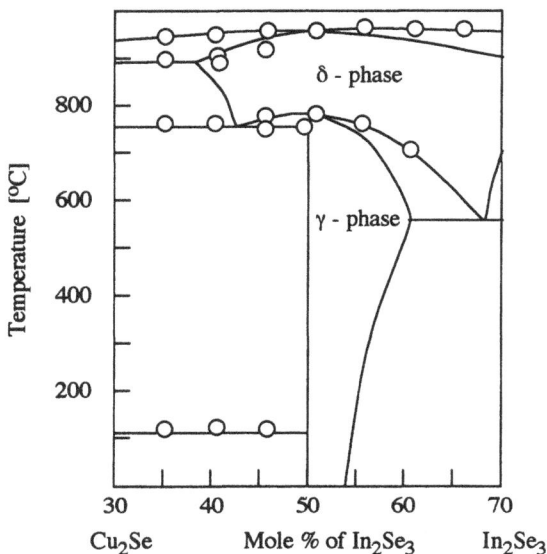

Figure 8.3 Pseudobinary phase diagram for CuInSe₂ based on Cu₂Se-In₂Se₃. Open circles represent experimentally determined transition points [10].

acterized by the molecularity Δm and the valence stoichiometry Δs, quantities that have been defined in Section 5.1 (equation (5.20)). The chalcopyrite structure within the composition range can be maintained by the incorporation of intrinsic point defects such as vacancies, interstitial atoms, and antisite defects. These defects are considered the main electrically active defects and have been related to the electrical properties of the material.

The first successful developments of copper ternary solar cells with efficiencies of about 12% were based on monocrystalline CuInSe₂ and they encouraged the further exploration of these materials. Several principal techniques have been used to grow single crystals of chalcopyrite I-III-VI₂ semiconductors [8].

1. The first experiences with the solidification of stoichiometric melts of CuInSe₂, CuInS₂, and CuInTe₂ uncovered a problem that has hindered the production of large crystals. The chalcopyrite-sphalerite transition and the anisotropic thermal expansion of the crystals along the a- and c-axes leads to a high density of cracks in the material during crystal growth and promotes the tendency to form polycrystals. The best quality and largest crystals have been produced by a high-pressure Bridgman process. The successful growth of n- and p-type crystals has been reported.

2. Direct growth from the melt is also possible from binary constituents (e.g., Cu_2Se-In_2Se_3 or $CuSe$-In_2Se_3 for $CuInSe_2$). Lower pressures are required here which minimizes the danger of explosion during growth.
3. Large platelets of $CuInSe_2$ single crystals have also been grown by chemical vapor transport, and thin layers (≈ 6 μm) have been deposited on ZnSe substrates using LPE. In the latter case, the solute was prepared from a stoichiometric melt and bismuth was used as a solvent.

Since the relevant copper ternaries have high absorption coefficients, they are primarily considered for thin-film solar cells, and the application and development of film growth techniques is therefore of greater importance. A variety of standard thin-film techniques, such as sputtering, PVD, electrodeposition, or spray pyrolysis, have been employed so far, but none is considered the definitive approach. The highest efficiencies have been obtained for $CuInSe_2$ films which are deposited by evaporation or selenization of metal films, described below:

1. Flash evaporation and (resistive heated) single boat deposition using single-phase powder has difficulty in controlling the loss of either one of the components if their vapor pressure is too different. This is the case for $CuInSe_2$, where deficiencies of selenium can result and the stoichiometry is difficult to control. Therefore, two- and three-source schemes have been developed in which either $CuInSe_2$ or selenium is evaporated from separate sources, or the films are deposited directly from the elemental constituents (Figure 8.4). The n- and p-type character of the films can be controlled by varying the different partial pressures.
2. A technologically simple process is the selenization of metal films (Figure 8.5). Metal layers of the different components are deposited on a substrate with a thickness according to the desired stoichiometry of the compound. The reaction of the films with selenium at elevated temperatures occurs when they are exposed to a selenium or H_2S atmosphere, or by depositing selenium directly on the metal layers [11]. Relatively large grains (>1 μm) develop under these conditions.

Polycrystalline films are often deposited on a metal layer (e.g., molybdenum for $CuInSe_2$), which forms a good ohmic contact. The deposition techniques produce films that can have grain sizes of about 1 μm or less, depending on the deposition conditions. It appears that, for optimal performance, grain sizes >1 μm are required in some cases [8].

8.2.2 Electro-Optical Properties and Lattice Defects

Absorption measurements for monocrystalline and polycrystalline $CuInSe_2$ show the enormously high absorption for this material in comparison with other semi-

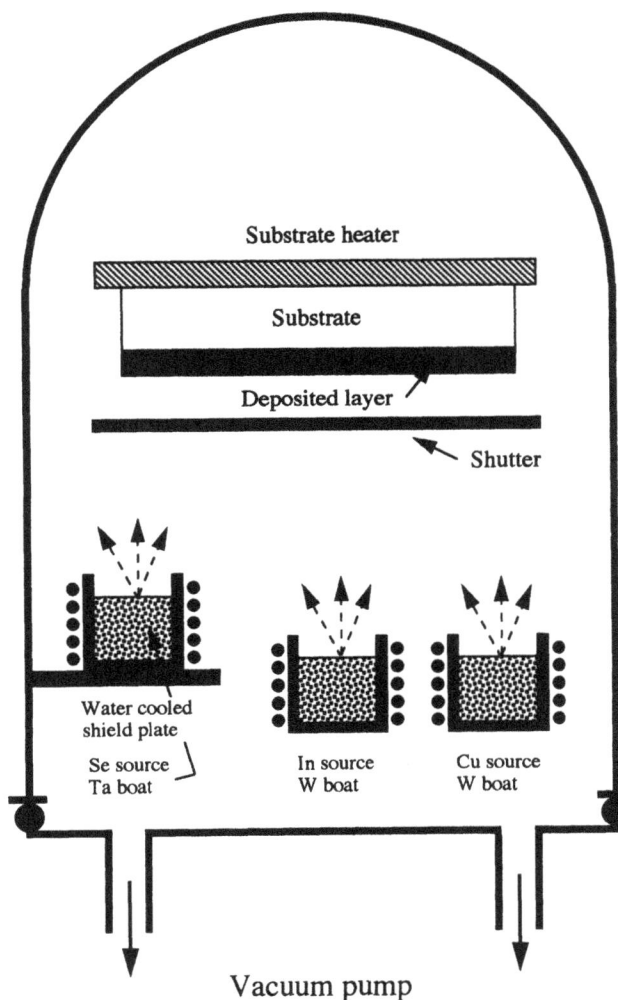

Figure 8.4 Schematic representation of the evaporation method for the fabrication of CuInSe$_2$ thin films using different sources for each element.

conductors (Figure 2.2) [8,12–14]. The fundamental absorption as a function of the photon energy can be described by (2.4) for $h\nu > 0.95$ eV, which confirms that it is an allowed direct band gap transition. Specimens prepared by different techniques and subjected to various annealing treatments show, however, that the empirical band gaps vary between 0.92 and 1.04 eV (Figure 8.6). It has been

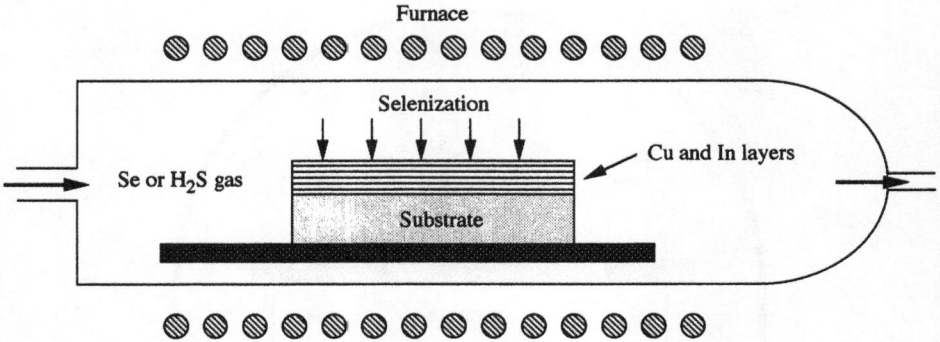

Figure 8.5 Schematic representation of the selenization method for the fabrication of CuInSe₂ thin films.

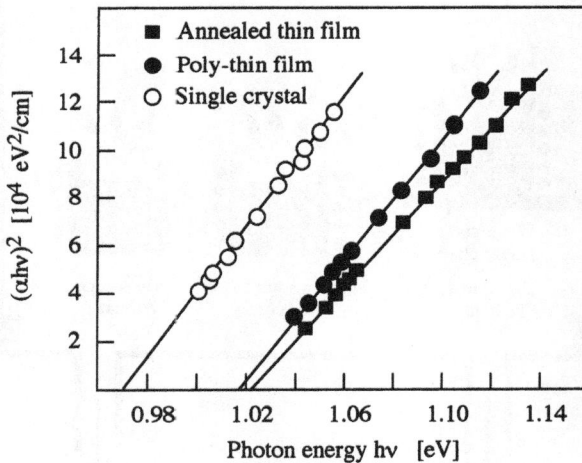

Figure 8.6 Plot of the absorption parameter $(\alpha h\nu)^2$ versus photon energy $h\nu$ for various CuInSe₂ samples showing the absorption behavior of a direct band gap semiconductor (see (2.4)). The variation of the band gap energy is observed for different microstructures of the semiconductor [13].

suggested that band gap narrowing due to high concentrations of point defects is responsible for the observed scatter of the band gaps. According to Section 2.2 (equation (2.7)), the band gap narrowing induced by high concentrations of shallow doping defects N is inversely proportional to the Debye screening length $L_D = (\epsilon\epsilon_0 KT/e^2 N)^{1/2}$. An analysis of the band gap energies as a function of the concentration of ionized point defects and free carriers in a variety of samples indeed confirms the $1/L_D$ dependence (Figure 8.7) [15]. For instance, the extrapolation to $1/L_D = 0$ in Figure 8.7 gives $E_g = 1.01$ eV, which is close to the currently accepted value for the undoped crystal $E_g = 1.04$ eV.

In the low-absorption-coefficient region ($h\nu < 0.95$ eV) of monocrystalline and polycrystalline $CuInS_2$ and $CuInSe_2$, deviations (tails) from the absorption behavior for a direct band gap material are observed. These tails, which have been attributed to defect- (copper vacancies, grain boundaries) related transitions from defect states into the conduction band [16], can obscure the direct band gap character of the absorption edge. The deviations in the low-energy region are also frequently observed for other ternary semiconductors and may be the reason, for instance, for some of the ambiguous results about the direct band gap character of $CuInTe_2$. Generally, however, the measurements support the direct band gap prediction for copper and silver ternaries, with some further explanation of the additional absorption near the band edges.

8.2.2.1 Monocrystalline CuInSe$_2$

An essential feature of the chalcopyrite semiconductors is the high density of point defects which determine not only the optical absorption to some extent, but also the conductivity and other electrical properties. The general observations about the electrical conductivity of monocrystalline $CuInSe_2$, $CuInS_2$, and $CuInTe_2$ are the following: $CuInSe_2$ and $CuInS_2$ can be made n- or p-type by changing the

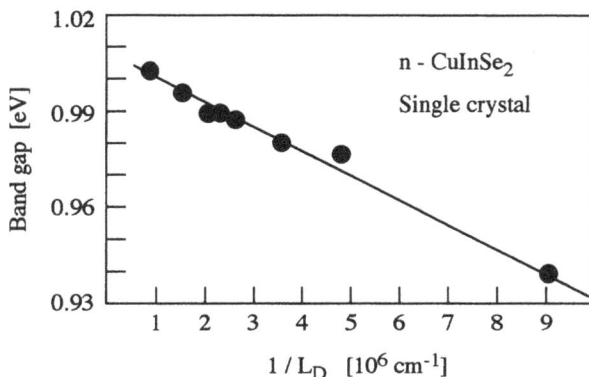

Figure 8.7 Band gap E_g as a function of the inverse Debye screening length $1/L_D$ for n-type CuInSe$_2$ single crystals. $L_D = (\epsilon\epsilon_0 KT/e^2N)^{1/2}$ depends on the concentration N of shallow doping defects. Extrapolation to $N = 0$ yields the band gap for an undoped crystal.

stoichiometry of the crystals, whereas $CuInTe_2$ can so far only be prepared as a *p*-type material, which prevents the fabrication of homojunction devices. Extrinsic doping by shallow impurities has been reported [8] but is usually not applied for device applications.

Typical values for the carrier concentrations in *n*- and *p*-type single-crystal $CuInSe_2$ are in the range 10^{16} to $10^{17} cm^{-3}$ (Table 8.3(c)) The mobilities of *n*-type samples can be as high as $\mu_n = 900$ cm^2/Vs and are generally lower for *p*-type material: $\mu_p \approx 10$ to 30 cm^2/Vs. Measured Hall mobilities as a function of temperature [17] have been analyzed assuming scattering by phonons and ionized point defects using (4.3) to (4.6). The results yielded ionized defect concentrations of about $N_a = 8.4 \times 10^{18}$ cm^{-3} (*p*-type) and $N_d = 7.9 \times 10^{17}$ cm^{-3} (*n*-type), which was considerably higher compared to the free carrier concentrations. This indicates a high degree of compensation and seems to be characteristic for copper ternary semiconductors.

Systematic investigation of the relationship between the conductivity and stoichiometry of the crystal is depicted in Figure 8.8 for $CuInSe_2$ [18,19]. It is generally accepted now that the results can be explained on the basis of a point defect model, which assumes that intrinsic point defects compensate for the deviation from the ideal stoichiometry and introduce shallow acceptor and donor levels into the band gap. The intrinsic point defects that have to be considered in general are vacancies, interstitials, and antisite defects. In a compound with the general formula ABX_2, twelve different native defects can occur, which are compiled for $CuInSe_2$ in Table 8.3 (see also Section 5.1). The main question is to identify the dominant doping defects for each composition.

The basis for the current interpretation is the assumption that chalcopyrite semiconductors are very similar to ionic crystals for which defect chemistry models

Table 8.3

Calculated Bulk Formation Energies H^F of all Intrinsic Defects in $CuInSe_2$ (H^{FB} are calculated formation energies for the same defects when formed at incorrect band sites at grain boundaries (e.g., in a polycrystalline material). The donor or acceptor character is determined assuming an ionic compound.)

Vacancies/Interstitials	$V_{Cu}^{0/-}$	$V_{In}^{0/-}$	$V_{Se}^{0/+}$	$Cu_i^{0/+}$	$In_i^{0/+}$	$Se_i^{0/-}$
ΔH_V [20]	2.6	2.8	2.4	4.4	9.1	22.4
ΔH_V [21]	3.2	2.4	2.6			
ΔH_{FB} [21]	1.6	1.1	2.4	2.3	2.8	2.7

Antisite defects	$Cu_{In}^{0/-}$	$In_{Cu}^{0/+}$	$Cu_{Se}^{0/-}$	$In_{Se}^{0/-}$	$Se_{Cu}^{0/+}$	$Se_{In}^{0/+}$
ΔH_A [20]	1.3	1.4	5.6	2.9	6.7	2.6
ΔH_A [21]	1.9	1.6	5.4	5.0	6.0	5.2
ΔH_{FB} [21]	1.6	1.1	2.4	2.3	2.8	2.7

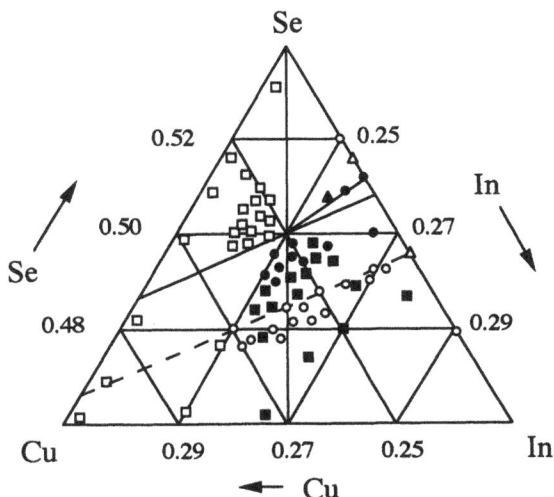

Figure 8.8 Schematic representation of the composition of *n*-type (closed symbols) and *p*-type (open symbols) CuInSe$_2$ single crystals in the ternary phase field near the ideal stoichiometry. Square symbols are from [18], circles from [27], triangles from [28].

are available. Theoretical calculations indeed confirm, at least for some compounds, the significant ionic contribution to the covalent bonding of the atoms. This is shown in a quantum mechanically based calculation of the total electron density for CuInSe$_2$ in Figure 8.9. Both the Cu-Se and In-Se bonds are polarized with a stronger polarization towards the anion site for the In-Se bond. It is thus reasonable to assume that the atoms carry a polarization charge and that the intrinsic defects can be charged as well. Considering that the chalcopyrites are also semiconductors, one has to assume that the intrinsic defects may introduce electronic levels in the band gap that can be occupied.

The underlying assumption for the defect chemistry in ionic crystals is that certain equilibria exist between the defects which can be described by a mass law of action. For instance, for the *A*-atoms in an *ABX*$_2$ compound, one can consider the following equilibria between vacancies, interstitials, and antisite defects

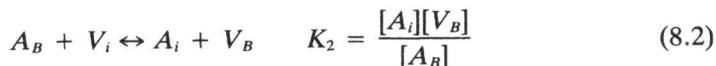

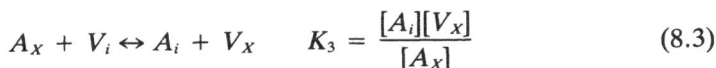

$$A_A + V_i \leftrightarrow A_i + V_A \qquad K_1 = [A_i][V_A] \tag{8.1}$$

$$A_B + V_i \leftrightarrow A_i + V_B \qquad K_2 = \frac{[A_i][V_B]}{[A_B]} \tag{8.2}$$

$$A_X + V_i \leftrightarrow A_i + V_X \qquad K_3 = \frac{[A_i][V_X]}{[A_X]} \tag{8.3}$$

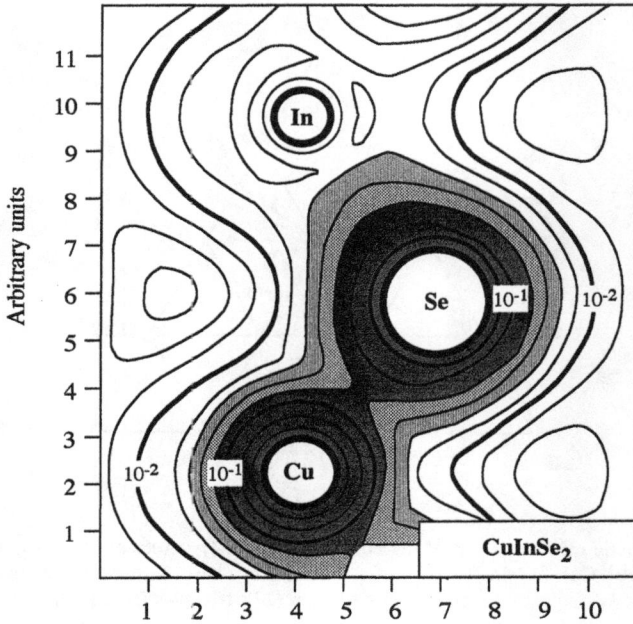

Figure 8.9 Calculated total valence charge density for next-neighbor atoms in CuInSe$_2$. Logarithmically spaced contour plot with shading highlighting the formation of a covalent lobe around the Cu-Se contact [29].

Corresponding equations for the B- and X-atoms can be easily developed. In thermodynamic equilibrium, the defect concentrations are thus not independent of each other, but are related by these internal equilibria. The equilibrium constants K_i are determined by the formation enthalpies ΔH_m and entropies ΔS_m of the different defects and are given by

$$K_i = \exp\left\{\sum_{m=1}^{M} v_m \Delta S_m / K\right\} \exp\left\{\sum_{m=1}^{M} v_m \Delta H_m / KT\right\} \tag{8.4}$$

where v_m are integers that describe the equilibrium reactions

$$\sum_{m=1}^{2} v_m [C_m] \leftrightarrow \sum_{m=3}^{4} v_m [D_m] \tag{8.5}$$

between corresponding defects. The integers on the left hand side are counted negative. The concentrations are also related to the stoichiometry of the crystal (for instance, given by the stoichiometry Δs and the molecularity Δm, which have been defined in (5.20)). For a given composition of the crystal, the concentrations of all defects are thus completely determined and can be calculated, provided that the formation enthalpies are known.

For $CuInSe_2$, the formation energies ΔH have been approximately calculated [20–22] and are included in Table 8.3. Assuming that the formation entropies are not too different, the results show that the defects with the lowest energies are the antisites In_{Cu} and Cu_{In}, and the vacancies V_{In}, V_{Cu}, and V_{Se}. Based on these formation energies and reasonable assumptions about the entropy of the defects, one can calculate the concentrations of all defects. The numerical results for $CuInSe_2$, depicted in Figure 8.10, show that, in general, only two or three defects dominate in a certain composition regime.

In the absence of doping elements of comparable concentrations, the electrical properties of the crystals are determined by the energy levels introduced by the intrinsic point defects. In most cases, the acceptor or donor character of the defects has been estimated from the charge they carry in the ionic compounds (Table 8.3). There are photoluminscence and other measurements consistent with the assump-

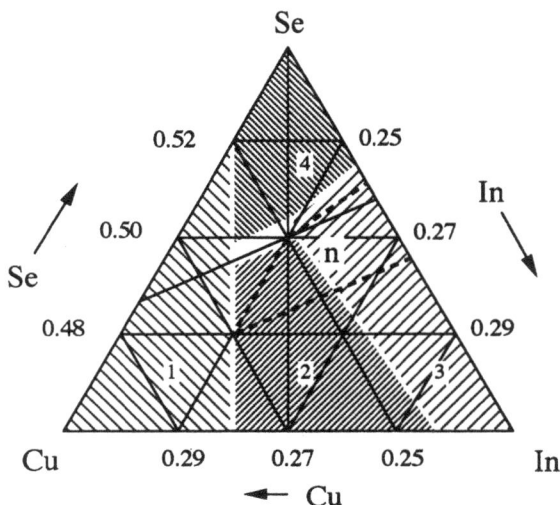

Figure 8.10 Calculated distribution of majority defect pairs in $CuInSe_2$ for large deviations from ideal stoichiometry and room temperature, depicted in a ternary phase diagram. The composition range bound by dashed lines shows n-type conductivity. For compositions close to the ideal stoichiometry other defect pairs occur which are not shown here. (1) $Cu_{Se}^{2+} - Cu_{In}^{2-}$, (2) $V_{Se}^{2+} - In_{Se}^{2-} - Cu_{Se}^{2-}$, (3) $In_{Cu}^{2+} - V_{Cu}^{-}$, (4) $In_{Cu}^{2+} - V_{In}^{2-} - V_{Cu}^{-}$.

tion that in CuInSe$_2$ the dominant defects introduce shallow levels only, and thus determine the doping of the material. Though the properties of none of the defects have been determined directly and unambiguously so far, there is some agreement among the results. Using the most likely set of trap levels for the dominant defect in CuInSe$_2$ (Table 8.4), one obtains the carrier concentrations and conductivity type as a function of the composition, as shown in Figure 8.11. The results are in fairly good agreement with experimental data, so a rather consistent picture has emerged now for the defect chemistry in monocrystalline CuInSe$_2$.

8.2.2.2 Polycrystalline Thin Films

The electronic properties of a polycrystalline compound and possibly the defect chemistry as well may differ considerably from the monocrystalline behavior, if the density of grain boundaries is high, which is usually the case in a thin-film material with grain sizes in the micron range. This general behavior is also observed for copper ternaries, although the electrical measurements from different groups appear to be critically dependent on the method of preparation and annealing treatments after fabrication, and a consistent picture has not emerged yet.

A systematic study of the resistivity and carrier concentration as a function of the composition of the films is depicted in Figure 8.12 [23]. The composition of the films during the evaporation process has been determined by changing the ratio [Cu]/[In] while maintaining the selenium concentration. The results show that in a narrow range of compositions the carrier concentration drops by several orders of magnitude and the conductivity changes from n- to p-type behavior.

A summary of the experimental data which show qualitatively the dependence of the conductivity on the composition is given in Figure 8.13. If the composition of the films is near the ideal stoichiometry, the material is always p-type, whereas a large deficiency of copper converts the conductivity into n-type behavior. The

Table 8.4
Summary of Experimental Results for the Energy Levels of Shallow-Level Defects and Their Assignment to Particular Intrinsic Defects in CuInSe$_2$

Defect	Donor Levels $E_C - E_T$ [meV]		Acceptor Levels $E_V + E_T$ [meV]	
V_{Cu}			70	
V_{In}			50	80
V_{Se}	60	80		
In_{Cu}	20	50		
Cu_{In}			50	100

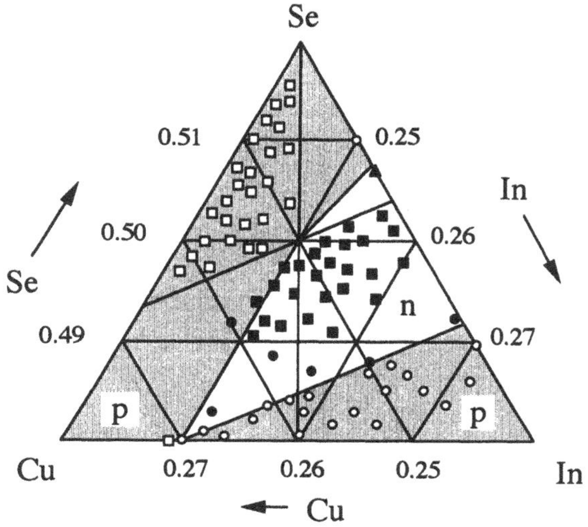

Figure 8.11 Calculated *n*- and *p*-type conductivity as a function of the stoichiometry for CuInSe₂ single crystals and a comparison with experimental results given in Figure 8.8. Calculated regions correspond to the defect chemistry given in Figure 8.10.

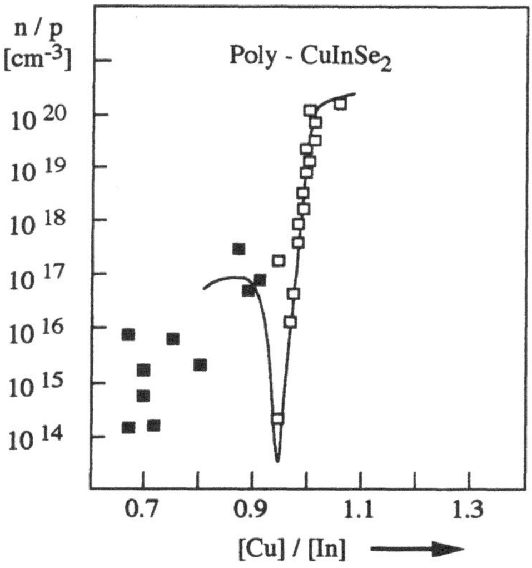

Figure 8.12 Carrier concentration of *p*- and *n*-type CuInSe₂ single crystals as a function of the composition [Cu]/[In]. Black symbols are *n*-type and white symbols are *p*-type [23]. The scatter of results for low copper concentrations is due to the formation of a second phase.

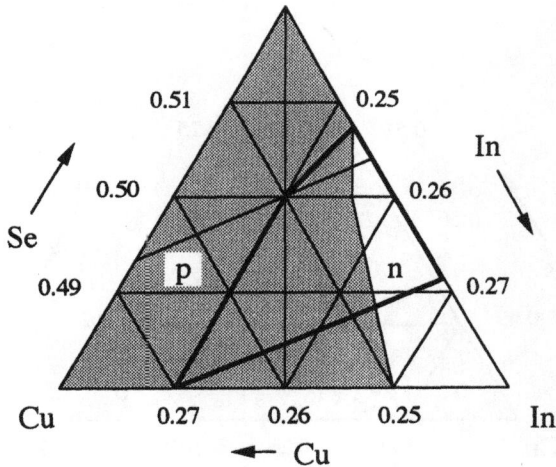

Figure 8.13 Summary of experimental results of the n- and p-type conductivity of polycrystalline CuInSe₂ thin films as a function of the stoichiometry, and comparison with experimental results for single-crystal CuInSe₂ given in Figure 8.11. The n-type region of mono-CuInSe₂ is bound by a solid line.

conversion is also correlated with a strong decrease in the carrier concentration and conductivity. The comparison with conductivity results for single crystals in Figure 8.11 shows that the correlation between conductivity and composition for selenium-deficient polycrystalline films ($\Delta s < 0$) differs significantly from the single-crystal behavior. This can be explained qualitatively by the influence of the grain boundaries, which become increasingly important when the grain sizes are in the micron range or below.

The experimental results for silicon and germanium and theoretical considerations in Section 5.5 have shown that a characteristic behavior of the conductivity in thin films can be a strong variation over several orders of magnitude in a narrow range of doping levels. This occurs when grain boundary traps are filled and the free charge carrier concentration is rapidly reduced. It has been shown that the important parameters determining the electrical behavior are the microstructure, such as the grain size and composition, the grain boundary trap distribution in the forbidden band gap, and the concentration of doping elements.

In ternary compounds, the interpretation of conductivity data is complicated because the doping is related to the composition. The previous discussion of the defect chemistry in monocrystalline CuInSe₂ has shown that the concentration of the intrinsic point defects that determine the doping level can be calculated in agreement with experimental results, but relies on a number of assumptions which have not been confirmed yet. Therefore, an independent determination of the concentration of the doping species in ternary polycrystalline films is not possible, contrary, for instance, to the situation for elemental semiconductors where the doping level is determined by extrinsically introduced impurities. An understanding of the electrical behavior of polycrystalline ternaries is thus even more complex than in single crystals, since the parameters which finally determine the majority carrier concentration and conductivity—the doping concentration, the concentrations of the compensating point defect, and the grain boundaries states—are basically unknown.

A comparison of electrical and structural measurements between single crystals and polycrystals is further complicated by the fact that the defect chemistry in semiconductors with a high density of grain boundaries can be considerably different from single crystals. It has been pointed out and discussed in Section 5.3 that grain boundaries in the chalcopyrite lattice are characterized by an atomic structure with a number of incorrect anion-anion or cation-cation sites which offer attractive places for the formation of intrinsic bulk defects. An approximate calculation of the formation energies of intrinsic point defects formed at particular grain boundary sites is included in Table 8.3. It shows that the formation energies of the major intrinsic point defects in monocrystalline and polycrystalline $CuInSe_2$ are different. The defects with the lowest formation energies are now In_{Cub} (1.1 eV), V_{Inb} (1.2 eV), and V_{Seb} (1.3 eV), compared to the bulk defects In_{Cu} (1.6 eV), Cu_{In} (1.9 eV), and V_{Se} (2.2 eV). The concentrations and electronic trap levels of vacancies, interstitials, and antisite defects in polycrystalline films may thus differ from those in a single crystal.

From the perspective of the solar cell application, the strong dependence of the conductivity on the composition makes the control of the fabrication of the polycrystalline films very difficult. In order to achieve high efficiency in a CdS-$CuInSe_2$ heterojunction cell, a p-type absorber with an optimal carrier concentration of about 10^{16} to 10^{17} cm^{-3} is required, which unfortunately lies in the concentration range where slight changes in the composition change the carrier concentration considerably.

8.2.2.3 Quaternary Compounds

The relatively small band gap of $CuInSe_2$ below the optimal value for photovoltaic applications of 1.5 eV limits the utilization of this material for single-junction devices. Nonetheless, most of the device structures so far have been based on the heterostructure concept with a wide band gap window material (e.g., CdS-$CuInSe_2$ cells), since the highest efficiencies have been obtained with this approach [8]. The research efforts were mainly directed towards the optimization of the device parameters. A further improvement of the performance can be expected if multijunction designs are employed in combination with a wider band gap material. Solar cells based on this concept, for instance, a-Si on CuInSe$_2$, have reached efficiencies of 14.6% [1].

Another approach to optimize the material is to change the band gap of the semiconductor by alloying. Since the band gap increases for the compound series CuInSe$_2$ (1.04 eV), CuGaSe$_2$ (1.68 eV), and CuAlSe$_2$ (2.67 eV), the electro-optical properties of the quaternary compounds CuIn$_{1-x}$Ga$_x$Se$_2$ and CuIn$_{1-x}$Al$_x$Se$_2$ compounds have recently attracted interest. In many cases, the band gap of a compound of two semiconductors with different band gaps can be calculated for the composition x from simple (linear or quadratic) relationships. For instance, for CuIn$_{1-x}$Ga$_x$Se$_2$ at room temperature, the dependence $E_g(x) = 1.018 + 0.575x + 0.108\,x^2$ has been found [11].

In contrast to CuInSe$_2$, which can be made p- and n-type, CuIn$_{1-x}$Ga$_x$Se$_2$ films are always p-type. Considering the conductivity diagram for polycrystalline CuInSe$_2$ (Figure 8.13), it is evident that with the substitution of indium by gallium, the n-type region shrinks and finally disappears. Since the n-type behavior is due to the presence of the donor In$_{Cu}$, one can assume that the corresponding Ga$_{Cu}$ antisite defect which gradually replaces In$_{Cu}$ in the alloy has a higher formation energy and thus a lower concentration. The concentration of holes and therefore the conductivity as a function of the composition x at room temperature shows a minimum at $x = 0.5$ and a strong increase towards the copper-rich side, most likely due to the formation of the secondary phase Cu$_2$Se (Figure 8.14). As with CuInSe$_2$, the compound CuIn$_{1-x}$Ga$_x$Se$_2$ has a very high absorption coefficient exceeding 2 $\times$ 10^4 cm^{-1} near the optical absorption edge.

The full potential of these and other quaternary compounds cannot be estimated at present. It is evident that the complexity of the material, such as composition, defect chemistry, and their relation to physical properties, increases. Currently, the main emphasis is still on the fabrication of thin films and the characterization of the electro-optical properties, and it is unclear in many cases if the material problems can always be solved.

8.2.3 Interface Properties and Oxidation Behavior

Heterostructure, multijunction, or Schottky barrier solar cells that are of appropriate cell design for polycrystalline compound semiconductors contain a variety of interfaces between different materials. In addition, the interface- (including surfaces) -to-volume ratio increases for thin-film solar cells because of thinner layers due to the higher absorption coefficients. The structural, chemical, and compositional characteristics of the interfaces are therefore particularly critical to the performance and operational lifetime of such solar cells.

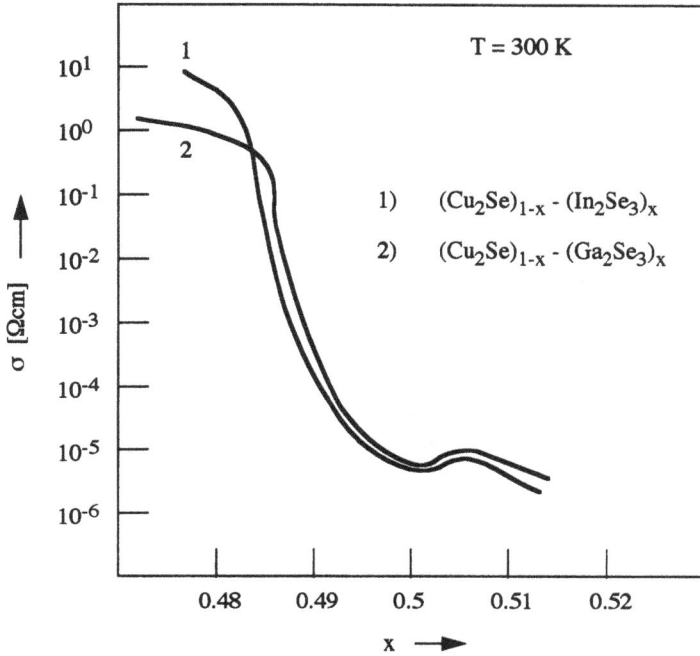

Figure 8.14 Conductivity of CuGaSe₂ thin films versus ratio of [Cu]/[Ga] and comparison with the conductivity of CuInSe₂ thin films versus ratio of [Cu]/[In] [11].

This has also been observed for the CuInSe₂ solar cells, where the correlation between interface properties and the device parameters has been studied intensely. The earlier concept depicted in Figure 3.11 showed low efficiencies, which could be related to undesirable properties of the CdS-CuInSe₂ interface. First, it was necessary to protect the light absorbing CuInSe₂ layer with a CdS or Cd(Zn)S layer from surface-environment interactions which easily degraded the efficiency. Second, it appeared that an important annealing step in air necessary for the improvement of the performance (discussed below) affects the properties of the cells only if CdS is on top. Therefore, efficient device structures are based on the Cd(Zn)S-CuInSe₂-heterojunction cell, for which two typical designs are shown in Figure 8.15.

These designs have been developed in response to the observation that for low resistivity material on either side of the junction, which is necessary to reduce the series resistance, copper precipitates form at the interface which protrude into the CdS layer and destroy the junction properties [8]. The first solution to the

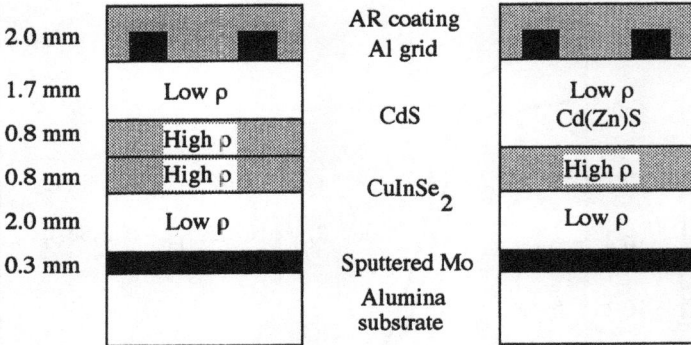

Figure 8.15 Cross sections of two high-efficiency CuInSe₂ solar cells [1]. The designs are developed to prevent the formation of copper precipitates at the heterojunction interface, which reduces the solar cell performance.

problem was to deposit thin layers of high resistivity (copper-poor CuInSe₂ and undoped CdS) on either side of the junction, which seems to prevent the formation of the copper phase. Another solution is the deposition of Cd(Zn)S with a concentration gradient towards the interface. Cd(Zn)S (15% to 30% at Zn) has also a wider band gap, which improves the efficiency slightly if good material quality can be maintained [24]. In both cases, the interface properties have been improved, although a complete understanding of the structure and chemistry of this interface has not emerged yet.

The defect structures of the interface, the grain boundaries, and the bulk are also important for understanding the effect of oxidation on the performance of CdS-CuInSe₂ solar cells. A major and critical step in the fabrication of high efficiency heterostructure CdS and Cd(Zn)S-CuInSe₂ solar cells is an annealing treatment (at 200°C) in air or oxygen after deposition. This step improves the open-circuit voltage and the short-circuit current. It has also been observed that this effect is to some extent reversible if the films are annealed in vacuum or soaked in a reducing solution [25]. Results from a variety of electrical and optical measurements show changes of the electrical activity of the interface and the bulk. Since oxygen diffusion in single crystals is slow but may be enhanced along grain boundaries, it is likely that grain boundary states are mainly responsible for the observed effect. An experimental result that supports this view is a dependence of the required annealing time on the grain size. The major effect of the oxidation seems to be an increase in the net acceptor density due to reduction of the donor density in the compensated thin films. Capacitance and Hall effect measurements on CuInSe₂ films have shown an increase in the acceptor concentration from about

10^{14} to 10^{16} cm^{-3}. IR measurements provide evidence for the formation of In-O bonds and scanning tunneling microscope investigations indicate the accumulation of oxygen at grain boundaries.

The annealing effect is less pronounced in selenium-rich films ($\Delta s > 0$), which points to the participation of the selenium vacancy, which has a low concentration in these films (Figure 8.10). These and other observations have led to a defect model that postulates the neutralization of selenium vacancies (V_{Seb}) formed at grain boundaries and are treated as donor defects. One can see from Table 8.3 that the selenium vacancy at the grain boundary is one of the possible donor defects because of the low formation energy. A schematic illustration of the atomic structure of a selenium vacancy in Figure 8.16 shows that an In-O bond can be formed if the oxygen atom occupies the vacant selenium site. This bond formation is supposed to neutralize the V_{Seb} donor and increase the net acceptor concentration in the material. Though the proposed process seems feasible, a better understanding of the defect chemistry in monocrystalline and polycrystalline CuInSe$_2$ films is certainly necessary to confirm the model.

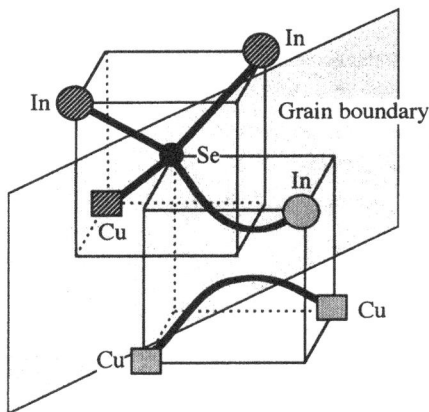

Figure 8.16 Structural atomic model of a vacancy V_{Se} at an incorrect Se-Se bond. Incorrect bonds occur, for instance, at grain boundaries in CuInSe$_2$ [21].

REFERENCES

[1] Kazmerski, L. L., *Intern. Mat. Rev.*, Vol. 34, 1989, p. 185.
[2] *Semiconducting Thin Films of II-VI Compounds*, S. Ignatowicz and A. Kobendza, eds., Chichester: Ellis Horwood, 1990.
[3] Kuriyama, K., and F. Nakamura, *Phys. Rev.*, Vol. B 36, 1987, p. 4449.
[4] Gibart, P., L. Goldstein, and J. L. Dormann, *Proc. 4th Int. Conf. on Ternary and Multinary Compounds*, Jpn. J. Appl. Phys., Vol. 19-3, 1980, p. 184.

[5] Wooley, J. C., R. G. Goodchild, O. H. Hughes, S. A. Lopez-Rivera, and B. R. Pamplin, *Proc. 4th Int. Conf. on Ternary and Multinary Compounds, Jpn. J. Appl. Phys.*, Vol. 19-3, 1980, p. 146.

[6] Jaffe, J. E., and A. Zunger, *Phys. Rev.*, Vol. B 29 (4), 1984, p. 1882.

[7] Romeo, N., *Proc. 4th Int. Conf. on Ternary and Multinary Compounds, Jpn. J. Appl. Phys.*, Vol. 19-3, 1980, p. 5.

[8] Kazmerski, L. L., and S. Wagner, *Current Topics in Photovoltaic*, T. J. Coutts and J. D. Meakin, eds., New York: Academic Press, 1985, p. 41.

[9] Wagner, S., *Proc. Symp. Mater. New Process. Tech. Photovolt. Electrochem. Soc.*, Vol. 83-11, 1983, p. 410.

[10] Palatnik, L. S., and E. J. Rogacheva, *Sov. Phys.-Dokl. (Engl. Transl.)*, Vol. 12, 1967, p. 503.

[11] H. W., Schock, *Polycrystalline Semiconductors, Springer Proceedings in Physics*, Vol. 35, H. J. Möller, H. Strunk, and H. Werner, eds., Berlin: Springer, 1989, p. 246.

[13] Shah, J. L., and J. H. Wernick, *Ternary Chalcopyrite Semiconductors*, New York: Pergamon, 1975.

[14] Kazmerski, L. L., *Ternary Compounds, Inst. Phys. Conf. Ser.*, E. D. Holah, ed., Vol. 35, 1977, p. 217.

[15] Neumann, H., *Solar Cells*, Vol. 16, 1986, p. 317.

[16] Rincon, C., J. Gonzalez, G. Sanchez Perez, and C. Bellabara, *Nuovo Cimento Soc. Ital. Fis.*, Vol. 2D, 1983, p. 1895.

[17] Irie, T., S. Endo, and S. Kimura, *Japn. J. Appl. Phys.*, Vol. 18, 1979, p. 1303.

[18] Neumann, H., and R. D. Tomlinson, *Solar Cells*, Vol. 28, 1990, p. 301.

[19] Abou-Elfotouh, F., D. J. Dunlavy, and T. J. Coutts, *Solar Cells*, Vol. 27, 1989, p. 237.

[20] Neumann, H., *Verbindungshalbleiter*, Leibzig: Akadem. Verlag, 1986, p. 392.

[21] Möller, H. J., *Solar Cells*, Vol. 31, 1991, p. 77.

[22] Kühn, G., and H. Neumann, *Z. Chem.*, Vol. 27, 1987, p. 197.

[23] Noufi, R., R. Axton, C. Herrington, and S. K. Deb, *Appl. Phys. Lett.*, Vol. 45 (6), 1984, p. 668.

[24] Poate, J. M., K. N. Tu, and J. W. Mayer, *Thin Films: Interdiffusion and Reactions*, New York: Wiley, 1978.

[25] Matson, R. J., R. Noufi, K. Bachmann, and D. Cahen, *Appl. Phys. Lett.*, Vol. 50, 1987, p. 158.

[26] Neumann, H., H. Sobotta, W. Kissinger, V. Riede, and G. Kühn, *Phys. Stat. Sol.*, Vol. B 108, 1981, p. 483.

[27] Yip, L. S., W. S., Weng, Z. A. Shukri, I. Shih, and C. H. Champness, *Proc. 10th E.C. Photovolt. Solar Energy Conf.*, Dordrecht: Kluwer Academic Publ., 1991.

[28] Rodak, E., and H. J. Möller, *Proc. 10th E.C. Photovolt. Solar Energy Conf.*, Dordrecht: Kluwer Academic Publ., 1991.

[29] J. E. Jaffe and A. Zunger, *Phys. Rev.*, Vol. B 28, p. 5822, 1983.

Chapter 9
Amorphous Thin-Film Semiconductors

The microstructure of thin films that can be deposited on a substrate by a variety of deposition techniques such as evaporation, sputtering, or chemical vapor deposition, depends on the substrate temperature. For instance, for silicon, a polycrystalline material (grain size ≈ 600Å to 1000Å) is obtained for a substrate temperature above 580°C, and a microcrystalline (≈ 50Å to 80Å) or an amorphous structure below this temperature, usually between 300° and 250°C. The low processing temperatures offer the possibility of depositing amorphous semiconductors on many different substrates, such as on polymers. Besides the fundamental scientific interest in the investigation of the amorphous state, there also exist favorable electro-optical properties that have stimulated the investigation of amorphous materials. Although the potential of amorphous silicon was recognized almost 20 years ago, the electrical and optical properties of evaporated or sputtered films were not encouraging enough at the beginning to use the films for technical applications. The conductivity could not be influenced by doping, nor did these films show any measurable photoconductivity. It soon became evident that the films had a high density of defect states, which prevented a significant and variable conductivity.

This situation changed when it was discovered that amorphous silicon, which was deposited by glow discharge decomposition of silane (SiH_4), exhibited a high photoconductivity and that the electrical conductivity was sensitive to small additions of PH_3 or B_2H_6 during deposition. It turned out that the enhanced electrical properties were related to hydrogen, which is released during the decomposition process and incorporated in the amorphous structure. This breakthrough was an important discovery for electronic and photovoltaic applications. Since then, many structural and electrical properties of amorphous silicon have been studied, which are summarized in many books and review articles [1–4]. At the same time, the technical development has produced solar cells for low-power applications such as watches and calculators. Amorphous silicon solar cells now account for about one-third of the world's photovoltaic market.

9.1 STRUCTURAL AND ELECTRO-OPTICAL PROPERTIES OF AMORPHOUS SILICON

Many of the the structural and physical concepts for the description of the crystalline solid state cannot be applied anymore to amorphous metals and semiconductors. For instance, the notion of a lattice defect, such as a dislocation or point defect, appears meaningless in a disordered structure. It was also unclear whether a band structure existed, which, in a crystalline structure, is closely related to the regular arrangement of atoms. The lack of adequate physical concepts at the beginning offered a great challenge for the investigation of the amorphous state of semiconductors. Although great progress has been made, many difficulties in the understanding of the physical behavior of amorphous materials still remain. For instance, a complete understanding of the important light-induced degradation of amorphous silicon has not been achieved yet. From the perspective of the photovoltaic behavior, the description of the amorphous state and the band structure are the most fundamental issues and shall be discussed first.

9.1.1 Band Gap States

Several deposition processes, such as CVD, reactive sputtering, sputter-assisted plasma CVD, and others, can be used to produce amorphous silicon for solar cells, although the (glow discharge) plasma deposition remains the major film deposition method. In these techniques, hydrogen is either added or released during the process and incorporated into the deposited film. Hydrogenated films contain about 4 to 40 at. % of hydrogen and are generally denoted by a-Si:H. The importance of incorporated hydrogen can be demonstrated, for instance, by annealing a hydrogenated film up to about 600°C. Figure 9.1 shows that above about 500°C, when the hydrogen is released from the film, the photoconductivity decreases abruptly by several orders of magnitude. Although it seems evident that hydrogen passivates defect levels in the amorphous films, great efforts have been taken to study the nature of these defects and their interaction with hydrogen. An essential observation was that the loss of hydrogen and the decrease in conductivity are accompanied by an increase in the density of unpaired electron spins, which can be measured by ESR [5]. According to (2.20), the photoconductivity is determined by the product of lifetime τ and mobility μ; therefore, the results indicate a relationship between $\mu\tau$ and the spin density given by $N_s\mu\tau$ = constant. Since it appears likely that, in the amorphous state of a semiconductor, unsaturated covalent bonds occur, it is assumed that the ESR signal measures the concentration of unsaturated *dangling* bonds. Therefore, the results indicate that hydrogen is trapped at defects with dangling bonds that are electrically active centers, and if unsaturated reduce

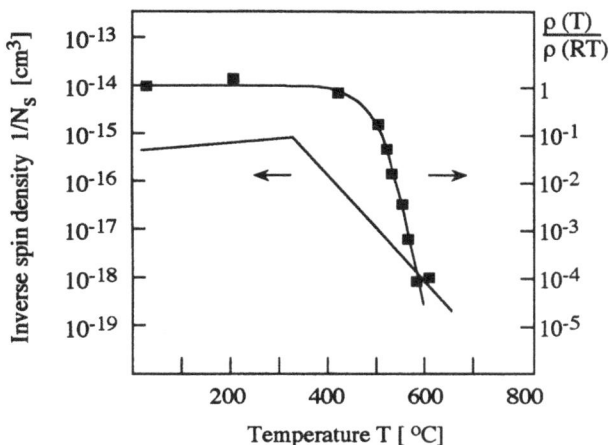

Figure 9.1 Normalized photoconductivity $\sigma(T)/\sigma(300K)$ of undoped a-Si:H after different annealing temperatures [4]. The corresponding dependence of the inverse of the spin density $1/N_s$ shows the correlation between the spin density and the conductivity.

the lifetime of free charge carriers. This concept turned out to be very useful for a description of the properties of the amorphous state.

In a perfect crystalline silicon structure, all four covalent bonds of an atom are saturated by next neighbors, and no unpaired spins occur. Structural models of an amorphous covalently bonded crystal show that the average number of next neighbors is lower than four, so a certain fraction of atoms are only coordinated with three or two nearest neighbors (Figure 9.2). Atoms with lower coordination occur because they locally release the strain in the disordered structure. The three-fold coordinated atoms possess an unpaired (dangling) bond that produces an electron spin signal in an ESR experiment. The amorphous structure is thus characterized by fourfold, threefold, and twofold coordinated atoms and the loss of the long-range order. Since to some extent the local next-neighbor environment of four coordinated atoms remains, there also exists a band structure resembling that of the crystalline state, but also showing fundamental differences. In particular, a band gap can be defined that determines the optical and electronic properties of the semiconductor. The structurally related dangling bond states introduce trap levels in the band gap and thus have to be considered a part of the band structure.

In order to describe the configuration of atoms in the structure, the notation T_z^q will be adopted, where T stands for the atom, q for the charge state, and z describes the coordination number. A neutral threefold coordinated atom is thus denoted by T_3^0. Twofold coordinated atoms T_2^0 can also occur and release the strain

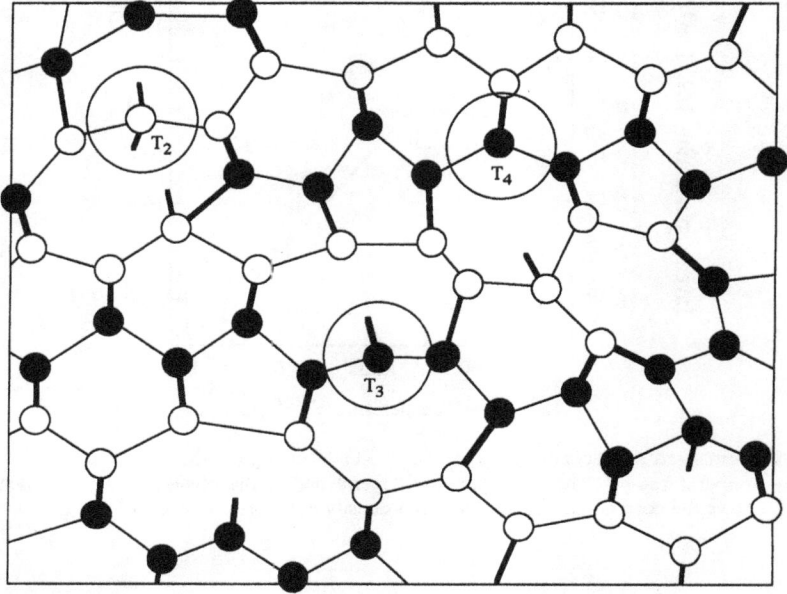

Figure 9.2 Structural atomic model of the covalent bonding in an amorphous semiconductor. Silicon atoms with two, three, and four next-neighbor atoms are indicated.

more effectively compared to a threefold coordinated atom [6]. Since the neutral T_2^0 center has only paired spins, it does not produce an ESR signal and cannot be detected. Nonetheless, it is assumed that these defects also occur. More complicated defects have also been postulated, such as pairs of charged twofold and threefold coordinated atoms $T_3^- - T_2^+$, complexes of several defects, or microvoids.

Hydrogen-free amorphous silicon has spin densities of up to 10^{20} cm^3, depending on the deposition conditions, and it is assumed that the spins are related to uncoordinated covalent bonds, mostly of the T_3^0 atom. A hydrogen atom with a single electron can easily supply an electron and complete an unsaturated covalent bond. This bond formation will remove the dangling bond level from the forbidden band gap (passivation) and trap the hydrogen. For instance, in a good (glow discharge) plasma deposited material, the spin density can be reduced down to 10^{15} cm^{-3} after passivation with hydrogen.

In order to obtain more insight into the electronic properties of amorphous silicon, it is necessary to determine the distribution of the trap states in the band gap. Figure 9.3 shows a result from experimental measurement in n-doped a-Si:H which depicts the main results [7]. Since the majority of atoms are still fourfold

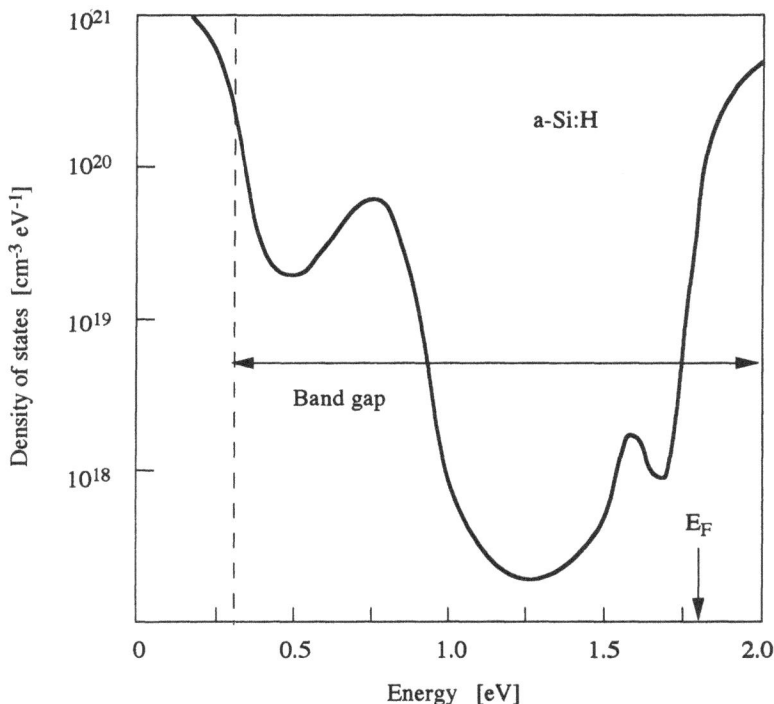

Figure 9.3 Density of states for an *n*-doped a-Si:H film determined by field-effect drift measurements [7]. The position of the Fermi level E_F is indicated by an arrow.

coordinated, the essential features of the crystalline band structure, the existence of extended valence and conduction band states, are still maintained. The lack of the long-range order and distortion of the covalent bonds of fourfold coordinated neutral silicon atoms T_4^0 in the amorphous structure gives rise to a gradual, approximately exponential decrease of the density of states (band tails) given by

$$N_{TC} = N_{C0} \exp \frac{E_c - E}{KT_{TC}} \qquad N_{TV} = N_{V0} \exp \frac{E - E_v}{KT_{TV}} \qquad (9.1)$$

KT_{TC} and KT_{TV} are the characteristic widths of the conduction and valence band tails, respectively. In silicon, typical values are about $N_{C0} = N_{V0} = 10^{21}$ to 10^{22} eV^{-1}cm^{-3}, and $KT_{TC} = 26$ meV and $KT_{TV} = 42$ meV, which means that the valence band tail extends deeper into the gap [8]. Because of the gradual transition of the density of the extended states, the band width is not clearly defined. Nonetheless, particular experimental techniques allow the determination of parameters that can be used to assign a certain value for the band width. For instance, the optical absorption, which will be discussed in the next section, can be analyzed analogously to a crystalline semiconductor and is usually used to determine the "optical band-width."

A fraction of tail states also originates from silicon atoms with three silicon neighbors and one hydrogen atom in tetrahedral configuration. These additional states from the silicon-hydrogen bond changes the optical band gap, depending on the hydrogen content. The optical gap increases from about 1.5 eV for pure amorphous silicon to about 2.0 eV when 30 at. % hydrogen is present. All states in the valence band tail $T_4^{+/0}$ are donor-like; that is, positive when empty and neutral when full. The states in the conduction band tail $T_4^{0/-}$ are acceptor-like; they are neutral when empty and negative when occupied.

Grossly distorted bonds form states deeper inside the gap, and there exists a critical limit for which the states have to be considered localized states. These critical energies are called the mobility edges (E_C, E_V) and are located below the optical band gap edges. The localized states inside the mobility edges essentially determine the Fermi energy and the electrical transport properties.

Superimposed on the band tail states are two deep-level bands in the lower and upper half of the band gap (at about $E_C - 0.4$ eV, $E_V + 0.6$ eV in amorphous silicon), which are thought to be related to dangling bonds of mainly threefold coordinated silicon atoms (see Figure 9.4). They are considered the defect bands of two different charge states of the neutral dangling bond T_3^0. The separation of the two levels is due to the electrostatic correlation energy of the two electrons. Since the local environment of the dangling bonds varies in the disordered structure, one can assume that the positions of their trap levels in the gap also vary slightly, which leads to the observed distribution of states. It is mostly assumed that the upper levels $E_{db}^{0/-}$ correspond to the negatively charged defect $T_3^- \Leftrightarrow T_3^0 + e$ when an electron is trapped, and the lower levels $E_{db}^{+/0}$ to the positive charge defect $T_3^+ \Leftrightarrow T_3^0 - e$.

In order to understand the effect of the Fermi energy on the occupation of the dangling bond levels, one can use (4.30) from Section 4.2. The numbers of dangling bonds in the different charge states N_{db}^+ and N_{db}^- are given by

$$N_{db}^+ = N_{db}^0 \exp\left\{\frac{E_{db}^{+/0} - E_F}{KT}\right\} \qquad N_{db}^- = N_{db}^0 \exp\left\{-\frac{E_{db}^{0/-} - E_F}{KT}\right\} \qquad (9.2)$$

with the total number given by $N_{db}^{tot} = N_{db}^0 + N_{db}^+ + N_{db}^-$. It is evident from these equations that a shift in the Fermi level changes the concentration of positively and negatively charged states, as indicated in Figure 9.4. In an undoped material,

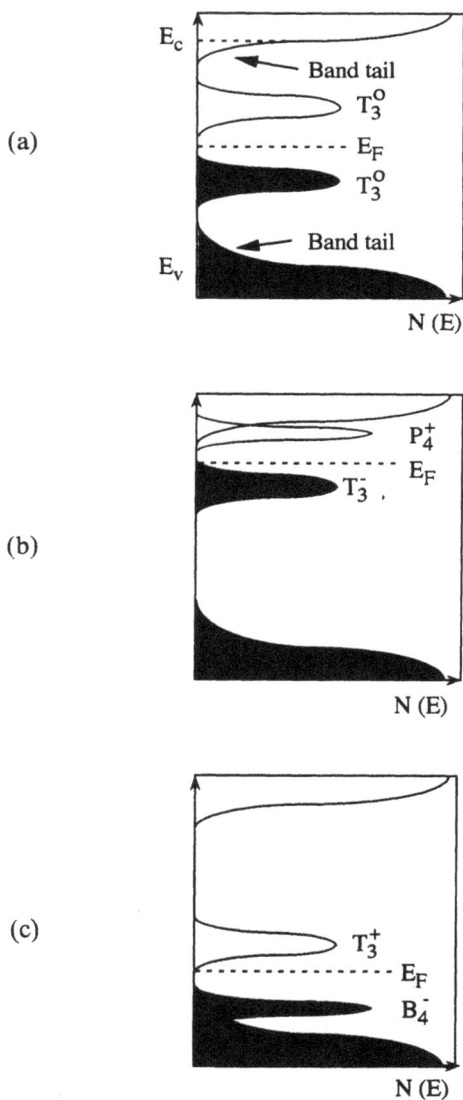

Figure 9.4 Simplified density of states model for dangling bond states in undoped and doped a-Si:H. E_C and E_V denote the conduction and valence band mobility edges: (a) undoped; (b) heavily P-doped; (c) heavily B-doped films. The energy levels of the doping elements are not single shallow levels, but are characterized by a distribution of states due to the variation in the local atomic environment of the doping atoms.

the bonds are neutral and occupied by one electron. In a p-type material, the lowe levels ($E_{db}^{+/0}$) begin to empty and the positively charged dangling bond T_3^+ domi nates, carrying no electrons. In n-type material, the upper levels ($E_{db}^{0/-}$) become partly (or totally) occupied, and the negatively charged dangling bond T_3^- with two electrons dominates. Therefore, the ESR signal which measures only the neu tral bond with an unpaired spin should decrease when the Fermi level shifts toward the band edges. This reduction of dangling bond T_3^0 centers is actually observed for low doping levels [9].

The capture cross sections for electrons σ_n and holes σ_p for the donor $T_3^{+/}$ and acceptor $T_3^{0/-}$ differ considerably: $\sigma_n^+ \gg \sigma_p^+$ and $\sigma_n^- \ll \sigma_p^-$, respectively, while the capture cross sections for electrons and holes for the neutral state are abou equal $\sigma_n^0 \approx \sigma_p^0$. These dangling bond states are effective recombination centers especially in undoped a-Si:H, where they can capture electrons and holes. A con sequence of the different charge states is therefore that the lifetime in undoped a Si:H is rather low and depends very sensitively on the position of the Fermi energy

Amorphous silicon usually also contains impurities that contribute to the density of states in the band gap. The most common impurities are oxygen, carbon and nitrogen with concentrations typically in the range of about 10^{18} to 10^{20} cm^{-3} These impurities have been shown to affect the diffusion length and thus the efficiency of solar cells (see, for instance, [10]).

9.1.2 Doping in a-Si:H

The problem of doping in amorphous silicon becomes evident if one considers tha atoms in the disordered structure are to some extent rather threefold (or less) coordinated compared to the fourfold coordination in crystalline silicon. Doping atoms in the crystalline structure supplies holes and electrons due to an excess of deficiency of electrons compared to the number of electrons necessary to form four covalent bonds (formulated as the 8-N rule). It has been argued that a doping atom on a substitutional threefold coordinated site will therefore not be able to supply holes or electrons according to this basic argument. Nonetheless, one can observe a doping effect in a-Si:H; however, only a fraction of the incorporated impurity atoms contributes to the free-carrier concentration compared to the doping situation in crystalline silicon. This behavior is shown in Figure 9.5, where the free-carrier concentration as a function of the impurity (phosphorus) concentration is presented for crystalline, microcrystalline and polycrystalline and amorphous silicon. The polycrystalline silicon (curve 2) shows the characteristic decrease of the carrier concentration below about 10^{18} cm^{-3} (for a grain size of about 0.1 μm) due to a trapping of charge carriers in the grain boundary states, which was discussed in Section 5.5. The hydrogen-free amorphous silicon deposited by a CVD process (curve 4) has a trap density of about 10^{19} cm^{-3} in the gap, which is indicated by

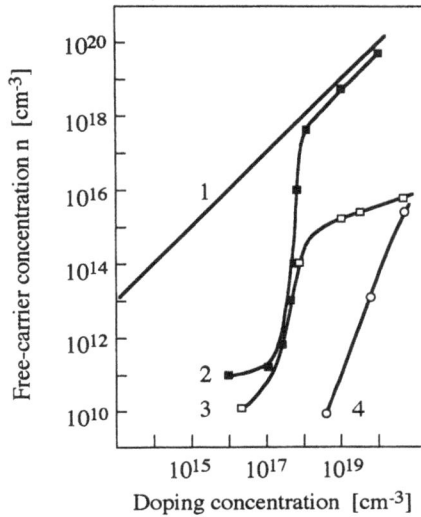

Figure 9.5 Free-carrier concentration n versus phosphorus concentration N_d for crystalline and amorphous silicon: (1) monocrystalline; (2) polycrystalline; (3) plasma-deposited a-Si:H; (4) CVD-deposited amorphous silicon [4].

the rapid decrease in free-carrier concentration below this value, when free carriers become trapped in these states [4]. The low carrier concentration could, however, also be due to an insufficient activation of the doping atoms when it is incorporated on a site where it is electrically not active. The behavior of the hydrogenated a-Si:H (curve 3) shows that both effects are important, namely, trapping of carriers in band gap states and low activation of the impurity atoms. Below about 10^{18} cm^{-3}, which corresponds to the trap density in this case, the free-carrier concentration decreases rapidly. However, even above about 10^{19} cm^{-3}, the free-carrier concentration remains several orders of magnitude below the impurity concentrations, which shows that only a fraction of the doping atoms is activated.

Phosphorus is the most common n-type dopant in a-Si:H. Normal structural bonding for phosphorus is threefold and one might expect that all P-atoms are preferentially incorporated in the P_3^0 configuration, where it is electrically inactive and merely a network relaxer. Experimental results show, however, that P-atoms are also incorporated in the fourfold configuration, where they act as a donor $P_4^{+/0}$. A possible explanation may be that phosphorus atoms in the fourfold configuration form a complex with another defect which is energetically comparable with the P_3^+ configuration. For instance, the $P_4^+ - T_3^-$ pair has been proposed as a possible complex configuration [6]. Although the configuration of the fourfold

P-atom may still be controversial, it is clear that a fraction of the incorporated atoms are electrically active and contribute impurity states to the band tails or just below. One can define the doping efficiency $\eta = [P_4]/[P_{tot}]$ of the electrically active component with respect to the total concentration of doping impurities. Experimental data for phosphorus and arsenic are given in Table 9.1. Despite the low activation, one can vary the Fermi energy over a wide range up to about 0.1 eV below the mobility edge for high concentrations of phosphorus.

Boron is the usual *p*-type dopant in a-Si:H. Because of its complex chemistry it is also difficult to predict the atomic configurations in the network where it becomes electrically active. Experimental data indicate that the vast majority of the boron atoms are in the neutral threefold coordinated configuration B_3^0 [6]. As with phosphorus, it has therefore been suggested that a defect pair must form to stabilize the electrically active fourfold configuration e.g., $B_4^- - T_3^+$. Doping efficiencies of about $\eta < 0.06$ have been reported, which indicates the low activation of boron [11]. For high boron concentrations the Fermi energy can also be shifted to about 0.1 eV above the mobility edge.

The doping behavior of phosphorus and boron in amorphous silicon differs also in another way from crystalline silicon. ESR measurements show that for slight doping levels the concentration of dangling bond centers decreases and finally completely disappears. This is in accordance with the conversion of the dangling bond center T_3^0 into the spin-paired centers T_3^- or T_3^+ when the Fermi energy shifts as discussed in the previous section. With the introduction of more impurity atoms, a new spin signal appears, however. The density of the dangling bond states N_{db} has been found to increase as $N_{db} \sim (N_d^{tot})^{1/2}$, where N_d^{tot} is the total concentration of the doping element [12]. The increase in the density of states for a doped film can be seen, for instance, in Figure 9.6 if one compares the increased defect concentration near midgap with the density of about 10^{16} cm^{-3} of undoped a-Si:H. Additionally, the distribution of the defect levels is also changed and varies with the deposition conditions. For heavily doped films, the ESR signal strength decreases again [13]. These doping-induced deep levels can also be observed in the

Table 9.1
Structural Configuration and Doping Efficiency of Some Doping Elements in Amorphous Silicon
(The Hall (μ_H) and drift (μ_d) mobilities of doped a-Si:H are included)

Doping Element	Doping Efficiency	Configuration	μ_H [cm²/Vs]	μ_d [cm²/Vs]
N-doping				
Phosphorus	<0.3	$P_4^+ - T_3^-$	0.06	13
Arsenic	0.2	$As_4^+ - T_3^-$		
P-doping				
Boron	<0.06	$B_4^- - T_3^+$	0.03	0.5

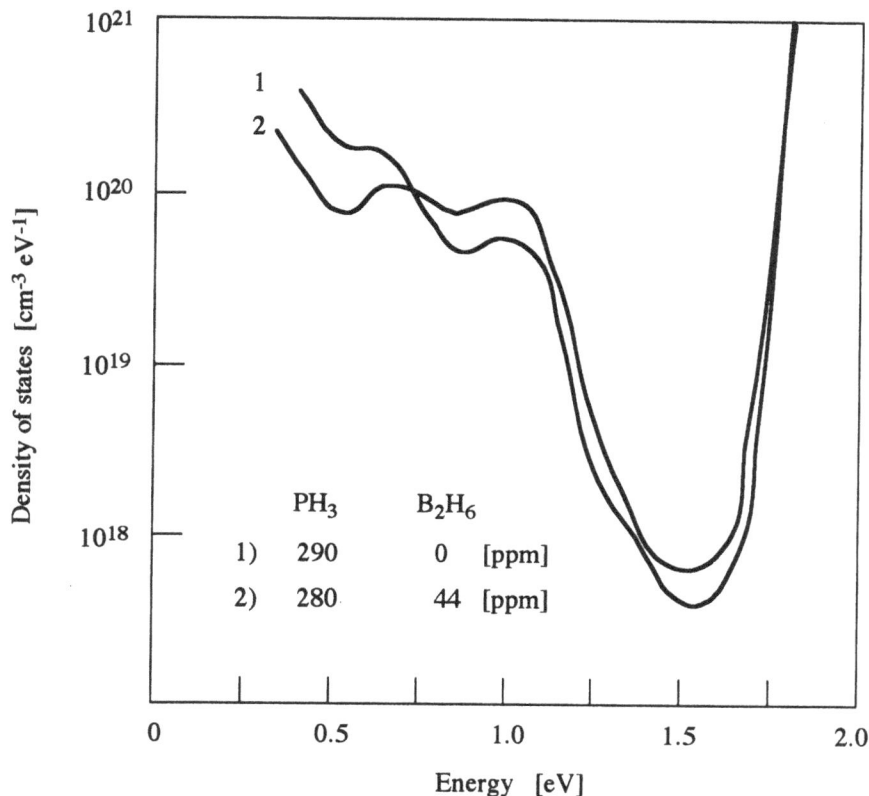

Figure 9.6 Density of states derived from DLTS measurements for phosphorus-doped a-Si:H containing various amounts of boron counterdoping. The comparison with the results of Figure 9.3 shows the increase of midgap states upon doping [27].

optical absorption as an increase of the subband gap absorption (see next section). Although several explanations have been proposed for the doping-induced deep trap centers, the nature of the dangling bond center remains controversial.

Summarizing the doping behavior, the important result is that the doping in amorphous silicon has two effects. First, it changes the Fermi energy and increases the concentration of free carriers. Second, since the doping impurity also introduces new deep-level centers, the doping effect is partly compensated for. Because of the increase of deep trap states the density of states of doped a-Si:H is changed compared to the intrinsic material. It is evident that the new defects also have a significant influence on the electrical conductivity and the lifetime in a-Si:H.

9.1.3 Recombination and Electrical Transport

The character and distribution of the band gap states determine the mobility and the lifetime in a-Si:H in different ways. While the shallow tail states essentially limit the *mobility* of charge carriers, the midgap states dominate the *recombination*. These midgap states change in concentration and distribution depending on the position of the Fermi level, as discussed in the previous section. In undoped material, the neutral dangling bond dominates, which is an especially active recombination center; therefore, their density has to be kept low to achieve a reasonable lifetime. In doped material with the Fermi energy shifted towards the mobility edges, most of the deep trap centers are occupied and therefore tend to act as trapping rather than recombination centers. However, since their density usually increases by several orders of magnitude ($>10^{17}$ cm^{-3}), they can also have a detrimental effect on the lifetime. Also important for the lifetime are band tail states, which can become recombination centers under illumination as the quasi-Fermi level approaches the band edges. This loss mechanism is not present in crystalline semiconductors; thus a-Si:H has a somewhat lower theoretical conversion efficiency than an ideal semiconductor with the same band gap. Generally, the lifetime in doped a-Si:H is shorter than in the intrinsic material, and most of the carrier collection in a solar cell has to be in undoped layers. Therefore, the *pin* configuration is usually used for amorphous solar cells, where the intrinsic region is wider and is the most active zone. Lifetimes have been estimated for undoped material with values ranging between 3 and 30 μs [14,15].

The band tail states have a strong influence on the *mobility* in amorphous silicon. Undoped a-Si:H is almost an insulator with a conductivity $\sigma < 10^{-10}$ Ω^{-1}cm^{-1}, and even doped amorphous silicon still has a very low conductivity of about $\sigma \approx 10^{-1}$ Ω^{-1}cm^{-1}. This is due to a small mobility with typical values below 0.1 cm^2/Vs at room temperature. The low mobility indicates that it is determined by mechanisms other than scattering of charge carriers in the extended states above the mobility edges. Because of the high density of localized band tail states, a significant fraction of carriers is trapped in these states. Conduction in the localized states is only possible by thermally activated hopping or tunneling between these states, which in both cases results in a very low mobility (below 0.01 cm^2/Vs) and a strong temperature dependence. Charge carriers can also be thermally activated into the extended states above the mobility edge. The mobility in amorphous silicon is thus generally a combination of carriers moving just above the mobility edge in the extended states and carriers hopping in the localized states. Since two different modes of transport occur, the mobility is described by the general expression (4.2) in Section 4.1.

The carrier Hall mobility μ_H determines the carrier transport by diffusion and thus the conductivity and diffusion length. The carrier mobility is measured, for instance, by the Hall effect, which is the usual method for conducting crystalline materials. In insulators and also in intrinsic amorphous semiconductors, the carrier transport in an electric field differs in general from the transport by diffusion and is described by the drift mobility μ_d. For strong applied fields, the carrier transport is mainly in the extended states, whereas for lower fields the drift mobility μ_d is due to capture and re-emission in the localized states. Typical values for the room temperature mobilities are summarized in Table 9.1. In solar cells, both μ_H and μ_d play an important role, depending on whether there are field-free regions in the structure. In that case, carrier transport is by diffusion and directly related to the magnitude of μ_H. In the presence of strong electric fields, for instance, in the junction region, the mobility is more determined by μ_d.

Contrary to crystalline semiconductors, the sign of the Hall voltage is opposite to that of the dominant charge carrier in amorphous semiconductors. This reflects the anomaly of the transport mechanism in disordered structures. The Hall mobility μ_H as a function of the temperature is shown in Figure 9.7 for a number of p- and n-type specimens [16]. The temperature dependence is exponential for boron, but shows a more complicated behavior for phosphorus. The analysis of the experimental data for phosphorus suggests that above 250K conduction is mainly through extended states, whereas below this temperature it takes place by hopping between localized states which, in doped material, would be mainly impurity levels. For high doping levels above 5×10^{18} cm^{-3}, impurity conduction becomes the dominant transport mechanism, even at room temperature [17].

The Hall mobility in undoped a-Si:H (with a slight excess of electrons) is also given in Figure 9.7 and shows an exponential increase above 360K, with an activation energy of about 0.13 eV. This corresponds to the activation energy for the drift mobility μ_d of electrons and indicates in this case that the mobility occurs through localized states above room temperature. The mobility at low temperatures is assumed to occur through tunneling.

The important parameters for the performance of the solar cell, mobility and lifetime, enter most of the equations describing the performance of a solar cell as a product $\mu\tau$. Therefore, experimental results are frequently presented for the mobility-lifetime product $\mu\tau$, which can be determined from photoconductivity and diffusion length measurements (see (2.20) in Section 2.3.1). A high-performance solar cell must efficiently collect all light-generated carriers in the active region of the device. This condition can only be met if the diffusion length L is greater than the thickness of the active region of the cell. The absorption behavior of amorphous silicon requires a thickness of about 0.5 μm to absorb most of the light. Using the expression $L = (KT\mu\tau/e)^{1/2}$, this translates into a mobility-lifetime product $\mu\tau > 10^{-7}$ cm^2/V. Considering the mobility values for electrons in undoped a-Si:H at room temperature from Figure 9.7 ($\mu_H \approx 0.1$ cm^2/Vs) and the lifetime values mentioned above (10 μs), one obtains $\mu\tau \approx 10^{-6}$ cm^2/V, which is sufficient for the operation of the cell. The mobility-lifetime product in undoped a-Si:H depends primarily on the density of deep trap states. Experimentally, it has been observed

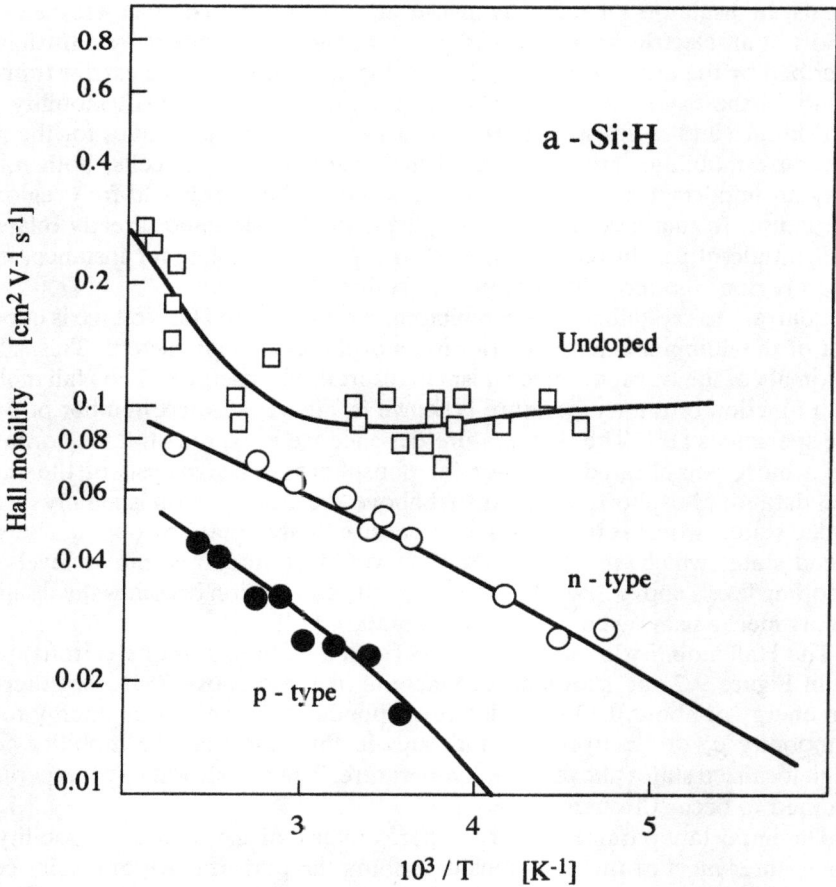

Figure 9.7 Temperature dependence of the Hall mobility for doped [16] and undoped a-Si:H [17]. For n-type samples, the volume gas ratio PH_3/SiH_4 is 3×10^4 ppm. For the p-type samples, the volume gas ratio B_2H_6/SiH_4 is 5×10^4 ppm. The solid line is a fit of (4.2) to the experimental data under the assumption that carrier transport occurs by drift and tunneling.

that for both electrons and holes the following relationship is valid for the product $\mu \tau N_{db}$ = const. (Figure 9.1) [18]. Since the dangling bond trap density N_{db} varies with the deposition technique, the diffusion length can be optimized by the deposition processes.

The doping dependence of the product $\mu \tau$ for electrons and holes is given in Figure 9.8 [19] for a boron-doped a-Si:H film. The low mobility values for the

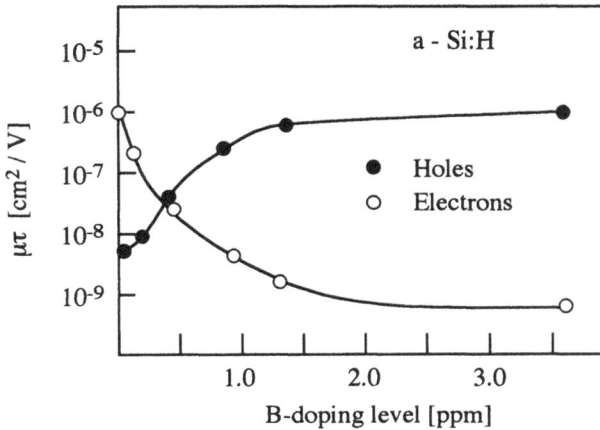

Figure 9.8 Mobility-lifetime product $\mu\tau$ as a function of the boron concentration for a-Si:H [19].

electrons (minority carriers here) at higher doping levels ($\mu\tau < 10^{-8}$ cm²/V) show that the collection efficiency in a p-doped region of a solar cell will be rather poor. It is therefore important to minimize the thickness of the doped regions usually below 10 nm (see, for instance, the typical configuration for a *pin* solar cell in Section 3.4, Figure 3.5). It should also be noted that, for undoped films, $\mu\tau$ is considerably smaller for holes than for electrons; thus, the diffusion length is mainly determined by electron transport. This difference suggests for the design of *pin* solar cells that the average travel path be minimized for the holes rather than for electrons. In fact, the more efficient cells are produced with a p^{+}-layer as the window for the incident light.

The conductivity of doped a-Si:H is determined by the mobility and concentration of majority carriers. The previous results have demonstrated that both parameters depend in a complex way on the doping level and the temperature. Experimental results for the room temperature conductivity for p- and n-type films as a function of the doping level are shown in Figure 9.9 [20]. Rather high doping levels are required to obtain the sufficient conductivity necessary for solar cell applications.

9.1.4 Optical Absorption

Crystalline silicon as an indirect semiconductor has a rather low absorption in the visible spectral range, since only a few of the electronic states near the band edge can take part in the absorption process because of the momentum conservation

328

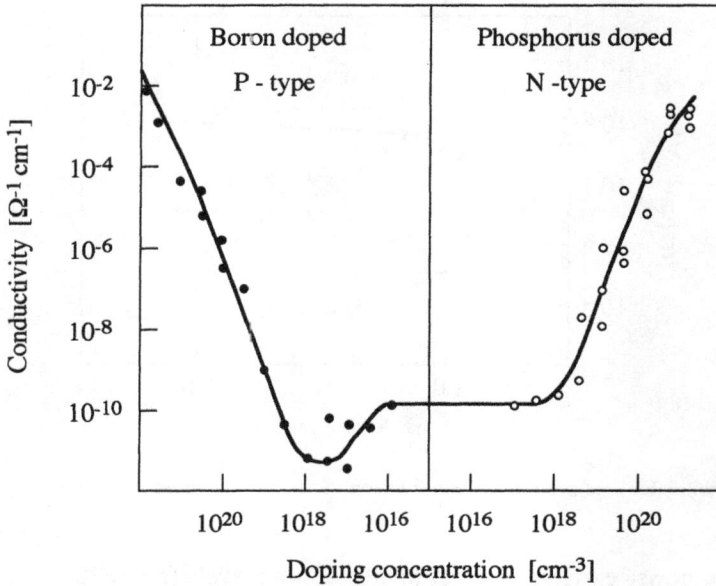

Figure 9.9 Room temperature conductivity of a-Si:H as a function of the implanted impurity concentration for *p*-type, boron-doped and *n*-type, phosphorus-doped films [20].

rule and the participation of phonons. This situation is different for amorphous silicon, since most of the electronic states near the band edges (tail states) are available for optical transitions. Correspondingly, the optical absorption coefficient is much higher compared to crystalline silicon, as can be seen in Figure 9.10. Over most of the visible light region (1.9 eV $< h\nu < 4.0$ eV), the absorption is greater than 10^4 cm^{-1}. For high photon energies ($h\nu > 1.6$ eV), the absorption can be described by the energy dependence of (2.3) for direct band transitions. The band gap E_g, which is obtained as a fitting parameter, is here called the *optical band gap* in order to distinguish it from the mobility band gap, which is inferred from transport measurements. For hydrogen-free a-Si:H, the optical gap is about 1.7 eV, and for photon energies above the gap the absorption coefficient is about ten times higher compared to crystalline silicon. This has the consequence that a film thickness of about 0.5 μm is already sufficient to absorb most of the light. It is interesting to note that it is the reduced optical penetration that compensates for the low carrier diffusion length and the poor conductivity in amorphous silicon.

For photon energies below the optical gap, the absorption drops off slower compared to the energy dependence extrapolated from higher energies. This

Wavelength [μm]

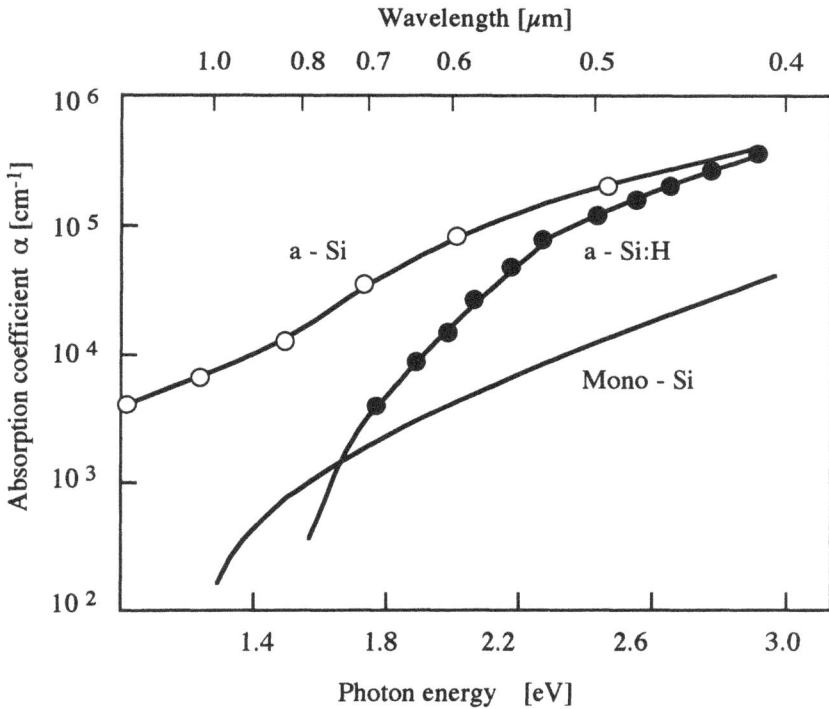

Figure 9.10 Optical absorption coefficient of monocrystalline silicon and hydrogenated and nonhydrogenated amorphous silicon as a function of the photon energy [4].

increased subband absorption is similar to the absorption in heavily doped crystalline semiconductors and can be described by the Urbach law (equation (2.8)):

$$\alpha_U(\nu) = C(\nu) \exp\left\{-\frac{E_g - h\nu}{E_0}\right\} \tag{9.3}$$

where E_0 is the Urbach energy which relates directly to the characteristic distribution width of the band tail states. An important consequence of the tail states is that nonradiative recombination can occur, which limits the lifetime and is thus inescapable for amorphous semiconductors. This is important for photovoltaic applications, since a limitation of the lifetime also limits the open-circuit voltage in solar cells, which can be seen from (2.47), where the reduction of lifetime increases the saturation current and thus reduces the open-circuit voltage. It has been estimated that the band tail states inferred from the transport measurements described above limit the open-circuit voltage of an amorphous silicon solar cell with an optical gap of 1.7 eV to about 1.0 eV [14]. In comparison, a crystalline semiconductor with the same band gap but only radiative recombination would have a maximum output voltage of 1.4 eV.

The deeper energy levels within the band gap have a strong influence on the absorption for lower photon energies. Figure 9.10 shows, for instance, that hydrogen-free amorphous silicon has a significantly larger absorption below the mobility gap of about 1.7 eV for lower photon energies, which is due to the much higher density of trapping states in the gap. Since the density of these band gap states depends strongly on the deposition conditions, the absorption strength varies accordingly. The passivation of these states in a-Si:H decreases the subband gap absorption significantly. The absorption in this range can be used to estimate the density of midgap states and to study their behavior. At the same time, the incorporation of hydrogen also changes the optical band gap, which varies between about 1.6 and 1.8 eV.

Doping has a similar effect: it generally increases the absorption in the entire visible range, but particularly at lower photon energies. This has important consequences for the device design, since doped regions also have high recombination losses and as much as 20% of the incident light can be lost, for instance, in the boron-doped p-layer, which is about 10 nm thick for a *pin* cell (Figure 3.5) [10].

9.1.5 Amorphous Silicon Alloys

The development of multi-junction cells seems to be the most promising device technology for improved efficiencies. Since multi-junction concepts require materials with different band gaps stacked upon each other in order to use the incident light more efficiently, it was necessary to develop amorphous films with different band gaps. This can be achieved by amorphous silicon alloys, where the band gap can be varied over a wide range, so the designer has substantial flexibility in constructing optimized devices. So far, amorphous silicon-germanium alloys appear to be the most promising semiconductors for the narrow band gap portion of double- or triple-junction devices. Since amorphous germanium has a lower band gap, it reduces the band gap of the silicon-germanium alloy with increasing germanium concentration.

In general, increasing the germanium content also decreases the mobility-lifetime product and hence the photoconductivity. Since at the same time no increase of the midgap density of states nor an increase in the tail density of states is observed, it has been suggested that the poor transport properties are an intrinsic property of the amorphous structure of the alloys. In particular, it is proposed that a preferential bonding between Ge-Ge occurs, which reduces $\mu\tau$ [21]. The situation can be improved by the addition of hydrogen, which seems to promote the Si-Ge bonding. In addition, alloys prepared with hydrogen also show less light degradation compared to the hydrogen-free amorphous alloys (see Section 9.2).

Amorphous silicon alloys with a wider band gap can be obtained by the addition of carbon and the formation of a-Si$_{1-x}$C$_x$:H alloys. However, many of the parameters relevant for the photovoltaic performance are not yet available and the development of these alloys is still at the exploration level. It shall also be mentioned that microcrystalline films of a-Si:H and a-SiC:H, which can be obtained at slightly higher deposition temperatures, are currently being investigated from the perspective of their utilization in multi-junction solar cells.

9.2 STABILITY AND LIGHT-INDUCED DEGRADATION

Hydrogenated amorphous silicon presents a serious problem which is less desirable for photovoltaic applications. It was first reported in 1977 that the photoconductivity of a-Si:H degrades when exposed to light. This phenomenon, which mainly affects undoped films, is known as the *Staebler-Wronski* (S-W) *effect*, named after the those who reported it first [22,23]. The main features of the effect are shown in Figure 9.11. First, it is observed that the photoconductivity of the films decreases during illumination and, secondly, that the dark conductivity decreases below its starting value. After heating the film in the dark at above 150°C, the photo-induced effect can be annealed and the conductivity returned to its original state. The effect is perfectly reversible and stable at room temperature. It can involve enormous changes in the conductivity, which has of course strong implications for the technical applications. Whereas the original state has a clearly defined conductivity, the metastable state can have a range of conductivities depending on the light exposure. The reversible S-W effect has been observed for a-Si:H films fabricated under a wide range of deposition conditions and also for a-Si-Ge alloys. The continuous transition from the stable to the metastable state depends, however, on the original properties of the amorphous silicon film and the level of illumination [24]. At room temperature and illumination of about 1 sun (typical for photovoltaic applications), it may take several hours to reach the metastable state.

ESR measurements show that during illumination the dangling bond density increases from a low starting value of about 10^{16} cm^{-3} up to values of about 10^{17} cm^{-3}. The photo-induced increase in the dangling bond density N_{db} increases the density of mid-gap states, which then can capture free carriers and thus reduce the dark conductivity. If one expresses the temperature dependence of the dark conductivity by

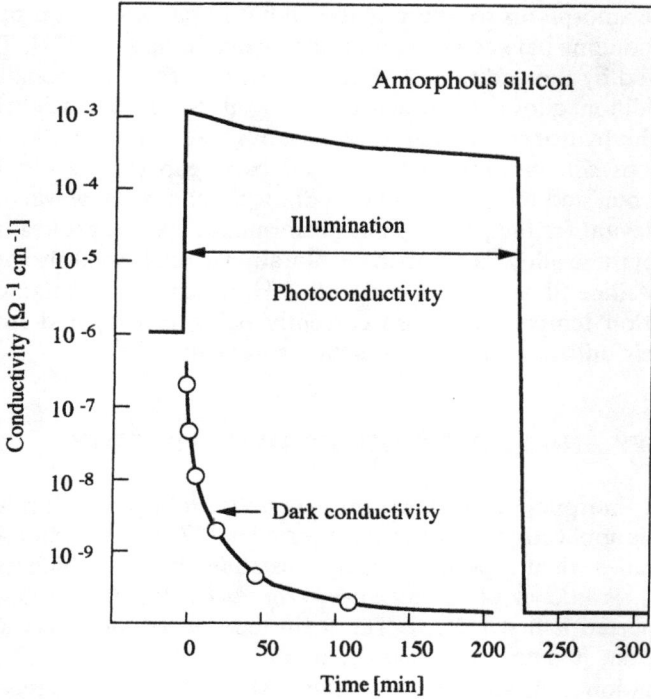

Figure 9.11 Staebler-Wronski effect for undoped a-Si:H films showing the decrease of the photoconductivity and dark conductivity during illumination with 2 kWm^{-2} filtered tungsten light [22].

$$\sigma = \sigma_0 \exp\left\{-\frac{E_a}{KT}\right\} \tag{9.4}$$

where $E_a = E_C - E_F$ is the activation energy for mobile electrons. Then experimental results show that for undoped films, for instance, E_a can change from about 0.55 to 0.9 eV, and for phosphorus-doped films between 0.3 to 0.55 eV corresponding to the shift of the Fermi energy towards the band center. The direct correlation of the change in conductivity and illumination with the corresponding movement of the Fermi level is, however, complicated by the fact that the prefactor σ_0 changes simultaneously, corresponding to the experimentally determined expression

$$\sigma_0 = C \exp(DE_a) \tag{9.5}$$

where C and D are constants. This exponential dependence of the prefactor on the activation energy is known as the Meyer-Nedel rule [25] and accounts for the observed saturation of the conductivity decrease. However, there is a large range of values for the constants C and D, which depend on the material and the deposition conditions.

The creation of dangling bonds by illumination is thus a self-limiting effect, which reaches a saturation value after several hours. The S-W effect is mainly observed for undoped films, but also occurs in both n- and p-type doped films. It has also been reported that for slight n-doping the dark conductivity can *increase* rather than decrease upon exposure of light [24]. For higher doping levels, above 10^{17} cm^{-3}, the Fermi level is essentially pinned by the acceptor or donor state and not affected any more by the formation of deep trap levels at a concentration between 10^{16} and 10^{17} cm^{-3}. Although there is general agreement that the formation of metastable compensating defects is responsible for the Fermi level shift, there is no satisfying explanation yet for the mechanisms that can cause the observed large changes in the prefactor.

The transition from the metastable state to the original state by annealing is a thermally activated process which has a unique activation energy of about 1.5 eV independent of the material and the conditions under which the metastable state has been reached. It has been observed that this value is close to the activation energy for the diffusion of hydrogen in amorphous silicon, which may suggest that hydrogen plays a role in the formation and/or passivation of the metastable defects.

The light-induced degradation problem has become one of the most investigated phenomenon in the photovoltaic research field, obviously because of its impact on this promising technology. Almost every conceivable technique has been applied to elucidate the problem, but a generally accepted model has not emerged yet to explain the degradation. It is evident now that the carrier recombination and/or capture induces the formation of metastable defects which return to their initial state by annealing. An important question for the explanation of the effect is whether the degradation is due to impurities or an intrinsic property of the disordered structure. Experiments performed on very pure a-Si:H films seem to rule out the participation of common impurities such as oxygen or nitrogen. Most models have therefore focused on processes that are related to the amorphous structure itself, particularly the weak (strongly distorted) silicon-silicon bonds in the band tails. It has been proposed that an elementary process can be the breaking of a weak bond, generating two adjacent dangling bonds which are stabilized by a relaxation of the network or the switching of the bonds (Figure 9.12). Numerous other models have been suggested, but none of these has emerged yet as a generally accepted theory; therefore, it appears that the problem of the Staebler-Wronski effect is not completely solved yet.

The degradation phenomena have important consequences for the operation and design of solar cells. In general, most of the degradation under illumination

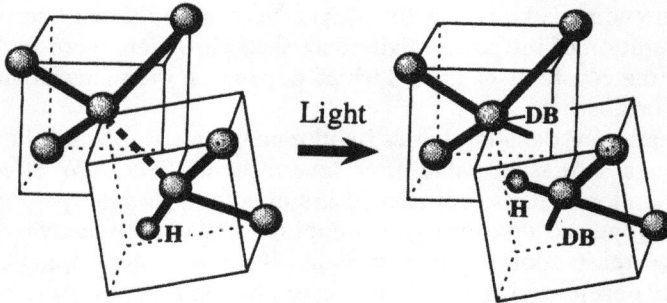

Figure 9.12 Schematic representation of the bond-breaking model in a-Si:H to account for the observation of metastable, light-induced dangling bonds as proposed in [26].

occurs in the first 100 hours of operation, and is accompanied by a drop in the efficiency of the solar cells by 10% to 30% (Figure 9.13) [10]. The degradation depends on the thickness of the intrinsic layer and the cell temperature: at higher operating temperatures (e.g., 80°C), the degradation is considerably lower compared to room temperature. The operation conditions also play a role: cyclic illumination (e.g., in a day-night rhythm) shows a lower degradation compared to a continuous illumination (with the same exposure time).

During illumination, both the photoconductivity and the dark conductivity decrease as discussed above and shown in Figure 9.11. In view of the previous results, the dynamics of the process and its impact on the efficiency are rather complex and cannot be evaluated easily. It is useful to consider the parameters that determine the photoconductivity (equation (2.20)), for instance, for undoped material:

$$\sigma_{ph} = \sigma + eG\mu_n\tau_n \tag{9.6}$$

where G is the generation rate of electron-hole pairs (per unit volume and time). Since $\mu_n\tau_n$ is higher for electrons in undoped material (see Figure 9.8), the hole contribution has been neglected here. During illumination, the dark conductivity σ (equation (9.4)) decreases as discussed above. The effect of the photo-induced defects on the second term is less evident from the previous results. The illumination can in principle affect the generation rate G, the mobility μ_n, and the lifetime τ_n. The generation rate G can be affected if the spectral response, which is mainly determined by the band structure near the band edges, is changed by light-generated defects. Similarly, one would expect a change in the mobility if the band tail states are greatly affected. Experimental results show, however, that both generation

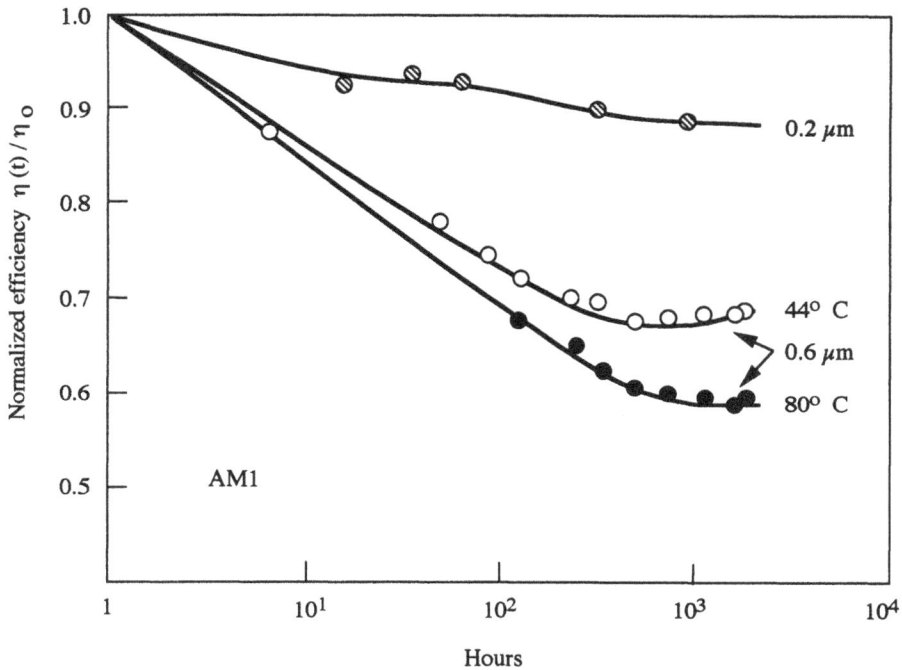

Figure 9.13 Normalized efficiency of a-Si:H solar cells with *pin* structures as a function of the exposure time for AM 1 illumination. Parameter is the thickness of the intrinsic layer of the cells [28].

efficiency and mobility are hardly changed, which indicates that the light-generated defect levels lie mainly in the center of the gap. Consequently, the light-induced changes in the photoconductivity are primarily due to the decrease of the lifetime resulting from the introduction of the metastable defects. The degradation of amorphous solar cells is therefore mainly due to changes in the depletion or undoped layer.

In general, the efficiency is determined by the change of the short-circuit current I_{sc}, the open-circuit voltage V_{oc}, and the fill factor FF (equations (2.47) to (2.50)). From the previous discussion, it is evident that I_{sc} is less affected, whereas the reduction of the minority carrier lifetime mainly increases the saturation current (equation (2.39)) and thus decreases the open circuit voltage. This is in qualitative agreement with experimental results obtained for amorphous solar cells. The simultaneous decrease of the dark conductivity increases the series resistance of the solar cell and therefore mainly affects the fill factor FF. Both factors reduce the efficiency of a solar with the duration of the illumination until the saturation of the defect formation has occurred.

Conventional solar cell designs for amorphous films use the *pin* structure (see Section 3.4, Figure 3.5) because of the lower recombination losses in the intrinsic layer compared to the doped layers, as discussed in the previous section. Since the degradation is primarily caused by changes in the properties of undoped a-Si:H layers, it is the intrinsic layer in a *pin* cell that is mostly affected. Evidently, a reduction in the thickness of the intrinsic layer improves the stability, as can be seen in Figure 9.13, though at the expense of the total efficiency if the layer becomes too thin. Slight doping of the intrinsic layer generally increases the stability, but also reduces the efficiency. Better stabilities are also reported for amorphous silicon alloys.

Technical solutions thus have to find a compromise between the fact that a certain thickness of the intrinsic layer is required to obtain a reasonable starting efficiency and that thinner undoped layers are less susceptible to degradation effects. The dependence on the thickness of the intrinsic layer for different cell structures (Figure 3.5) is shown in Figure 9.14. Layer thicknesses below 0.4 μm are usually necessary to obtain a reasonable stability of the cell. Amorphous silicon solar cells with rather thin intrinsic layers are usually not thick enough anymore to absorb enough light; therefore, multijunction structures are required to compensate for the optical loss. The best results for unstabilized cells (12.4% efficiency) have been reported for a-Si:H with a tandem *pin* layer structure, which consists of two *pn* junctions (see Figure 3.12). This result is still well below the theoretical limit of about 20% for a single-junction cell, which follows from the upper limits for I_{sc} from the optical absorption data, for the maximum open-circuit voltage of about 1.05 eV, and for the fill factor $FF = 0.86$ [14].

The efficiencies of large-area modules are considerably lower compared to those of a single cell. This is mainly due to series resistance losses (see Section 2.4.3) and the fact that the deposition of a uniform layer amorphous silicon on larger substrates ($>10 \times 10$ cm^2) is difficult and leads to "dead" areas, which do not contribute to the light conversion. Typical for a module are starting efficiencies of about 7% to 10% and 5% to 7% after stabilization.

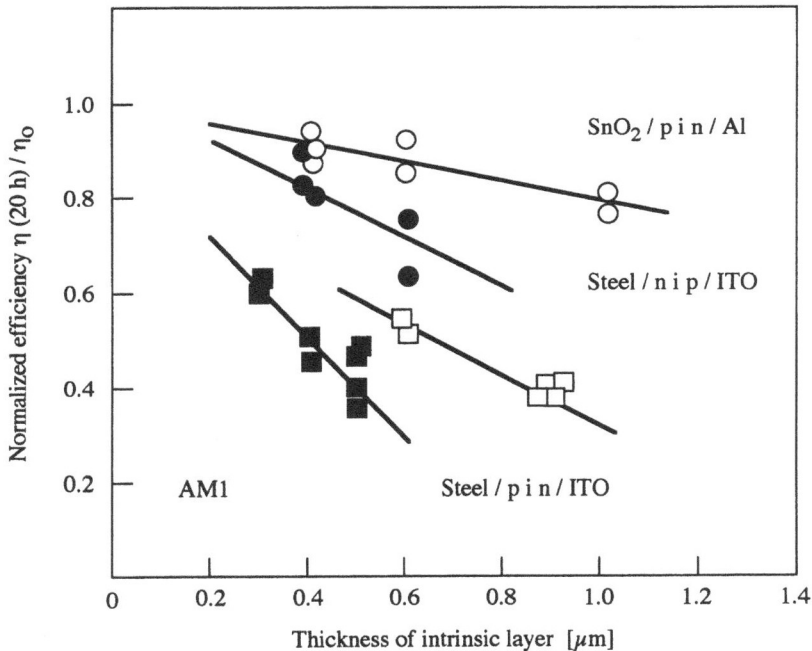

Figure 9.14 Degradation behavior of a-Si:H solar cells with different *pin* structures. The normalized efficiency after 20 hours of illumination (AM1 spectrum) is plotted as a function of the thickness of the intrinsic layer [29].

REFERENCES

[1] Willardson, R. K., and A. C. Beer, eds., *Semiconductors and Semimetals*, Vol. 21A-D, New York: Academic Press, 1984.
[2] Joannopoulos, J. D., and G. Lucovsky, eds., *Hydrogenated Amorphous Silicon I and II*, Berlin, New York: Springer, 1984.
[3] Adler, D., Y. Hamakawa, and A. Madan, *Mater. Res. Soc. Symp. Proc.*, Vol. 70.
[4] Winterling, G., and G. Müller, *Physica Scripta*, Vol. T13, 1986, p. 45.
[5] Wolford, D. J., J. A. Reimer, and B. A. Scott, *Appl. Phys. Lett.*, Vol. 42, 1983, p. 369.
[6] Adler, D., *Semiconductors and Semimetals*, Vol. 21A, R. K. Willardson and A. C. Beer, eds., New York: Academic Press, 1984, p. 291.
[7] Spear, W. E., P. E. Lecomber, and A. J. Snell, *Phil. Mag.*, Vol. 33, 1976, p. 935.
[8] Tiedje, T., *Semiconductors and Semimetals*, Vol. 21C, R. K. Willardson, and A. C. Beer, eds., New York: Academic Press, 1984, p. 207.
[9] Margarino, J., D. Kaplan, A. Friedrich, and A. Deneuville, *Phil. Mag.*, Vol. B 45, 1982, p. 285.
[10] Carlson, D. E., *Semiconductors and Semimetals*, Vol. 21D, R. K. Willardson and A. C. Beer, eds., New York: Academic Press, 1984, p. 7.
[11] Street, R. A., *Phys. Rev. Lett.*, Vol. 49, 1982, p. 1187.

[12] Müller, G,. H. Mannsberger, and S. Kalbitzer, *Phil. Mag.*, Vol. B 53, 1986, p. 257.

[13] Street, R. A., D. K. Biegelsen, and J. C. Knights, *Phys. Rev.*, Vol. B 24, 1981, p. 969.

[14] Tiedje, T., *Appl. Phys. Lett.*, Vol. 40, 1982, p. 627.

[15] Snell, A. J., W. E. Spear, and P. E. Lecomber, *Phil. Mag.*, Vol. 43, 1981, p. 407.

[16] Lecomber, P. E., D. I. Jones, and W. E. Spear, *Phil. Mag.*, Vol. 35, 1977, p. 1173.

[17] Dresner, J., *Appl. Phys. Lett.*, Vol. 37, 1980, p. 742.

[18] Tiedje, T., J. M. Cebulka, D. L. Morel, and B. Abeles, *Phys. Rev. Lett.*, Vol. 46, 1980, p. 1425.

[19] Catalano, A., R. R. Arya, M. Bennett, L. Yang, J. Morris, B. Goldstein, B. Fieselmann, J. Newton, and S. Wiedemann, *Solar Cells*, Vol. 27, 1988, p. 25.

[20] Kalbitzer, S., G. Müller, P. E. Lecomber, and W. E. Spear, *Phil. Mag.*, Vol. B 41, 1980, p. 439.

[21] Cody, G. D., C. R. Wronski, B. Abeles, R. B. Stephens, and B. Brooks, *Solar Cells*, Vol. 2, 1980, p. 227.

[22] Wronski, C., and D. Staebler, *J. Appl. Phys.*, Vol. 51, 1980, p. 3262.

[23] Wronski, C., and D. Staebler, *Appl. Phys. Lett.*, Vol. 31, 1977, p. 292.

[24] Wronski, C., *Semiconductors and Semimetals*, Vol. 21C, R. K. Willardson and A. C. Beer, eds., New York: Academic Press, 1984, p. 347.

[25] Meyer, W., and H. Z. Nedel, *Tech. Phys.*, Vol. 12, 1937, p. 588.

[26] Dersch, H., M. Stuke, and J. Beichler, *Appl. Phys. Lett.*, Vol. 38, 1981, p. 456.

[27] Cullen, P. J., P. Harbison, D. V. Lang, and D. Adler, *J. Non-Cryst. Solids*, Vol. 59/60, 1983, p. 261.

[28] Fortmann, C. M., J. O'Dowd, J. Newton, and J. Fischer, *Proc. Int. Conf. on Stability of Amorphous Alloy Materials*, 1987.

[29] Kniffler, N., G. Mück, G. Müller, M. Simon, and G. Winterling, *BMFT Report Photovoltaic*, German Society for Solar Energy, 1984.

Index

Absorption, in semiconductors, 12–19
Absorption coefficient, 12–15
Acid leaching, 234
Air mass one (AM1) radiation, 10
Air mass zero (AM0) radiation, 10
    dislocation, 161
Aluminothermal reduction, 233
Ambipolar diffusion coefficient, 20
Ambipolar mobility, 20
AM1 (air mass one) radiation, 10
Amorphous silicon, 5, 313
    alloys, 330–331
    band gap states in, 314–320
    doping in, 320–323
    optical absorption in, 327–330
    recombination and electrical transport in, 324–327
    stability and light-induced degradation of, 331–337
    technology of, 67–68
AM0 (air mass zero) radiation, 10
Anti-phase boundaries, 278, 279
Antireflection (AR) coatings, 42–43
Antisite defects, 108, 118–120
A-swirls, 135
Auger recombination, 47–48, 85

Back-side passivated emitter solar cell, 54, 55
Back surface field (BSF), 52
Band-band recombination, 84–86
Band gap, 6
Band gap energies, 58
Band gap states, in amorphous silicon, 314–320
Base current, 38
Base width, 37–38
    dislocation, 161
Binary compounds, point defects in, 108, 118–120

Boron, as dopant of amorphous silicon, 322
Bridgman-Stockbarger process, 235–237
Bridgman technique, 270
Broken bonds. *See* Dangling bonds
BSF. *See* back surface field
B-swirls, 135
Bulk precipitation, 197–200
Buried contact cell with laser grooved
    surface, 54, 55

Carbon
    interaction with oxygen, 135–136
    in silicon, 125–126, 131–138
Carbon-related defects, electrical activity
    of, 137–138
Carrier concentrations, 19–20
CdS cell, 287
CdTe cell, 287–290, 291
Chalcopyrite compounds, 290–293
    crystal and film growth techniques for, 293–296
    electro-optical properties and lattice defects
      in, 296–308
    interface properties and oxidation behavior
      of, 308–311
    intrinsic defects in, 120–125
    monocrystalline, 299–304
    polycrystalline, 304–307
    quaternary, 292, 307–308
Charge carriers
    mobility of, 78–83
    recombination of, 83–105
Coincidence site lattice (CSL), 173–175
Compound parabolic concentrator (CPC), 74
Compound semiconductors, 267–269
    epitaxial, 268, 277–283